Advanced Analytic Methods
in Applied Mathematics, Science, and Engineering

10 09 08 07 10 9 8 7 6 5 4 3 2 1

Published by LuBan Press
25 West Street
Boston, MA 02111-1213 USA
Fax (617) 426-3669
"Opening Minds, Expanding Worlds"™

Library of Congress Cataloging-in-Publication Data

Cheng, Hung, 1937-
Advanced analytic methods in applied mathematics, science, and engineering / by Hung Cheng.
p. cm.
Includes bibliographical references and index.
ISBN-13: 978-0-9758625-1-3
ISBN-10: 0-9758625-1-0
1. Mathematical analysis–Textbooks. I. Title.

QA300.C4438 2007
515–dc22

2006048987

Printed in the United States of America

Contents

Preface

For almost four decades, the Department of Mathematics at MIT has offered a course called "Advanced Analytic Methods in Science and Engineering."[1] The purpose of this course is to strengthen the mathematical background of all entering graduate students, so they may be better prepared for their respective coursework and specialties.

During the past ten years I've been teaching this course I occasionally wrote notes for the students in my class, intended only to supplement the course's textbooks. However, at the end of one recent semester several students suggested I make my materials, which were by that time more than simply "notes," accessible to students beyond MIT. This textbook is the result of expansion of and revisions to that material.

The background of students taking the course is usually fairly diverse. Many of them lack some of the fundamentals that would prepare them for a graduate math course. The first five chapters owe their origin to the need for helping such students, bringing them up to speed. The last five chapters contain more advanced materials.

Teachers and students will thus find that this book's content is flexible enough to meet the needs of a variety of course structures. For a one-semester course with emphasis on approximate methods, a teacher may just skim over the first five chapters, leaving the students to read in more detail the parts they need most. Such a plan would be especially useful for graduate students entering a Ph.D. program in engineering, science, or applied mathematics. But if this book is adopted for a course in advanced calculus for undergraduate engineering, science, or applied mathematics students, then Chapters 1–6 should be emphasized. Chapters 1–5, plus a few selected later chapters, would be suitable for a graduate course for Master's degree students. In addition, Chapters 3, 4, and 5 may be used as part of the materials for a course on partial differential equations.

[1] The course was created by Professor Harvey Greenspan. In 1978 Carl M. Bender and Steven A. Orszag, two lecturers of this course, authored a textbook, *Advanced Mathematical Methods for Scientists and Engineers* (New York, McGraw-Hill, Inc.; reprinted by Springer-Verlag New York, Inc., 1998).

While most graduate students and upper-class undergraduate students have already had a full semester of ordinary differential equations, some of them may need a refresher. Therefore, Chapter 1 includes a very brief summary of ordinary differential equations. This first chapter is also a convenient place to reintroduce the elementary but powerful operator method to students. The operator method enables one to more quickly produce the particular solutions of certain linear ordinary differential equations as well as partial differential equations, and it also facilitates many other calculations.[2]

Chapter 2 is for students who need a quick summary of some of the relevant materials in complex analysis. The important but often neglected subjects of branch points and branch cuts are included, as well as a short discussion of the Fourier integral, the Fourier series, and the Laplace transform.

Many of the analytic methods discussed in this book arose from the need to solve partial differential equations. To help the reader see that connection, Chapters 3, 4, and 5 address partial differential equations.

Because many problems encountered in real life are often not solvable in a closed form, it will benefit a student to learn how to do approximations. Chapter 6 presents the methods of series solutions. A few well-known special functions are used as examples in order to help students gain some familiarity with these functions while learning the methods of series solutions. I will address the topic of irregular singular points of an ordinary differential equation, which is not usually covered in standard textbooks on advanced calculus, such as F. B. Hildebrand, *Advanced Calculus for Applications*, Prentice Hall, 1976. The series solution expanded around an irregular singular point of an integral rank is generally divergent and leads naturally to the concept of asymptotic series, which we'll cover in subsequent chapters.

Chapter 7 discusses the WKB method. This method gives good approximate solutions to many linear ordinary differential equations with a large parameter or those with coefficients that are slowly varying. It is also helpful for yielding solutions near an irregular singular point of a linear differential equation. While the lowest-order WKB solutions are obtained by solving nonlinear differential equations, the higher-order WKB approximations are obtained by iterating linear differential equations. The last section of this chapter discusses the solutions near a turning point.

Chapter 8 addresses the Laplace method, the method of stationary phase, and the saddle point method, which are useful for finding the asymptotic series of

[2]While this method has been routinely used in field theories, particularly with the derivation of various Green functions, it has not been adequately covered in most undergraduate textbooks, with the notable exception of *Differential Equations* by H. T. H. Piaggio, G. Bell and Sons, Ltd., London, 1946 (reprinted in the U.S. by Open Court Publishing Company, LaSalle, Illinois, 1948).

integrals with a large parameter. In the saddle point method, we deviate from the rigorous approach of finding the path of steepest descent. Instead, we advocate finding just a path of descent, as this may somewhat reduce the solution's chores.

Chapters 9 and 10 address the subjects of regular perturbation and singular perturbation. Chapter 9 is devoted to the topic of boundary layers, and Chapter 10 covers the topic of small nonlinear oscillations.

Throughout this book I emphasize a central theme rather than peripheral details. For instance, before discussing how to solve a class of advanced problems, I relate it to the basics and, when possible, make comparisons with similar but more elementary problems. As I demonstrate a method to solve a certain class of problems, I start with a simple example before presenting more difficult examples to challenge the minds of the students. This process gives students a firmer grasp of the subject, enabling them to acquire the key idea more easily. Hopefully, ours will make it possible for them to do mathematics without the need of memorizing a large number of formulae. In the end I hope that they will know how to approach a general problem; this is a skill that leaves students better prepared to treat problems unrelated to the ones given in this book, which they'll likely encounter in their future academic or professional lives.

During my classroom lectures, I emphasize interaction with the students. I often stop lecturing for a few minutes to pose a question and ask everyone to work through it. I believe this method helps to encourage students to learn in a more thorough way and to absorb concepts more effectively, and this book reflects that interactive approach; many "Problems for the Reader" are found throughout the text. To deepen their understanding of the themes that they're learning, students are encouraged to stop and work on these problems *before* looking at the solutions that follow.

This book also passes on to learners some of the problem-solving methods I've developed through the years. In particular, parts of Chapters 9 and 10 offer techniques, which I hope will benefit students and researchers alike. Indeed, I believe that the renormalization methods given in Chapter 10 are more powerful than other methods treating problems of non-linear oscillations so far available.

I am indebted to the group of students who encouraged me to publish this book. Several students have read the field test version of this book and have given me their very helpful suggestions. They include Michael Demkowicz, Jung Hung Lee, Robin Prince, and Mindy Teo. Also, Dr. George Johnston read Chapter 7 and gave me very useful comments. I want to thank Professor T. T. Wu of Harvard University, who introduced me to the saddle point method several decades ago with a depth I had never fathomed as a graduate student. I thank Mr. David Hu for the graphs in Chapters 2 and 8. Special thanks are due to Dr. Dionisios Margetis for graphs in Chapter 9 and the compilation of an extensive bibliography, and to Mr. Nikos Savva for graphs in Chapters 3 and 9. I also am truly grateful to Professor John Strain for his inexhaustible efforts in reading through all of the chapters in the first draft. I am greatly indebted to Dr. H. L. Hu for the many graphs he tirelessly drew for this book.

Chapter 1

The Algebra of Operators

In the first five chapters of this book, we will review some of the fundamentals for the beginning graduate student in science or engineering. This review focuses on a few mathematical topics that are essential and yet are sometimes inadequately covered in undergraduate programs. The last five chapters will deal with more advanced mathematical topics.

A. Calculus

We'll begin with learning how to do differentiation more efficiently. Your reaction upon hearing this is probably that you have learned to do differentiation very well in high school and in college, and you need no more lessons on this elementary topic. But many of you probably do not yet know the information explained below, and learning it may make it more enjoyable for you to perform certain kinds of calculations in the future.

Let us first introduce a shorthand notation for the differential operator:

$$D \equiv \frac{d}{dx}. \tag{1.1}$$

We note that

$$De^{mx} = me^{mx}.$$

This means that when it operates on e^{mx}, the differential operator D is simply equal to the constant m:

$$D \to m.$$

Indeed,

$$D^2 e^{mx} = m^2 e^{mx}.$$

Thus, when it operates on e^{mx},

$$D^2 \to m^2.$$

Similarly, when it operates on e^{mx},

$$D^n \to m^n, \quad n = 2, 3 \cdots.$$

Therefore, when it operates on e^{mx}, the operator $P(D)$ can be replaced by

$$P(D) \to P(m), \tag{1.2}$$

whenever $P(D)$ is a polynomial in D.

Next we note that

$$(e^{mx}v)' = e^{mx}v' + me^{mx}v.$$

Using the notation D, we have

$$D(e^{mx}v) = e^{mx}Dv + me^{mx}v. \tag{1.3}$$

By factoring out the common exponential, we get

$$D(e^{mx}v) = e^{mx}(D+m)v. \tag{1.4}$$

While (1.4) is just another way to express (1.3), we may look at (1.4) in the following light: eq. (1.4) says that we are allowed to move e^{mx} from the right of D to the left of D, provided that we compensate by replacing D with $(D+m)$.

By applying (1.4) twice in succession, we have

$$D^2 e^{mx}v = D(De^{mx}v) = De^{mx}(D+m)v = e^{mx}(D+m)^2 v.$$

More generally,

$$P(D)e^{mx}v = e^{mx}P(D+m)v, \tag{1.5}$$

where $P(D)$ is a polynomial of D. Thus we are allowed to move e^{mx} from the right of $P(D)$ to the left of $P(D)$, provided we replace D by $D+m$. It's possible to go a little further. We have

$$(uv)' = uv' + u'v,$$

which can be written as

$$D(uv) = u\left(D + \frac{u'}{u}\right)v.$$

The function $\frac{u'}{u}$ is equal to the derivative of $\ln u$, and is hence called the logarithmic derivative of u. The equation above says that we are allowed to move u from the right of D to the left of D if we compensate by replacing D with $D + \frac{u'}{u}$. This also works for the second derivative of uv. We have, by moving u to the leftmost position in two successive steps,

$$D^2uv = D(Duv) = Du\left(D + \frac{u'}{u}\right)v = u\left(D + \frac{u'}{u}\right)^2 v,$$

which means that we are allowed to move u from the right of D^2 to the left of D^2 if we compensate by replacing each factor D in D^2 by $D + \frac{u'}{u}$.

This is obviously true for the operators $D^3,\ D^4\ \cdots$ as well. Thus we have

$$P(D)uv = uP\left(D + \frac{u'}{u}\right)v. \tag{1.6}$$

Let

$$u = e^f;$$

then (1.6) becomes

$$P(D)e^{f(x)}Y = e^{f(x)}P(D + f'(x))Y. \tag{1.7}$$

Equation (1.5) is a special case of (1.7) with $f(x) = mx$.

Problem for the Reader

Find the logarithmic derivative of x^a and e^{bx}.

Solution

Since $\ln x^a = a \ln x$, the logarithmic derivative of x^a is $\frac{a}{x}$.
Since $\ln e^{bx} = bx$, the logarithmic derivative of e^{bx} is b.

There is an equation, called the Bessel equation, that comes up in many problems in science and engineering. This equation is given by

$$y'' + \frac{1}{x}y' + \left(1 - \frac{p^2}{x^2}\right) y = 0,$$

where p is a constant, and is called the order of the Bessel equation.

Problem for the Reader

Make a change of variable for the Bessel equation so that the resulting equation has no term proportional to y'.

Solution

Let us write

$$y'' + \frac{1}{x}y' = \left(D + \frac{1}{x}\right) Dy.$$

We make the change of the dependent variable

$$y = x^s Y,$$

where s is a constant. With the use of (1.6), we find that

$$\left(D + \frac{1}{x}\right) Dx^s Y = x^s \left(D + \frac{1+s}{x}\right) \left(D + \frac{s}{x}\right) Y.$$

The above expression has no term linear in D if we choose

$$1 + s = -s,$$

or

$$s = -\frac{1}{2}.$$

This is because

$$(D-a)(D+a) = D^2 - a^2 + (Da - aD).$$

We note that had D been a number, $(Da - aD)$ would be equal to zero, and the expression above would have been

$$(D-a)(D+a) = D^2 - a^2,$$

an elementary formula. But since D is a differential operator, we have

$$(Da - aD)Y = \frac{d}{dx}(aY) - a\frac{d}{dx}Y = a'Y$$

or, more compactly,

$$(Da - aD) = a'.$$

Thus we have

$$(D-a)(D+a) = D^2 - a^2 + a'. \tag{1.8}$$

Note that there is no term linear in D in (1.8). Thus, to eliminate the first derivative term in the Bessel equation, we choose

$$y = x^{-\frac{1}{2}}Y,$$

and the Bessel equation becomes

$$Y'' + \left(1 - \frac{p^2 - \frac{1}{4}}{x^2}\right)Y = 0.$$

The logarithmic derivative also facilitates the calculation of the derivative of a product of a large number of factors. Let

$$g(x) = a_1(x)a_2(x)\cdots a_n(x);$$

by applying (1.6) to each of the factors successively, we get

$$\begin{aligned} Dg(x) &= a_1(x)a_2(x)\cdots a_n(x)\left(D + \sum_1^n \frac{a_m'(x)}{a_m(x)}\right)1 \\ &= a_1(x)a_2(x)\cdots a_n(x)\sum_1^n \frac{a_m'(x)}{a_m(x)}. \end{aligned}$$

Problem for the Reader

Find the derivative of the function

$$g(x) = \left(1+\frac{x}{1}\right)\left(1+\frac{x}{2}\right)^2\left(1+\frac{x}{3}\right)^3\left(1+\frac{x}{4}\right)^4\left(1+\frac{x}{5}\right)^5.$$

Solution

If we calculate the derivative of $g(x)$ defined above in a straightforward way, we will get five terms, each of which has five factors, a horrendous expression.

Since the logarithmic derivative of

$$\left(1+\frac{x}{n}\right)^n$$

is

$$\frac{n}{n+x},$$

we get

$$g'(x) = g(x)\left(\frac{1}{1+x}+\frac{2}{2+x}+\frac{3}{3+x}+\frac{4}{4+x}+\frac{5}{5+x}\right),$$

which is more manageable.

Obviously, the advantage of using logarithmic derivatives is pronounced as we differentiate the product of a large number of factors.

Problem for the Reader

Calculate the derivative of

$$g(x) = \prod_{n=1}^{\infty}\left(1+\frac{x}{n}\right)e^{-\frac{x}{n}},$$

which is a product of an infinite number of factors.

Solution

The logarithmic derivative of $\left(1+\frac{x}{n}\right)e^{-\frac{x}{n}}$ is

$$\frac{1}{x+n}-\frac{1}{n}=-\frac{x}{n(n+x)};$$

thus we have

$$g'(x)=-g(x)\sum_{n=1}^{\infty}\frac{x}{n(n+x)}.$$

An aside: The function $g(x)$ given above is proportional to the inverse of the Gamma function $\Gamma(x)$, which we will discuss in Chapter 6. Indeed,

$$\frac{1}{\Gamma(x)}=xe^{\gamma x}g(x),$$

where γ is the Euler constant, which is equal to $0.5772156649\cdots$.

B. Ordinary Differential Equations

Next we'll do a brief review of ordinary differential equations (ODEs).

We may classify an ODE by its order and by whether it is linear or nonlinear.

Generally, it is much more difficult to solve a nonlinear ODE than a linear ODE in a closed form. And it is much more difficult to solve a high-order ODE than a low-order ODE in a closed form.

Consider the simplest of differential equations:

$$\frac{dy}{dx}+p(x)y=q(x). \tag{1.9}$$

In (1.9), $p(x)$ and $q(x)$ are functions of x. Since the highest-order derivative in the equation above is the first derivative, the equation is called an ODE of the first order. And since the equation is linear in y ($\frac{dy}{dx}$ is regarded as linear in y), the equation is called linear. Being linear and of the lowest order possible, eq. (1.9) can be solved in a closed form, as we shall show.

If $q=0$, the equation is called homogeneous.

If $q\neq 0$, the equation is called inhomogeneous.

Problem for the Reader

Classify the following ordinary differential equations:

$$\textbf{a. } y'' - xy = 1. \tag{1.10}$$

$$\textbf{b. } y' = e^{x+y}. \tag{1.11}$$

Solution

Equation (1.10) is linear, inhomogeneous, and of the second order.

Equation (1.11) is nonlinear and of the first order.

Let us return to (1.9) with $q = 0$. We have

$$\frac{dy}{dx} = -p(x)y. \tag{1.12}$$

This equation is exactly solvable for any $p(x)$, no matter how ugly $p(x)$ looks. We have

$$\frac{dy}{y} = -p(x)dx;$$

thus

$$\ln y = -\int_0^x p(x')dx' + c',$$

or

$$y = ce^{-P(x)},$$

where

$$P(x) \equiv \int_0^x p(x')dx', \tag{1.13}$$

and $c = e^{c'}$ is an arbitrary constant.

We note that the solution of a first-order ODE such as (1.12) has one arbitrary constant. Consequently, the solution of (1.12) satisfying an initial condition is unique. For example, the solution of (1.12) satisfying $y(0) = 1$ is

$$y(x) = e^{-P(x)}.$$

I qualify this statement by mentioning that this is true only if $p(x)$ does not have a singularity at the initial point. Let $p(x)$ be equal to $1/x$, which blows up at $x = 0$. Let the initial point be the origin and let us require that $y(0) = 0$. The solution of the differential equation

$$y' = \frac{1}{x}y$$

is

$$y(x) = cx,$$

where c is a constant. Any choice of c will satisfy $y(0) = 0$, and hence there are infinitely many solutions.

The point $x = 0$ is known as a singular point of the differential equation $y' = y/x$. A condition on the solution set at a singular point may not be meaningful.

As another example, let us change the sign of the right side of the equation we just discussed. We have

$$y' = -\frac{1}{x}y.$$

Let the initial condition be $y(0) = 1$. The solution of the differential equation is

$$y = \frac{c}{x},$$

where c is a constant. There is no value of c that enables us to satisfy the initial condition $y(0) = 1$.

When we deal with a linear differential equation, especially one of the second order or an even higher order, it is often helpful to spot the singular points of this equation before we try to solve it. This will be discussed more fully in Chapter 6. We shall also quote without proof Picard's theorem, which applies to all first order differential equations, be they linear or nonlinear: Let

$$y' = f(x, y).$$

If f and $\partial f/\partial y$ are continuous in a neighborhood of the point (x_0, y_0), then there exists a continuous solution $y(x)$ of the differential equation that satisfies $y(x_0) = y_0$.

Next we shall show that (1.9) with $q(x) \neq 0$ can also be solved in a closed form.

Let us rewrite (1.9) as

$$(D + p(x))y = q(x).$$

Let

$$y(x) \equiv e^{f(x)} Y(x).$$

With (1.7) in mind, we require that $f(x)$ satisfies

$$\frac{d}{dx} f(x) = -p(x),$$

or

$$f(x) = -P(x).$$

With such a choice we get

$$e^{-P(x)} DY = q(x),$$

or

$$DY = e^{P(x)} q(x).$$

Thus we have

$$Y(x) = \int_0^x e^{P(x')} q(x') dx' + c.$$

The solution of (1.9) is then

$$y(x) = e^{-P(x)} \left(\int_0^x e^{P(x')} q(x') dx' + c \right).$$

Thus we have shown that a first-order linear ODE can be solved in a closed form, be it homogeneous or inhomogeneous, and that the solution has one arbitrary constant.

Some nonlinear first-order differential equations are also solvable. Here is one example.

Example 1:

Solve

$$y' = e^{x+y}.$$

Solution

We have from the ODE that

$$e^{-y}dy = e^x dx.$$

The integrand of the left side is a function of y only, while that of the right side is a function of x only. Such an equation is called separable, and is solved by one integration. We get

$$-e^{-y} = e^x + c.$$

On the other hand, a second-order ODE like (1.10), which has nonconstant coefficients (the coefficient of y in (1.10) is not a constant), is much more difficult to solve, even though it is linear.

While it is usually difficult to obtain closed form solutions of linear equations like (1.10), some general properties of its solution can be deduced.

Consider the nth-order ODE

$$\frac{d^n y}{dx^n} + a_{n-1}(x)\frac{d^{n-1}y}{dx^{n-1}} + \cdots + a_0(x)y = b(x), \tag{1.14}$$

which is linear and inhomogeneous. Let Y be the solution of (1.14) with $b = 0$:

$$\frac{d^n Y}{dx^n} + a_{n-1}(x)\frac{d^{n-1}Y}{dx^{n-1}} + \cdots + a_0(x)Y = 0. \tag{1.15}$$

It is easy to show that if Y_1 is a solution of (1.15), so is cY_1. This can be proved by multiplying (1.15) by c. More generally, it is easy to prove that if $Y_1, Y_2 \cdots Y_n$ are solutions of (1.15), so is

$$c_1Y_1 + c_2Y_2 + \cdots + c_nY_n. \tag{1.16}$$

Therefore, once we find n independent solutions of (1.15), we may construct a solution of (1.15) with n arbitrary constants, which is the most general solution of (1.15).

It is also easy to prove that if y_P is a solution of (1.14), i.e.,

$$\frac{d^n y_P}{dx^n} + a_{n-1}(x)\frac{d^{n-1}y_P}{dx^{n-1}} + \cdots + a_0(x)y_P = b(x),$$

and if y satisfies (1.14), then $(y - y_P)$ satisfies (1.15). Thus we have

$$y = y_P + Y. \tag{1.17}$$

This tells us that an efficient way to find the most general solution of (1.14) is to do it in two steps:

Step 1: We first set $b = 0$ and find the most general solution of (1.15). This solution is called the complementary solution of (1.14).

Step 2: We find just one particular solution (any solution) of (1.14), which we call y_P.

The most general solution of (1.14) is (1.17).

Note that there is another class of ODE that can be solved in closed forms. These equations are linear ODEs with constant coefficients. They can be either homogeneous or inhomogeneous.

Problem for the Reader

Find a particular solution of the following differential equation:

$$(D^{100} + 1)y = e^x.$$

Solution

There is more than one way to find a particular solution of this equation, but we advocate doing it in the following way. We treat D as if it were a number and move the operator $(D^{100} + 1)$ to the right-hand side of the differential equation obtaining

$$y_P = \frac{1}{D^{100} + 1} e^x,$$

where y_P is a particular solution of the differential equation. By using (1.2) with $m = 1$, we get

$$y_P = \frac{1}{1^{100} + 1} e^x = \frac{1}{2} e^x.$$

You may worry that since $(1 + D^2)^{-1}$ is not a polynomial of D, (1.2) does not apply. So let us try to understand what we've done.

Problem for the Reader

Verify that the solution given above satisfies the differential equation.

Solution

$$(D^{100}+1)\frac{e^x}{2}=(1+1)\frac{e^x}{2}=e^x.$$

I mention that the conventional method of finding a particular solution of the differential equation above is that of undetermined coefficients, which is based on an ansatz. (An ansatz is an assumed form of mathematical expression. Sometimes a difficult problem can be completely solved with an ansatz. See problem 6 in this chapter.) For this differential equation, the ansatz of the method of undetermined coefficients is that $y_p(x)$ is of the form ce^x. The coefficient c is determined by plugging this solution into the differential equation and requiring that the equation be satisfied. While it is not being stated explicitly, the method we advocate here also employs exactly the same ansatz: that the particular solution is of the form ce^x. As we have seen, c is equal to the inverse of $(D^{100}+1)$ with D replaced by unity. Thus our method here is just a shorthand version of the conventional method of undetermined coefficients.

Problem for the Reader

Find a particular solution of the following differential equation:

$$y''-y=e^x.$$

Solution

A particular solution of the differential equation above is

$$y_P=\frac{1}{D^2-1}e^x.$$

If we apply (1.2), then the denominator in the above expression is zero and we get $y_P=\infty$, not a meaningful result.

Instead, we first write

$$y_P=\frac{1}{(D-1)}\frac{1}{(D+1)}e^x.$$

By applying (1.5), we get

$$y_P = e^x \frac{1}{D} \frac{1}{(D+2)} 1. \tag{1.18}$$

By setting

$$\frac{1}{(D+2)} 1 = \frac{1}{2},$$

(1.18) becomes

$$y_P = e^x \frac{1}{D} \frac{1}{2}.$$

Since the inverse of differentiation is integration, we assert that

$$\frac{1}{D} 1 = \int dx = x + c.$$

This assertion can be proved by calling

$$\frac{1}{D} 1 \equiv y.$$

We have

$$Dy = 1,$$

which tells us that y is equal to $x + c$. We may set c to zero as all we need is one particular solution. Then we have

$$y_P = \frac{e^x x}{2}.$$

Some readers may feel uneasy about applying (1.5) with $(D^2 - 1)^{-1}$, which is not a polynomial. So let us justify it. The particular solution $y_p(x)$ we seek satisfies

$$(D-1)(D+1) y_p(x) = e^x.$$

Let

$$y_p(x) \equiv e^x Y(x)$$

and substitute it into the equation for $y_p(x)$. Since $(D-1)(D+1)$ is a polynomial of D, it is justified to apply (1.5), and we get

$$D(D+2) Y(x) = 1,$$

which leads to (1.18).

Example 2:

Find the general solution of

$$y^{(100)} + y = \cosh x.$$

Solution

Step 1: We solve the corresponding homogeneous equation:

$$Y^{(100)} + Y = 0.$$

Try

$$Y = e^{mx};$$

then m must satisfy

$$m^{100} = -1. \tag{1.19}$$

Equation (1.19) is a polynomial equation with 100 roots, which we will denote as m_n, $n = 1, \cdots 100$, with

$$m_n = e^{\frac{\pi(2n+1)i}{100}}.$$

If you have forgotten how to solve (1.19), read section A of Chapter 2. Thus the complementary solution of the present example is

$$Y = c_1 e^{m_1 x} + \cdots + c_{100} e^{m_{100} x}. \tag{1.20}$$

Step 2: We find one particular solution of the ODE. We have

$$y_P = \frac{1}{D^{100}+1} \cosh x = \frac{1}{D^{100}+1} \frac{1}{2}(e^x + e^{-x}).$$

Making use of (1.2), we have

$$y_P = \frac{1}{1+1} \frac{1}{2}(e^x + e^{-x}) = \frac{1}{2} \cosh x. \tag{1.21}$$

The most general solution is

$$y = y_P + Y. \tag{1.22}$$

We note that we first split $\cosh x$ into the sum of two exponentials, then it turned out that we made the same replacement $D^{100} \to 1$ for both exponentials. Thus we were wasting our time doing the splitting. Why is it so?

Let us note that

$$D \cosh mx = m \sinh mx.$$

In other words, D operating on $\cosh mx$ is not equal to a constant times $\cosh mx$. Thus it is true that the replacement (1.2) cannot be made when D operates on $\cosh mx$. However, by differentiating the equation above one more time, we get

$$D^2 \cosh mx = m^2 \cosh mx.$$

This means that we may make the replacement

$$D^2 \to m^2 \tag{1.23}$$

when D^2 operates on $\cosh mx$. With this understanding, y_P in Example 2 can be calculated without the need of splitting $\cosh x$ into the sum of two exponentials.

It's easy to prove that we may also make the replacement (1.23) if D^2 operates on $\sinh mx$.

Also note that when applying to $\cos mx$ or $\sin x$, we may make the replacement

$$D^2 \to -m^2.$$

Example 3:

Find a particular solution for

$$y'' - y = \cosh x.$$

Solution

We have

$$y_p(x) = \frac{1}{D^2 - 1} \cosh x.$$

If we replace D^2 by unity, the right side blows up. Thus we write

$$y_p(x) = \lim_{m \to 1} \frac{1}{D^2 - 1} \cosh mx = \lim_{m \to 1} \frac{\cosh mx}{m^2 - 1}.$$

The solution $y_p(x)$ is actually the same as before and is equal to infinity, as the denominator $(m^2 - 1)$ vanishes in the limit $m \to 1$. We will show that it is possible to add to y_p a solution of the corresponding homogeneous equation and make the sum finite. The term we add to y_p will be chosen to be $-\cosh x/(m^2-1)$, which is a solution of the equation $Y''-Y=0$. The resulting solution is

$$\lim_{m\to 1}\frac{\cosh mx-\cosh x}{m^2-1}.$$

This is allowed, as adding to a particular solution a solution Y of the homogeneous equation gives another particular solution. The advantage of so doing is that the numerator of the latter particular solution vanishes as $m\to 1$, and hence the latter particular solution is in the form of a limit of zero divided by zero. This limit can be found by using l'Hopital's rule, which allows us to differentiate the numerator and the denominator with respect to m before taking the limit $m\to 1$. We get

$$y_p(x)=\frac{x\sinh x}{2}.$$

Example 4:

Find the general solution of

$$y''+2y'+y=2\sinh 3x+x+3e^{-x}.$$

Solution

Step 1: We find the complementary solution by solving

$$m^2+2m+1=0,$$

which has the double root $m=-1$. Hence we have found only one of the independent complementary solutions.

To find the other independent complementary solution, let

$$Y\equiv e^{-x}v,$$

and substitute the expression above into the equation

$$(D+1)^2Y=0.$$

We get

$$D^2v=0,$$

or

$$v=c_1+c_2x.$$

Thus the complementary solution is

$$Y = e^{-x}(c_1 + c_2 x).$$

A particular solution is

$$y_P = \frac{1}{D^2 + 2D + 1}(2\sinh 3x + x + 3e^{-x}).$$

Now

$$y_{P1} \equiv \frac{1}{D^2 + 2D + 1} 2\sinh 3x = \frac{1}{D+5}\sinh 3x,$$

where (1.23) has been used. We reduce the above equation further by multiplying the numerator and the denominator by $(D - 5)$. Thus we have

$$y_{P1} = \frac{D-5}{D^2 - 25}\sinh 3x = \frac{D-5}{9-25}\sinh 3x = -\frac{3\cosh 3x - 5\sinh 3x}{16}.$$

Next we have

$$y_{P2} \equiv \frac{1}{(1+D)^2} x = (1 - 2D + \cdots)x = x - 2,$$

as the Maclaurin series for $(1 + D)^{-2}$ is $(1 - 2D + 3D^2 - \cdots)$, and as

$$D^2 x = D^3 x = \cdots = 0.$$

Next we have

$$y_{P3} \equiv \frac{1}{(D+1)^2}(3e^{-x}) = 3e^{-x}\frac{1}{D^2}1.$$

But the inverse of D is integration. Thus we have

$$y_{P3} = 3e^{-x}\int_0^x dx' \int_0^{x'} dx'' = \frac{3}{2}e^{-x}x^2.$$

The most general solution of the ODE in this example is the sum of y_{P1}, y_{P2}, y_{P3}, and Y.

Below is a formula that will enable us to handle the general case in which the inhomogeneous term e^{mx} is a solution of the corresponding homogeneous equation. Let

$$P(D)y_P = e^{mx},$$

where

$$P(D) \equiv (D-m)^n p(D),$$

with

$$p(m) \neq 0.$$

We have

$$\begin{aligned} y_P(x) &= \frac{1}{(D-m)^n p(D)} e^{mx} \\ &= \frac{1}{(D-m)^n p(m)} e^{mx} \\ &= \frac{e^{mx}}{p(m)} \frac{1}{D^n} 1 = \frac{e^{mx}}{p(m)} \frac{x^n}{n!}. \end{aligned}$$

Example 5:

Solve

$$y'' + y = f(x),$$

where $f(x)$ is an awful-looking function.

Solution

The complementary solution is

$$Y = a \cos x + b \sin x.$$

A particular solution is

$$y_P = \frac{1}{D^2+1} f(x) = \frac{1}{2i} \left(\frac{1}{D-i} - \frac{1}{D+i} \right) f(x).$$

We have

$$\frac{1}{D-i} f(x) = \frac{1}{D-i} e^{ix} (e^{-ix} f(x)) = e^{ix} \frac{1}{D} e^{-ix} f(x),$$

where we have moved the factor e^{ix} all the way to the left and compensate by replacing D with $D+i$. By making use of the fact that the inverse of differentiation is integration, we get from the expression above that

$$\frac{1}{D-i} f(x) = e^{ix} \int_0^x dx' e^{-ix'} f(x').$$

Similarly

$$\frac{1}{D+i}f(x) = e^{-ix}\int_0^x dx' e^{ix'} f(x').$$

Thus

$$y_P = \int_0^x dx' \frac{e^{i(x-x')} - e^{-i(x-x')}}{2i} f(x') = \int_0^x dx' \sin(x-x') f(x').$$

Example 6:

Solve the boundary-value problem

$$\frac{d^2y}{dx^2} - y = g(x), \ -\infty < x < \infty$$

with the boundary conditions

$$y(\infty) = y(-\infty) = 0.$$

It is assumed that the function $g(x)$ vanishes at infinity and has a Fourier transform.

Solution

Let us express $g(x)$ by the Fourier integral (see Section F in Chapter 2 for a brief discussion of Fourier transform)

$$g(x) = \int_{-\infty}^{\infty} \tilde{g}(k) e^{ikx} \frac{dk}{2\pi},$$

where $\tilde{g}(k)$ is the Fourier transform of $g(x)$. Then a particular solution is

$$y_p(x) = \frac{1}{D^2-1}\int_{-\infty}^{\infty} \tilde{g}(k) e^{ikx} \frac{dk}{2\pi} = -\int_{-\infty}^{\infty} \frac{\tilde{g}(k)}{k^2+1} e^{ikx} \frac{dk}{2\pi}.$$

The solution $y_P(x)$ vanishes as $x \to \pm\infty$, satisfying the boundary conditions. To prove this, we note that when $|x| \gg 1$, the factor e^{ikx}, considered as a function of k, oscillates rapidly, resulting in great cancellations as one integrates over k. Indeed, the Riemann-Lebesgue lemma asserts that $y_p(x)$ vanishes as $|x| \to \infty$. (We will discuss this more extensively in Chapter 8.)

The complementary solution of the differential equation is

$$ae^x + be^{-x},$$

where a and b are constants. Since e^x (e^{-x}) blows up as $x \to \infty$ ($x \to -\infty$), the boundary conditions at $x = \pm\infty$ are satisfied only if both a and b are zero.

Next we consider the differential equation

$$x^n \frac{d^n y}{dx^n} + a_{n-1} x^{n-1} \frac{d^{n-1} y}{dx^{n-1}} + \cdots + a_1 x \frac{dy}{dx} + a_0 y = 0.$$

The coefficient of the mth derivative of y in the ODE is equal to the constant a_m times x^m, and is not a constant. This equation is unchanged if we replace x by cx; thus it is called equidimensional. This ODE can be solved in a closed form by the change of variable

$$X \equiv \ln x, \text{ or } x = e^X.$$

Problem for the Reader

Express $x\frac{d}{dx}$ in terms of the variable X and D, here $D \equiv \frac{d}{dX}$. Also, do the same for $x^n \frac{d}{dx^n}$.

Solution

We have

$$\frac{d}{dx} = \left(\frac{dx}{dX}\right)^{-1} \frac{d}{dX} = e^{-X} \frac{d}{dX} = e^{-X} D;$$

thus

$$x \frac{d}{dx} = D.$$

We also have

$$\frac{d^2}{dx^2} = (e^{-X} D)(e^{-X} D) = e^{-2X}(D-1)D;$$

thus

$$x^2 \frac{d^2}{dx^2} = D(D-1).$$

More generally, we have

$$\frac{d^n}{dx^n} = e^{-nX} D(D-1)\cdots(D-n+1).$$

Thus

$$x^n \frac{d^n}{dx^n} = D(D-1)\cdots(D-n+1), \; n = 1, 2, 3 \cdots . \tag{1.24}$$

Therefore, a homogeneous and equidimensional ODE can be transformed into an ODE with constant coefficients and is thus exactly solvable. The solutions are of the form $e^{mX} = x^m$. Indeed, when it operates on x^m, $x^n \frac{d^n}{dx^n}$ can be replaced by

$$m(m-1)\cdots(m-n+1).$$

We shall next consider a nonlinear ODE.

Example 7:

Solve

$$y' = a(x) + b(x)y + c(x)y^2.$$

Note that the right side of the equation above is quadratic, not linear, in y. If $c(x) = 0$, the equation above is linear and is exactly solvable, as we have discussed. If $a(x) = 0$, this equation can be linearized by the change of the dependent variable $u = y^{-1}$. (See homework problem 1c.) If $a(x)c(x) \neq 0$, this is a first-order nonlinear equation known as Riccati's equation.

Solution

Let

$$y(x) \equiv -\frac{u'(x)}{c(x)u(x)}, \tag{1.25}$$

and substitute it into the Riccati equation. We find, after a little algebra, that

$$u''(x) - \left[\frac{c'(x)}{c(x)} + b(x)\right] u'(x) + a(x)c(x)u(x) = 0. \tag{1.26}$$

We have therefore succeeded in linearizing the Riccati equation, but at the cost of increasing the order of the equation by one, from a first-order equation to a second-order equation. The resulting second-order equation is generally of nonconstant coefficients unless $a(x)c(x)$ and

$$\frac{c'(x)}{c(x)} + b(x)$$

are both equal to constants, and may or may not be solved in a closed form.

While the solution $u(x)$ of (1.26) has two arbitrary constants, the solution $y(x)$, expressed in terms of $u(x)$ by (1.25), is unchanged if u is replaced by a constant times u. Thus the solution of the Riccati equation has only one arbitrary constant, which it should since the Riccati equation is a first-order ODE.

Generally, a second-order ODE with nonconstant coefficients is difficult to solve in a closed form. But a bilinear form of its solutions turns out to have a closed form. Let y_1 and y_2 be two independent solutions of a linear and homogeneous second-order ODE, i.e.,

$$y_1'' + p(x)y_1' + q(x)y_1 = 0 \tag{1.27a}$$

and

$$y_2'' + p(x)y_2' + q(x)y_2 = 0. \tag{1.27b}$$

The Wronskian of y_1 and y_2 is defined to be

$$W(y_1, y_2) \equiv y_1(x)y_2'(x) - y_1'(x)y_2(x), \tag{1.28}$$

which is a bilinear form of y_1 and y_2. Let us take the difference of (1.27a) multiplied by y_2 and (1.27b) multiplied by y_1. We get

$$W' + p(x)W = 0.$$

Thus we have

$$W = ce^{-P(x)}, \tag{1.29}$$

where $P(x)$ is given by (1.13). Hence the Wronskian of two independent solutions of a second-order, linear, and homogeneous equation can always be obtained in closed form. We note that W is a constant if $p = 0$.

When the exact solutions are out of our grasp, knowing the exact form of the Wronskian sometimes gives us insights into the behavior of the solutions of the differential equation. And if by luck or otherwise we know one of the solutions of the equation, say y_2, then (1.29) forms a linear first-order differential equation for y_1:

$$y_1' - \frac{y_2'}{y_2} y_1 = \frac{e^{-P}}{y_2},$$

where we have chosen c to be equal to -1. From this equation we solve for y_1 and get

$$y_1(x) = y_2(x) \int^x \frac{e^{-P(x')}}{y_2^2(x')} dx'.$$

The general solution of the second-order differential equation is then obtained as a linear superposition of y_1 and y_2.

Finally, we consider the inhomogeneous linear ODE

$$y'' + p(x)y' + q(x)y = f(x). \tag{1.30}$$

We mention without proof that a particular solution of (1.30) is

$$y_1(x) \int \frac{-y_2 f(x)}{W(y_1, y_2)} dx + y_2(x) \int \frac{y_1 f(x)}{W(y_1, y_2)} dx, \tag{1.31}$$

where $y_1(x)$ and $y_2(x)$ satisfy (1.27). Thus the general solution of the corresponding homogeneous equation enables us to obtain the solution of the inhomogeneous equation (1.30) in the form of integrals. Equation (1.31) can be proven with the use of the method of variation of parameters.

Reference

The use of D can be found in H. T. H. Piaggio, *Differential Equations* (John S. Swift Co., Inc., Chicago, 1920).

Homework Problems for This Chapter

Solutions to the Homework Problems can be found at www.lubanpress.com.

1. Solve the following first-order differential equations:

a. $x(1+x)y' + \frac{1+x}{2} y = \sqrt{x}.$

b. $y' + \ln y + 1 = 0.$

Hint: The solution cannot be expressed by elementary functions. It is an incomplete Gamma function. (See eq. (21) on p. 266 of Chapter 6 in *Higher Transcendental Functions* (Bateman Manuscript Project, Volume 1), by Erdelyi, Magnus, Oberhettinger and Tricomi (McGraw-Hill, Inc., 1953).)

c. $y' = \frac{1}{x}y + xy^2.$

Hint: Let $u \equiv \frac{1}{y}$.

d. Transform the following equation into a linear one: $y' = \frac{1}{y} + 3x.$

Hint: Let

$$y \equiv u + \frac{3x^2}{2}.$$

Then consider u as the independent variable and x as the dependent variable. The resulting equation is a Riccati equation that can be linearized into the Airy equation to be discussed in Chapter 6.

2. Solve the following higher-order equations:

a. $y'' + 5y' + 6y = e^{-2x} + 4e^{2x},$

b. $y'' + y' + y = \cos x,$

c. $y'' + y = x^2 + 3\cos x,$

d. $x^3y''' + 3x^2y'' + 3xy' + y = \ln x,$

e. $y'' + 2y' + y = 5xe^{-x} + \cosh x,$

f. $y''' + y'' - y' - y = xe^x + 7\cosh x.$

3. Show that if y_1 is any solution of the Riccati equation in Example 7, then by making the substitution

$$y = y_1 + \frac{1}{v},$$

we get

$$v' + (b + 2y_1c)v = -c.$$

4. Let y_1 and y_2 both satisfy the equation

$$y'' + p(x)y' + q(x)y = 0.$$

Show that if

$$\frac{y_1'(x_0)}{y_1(x_0)} = \frac{y_2'(x_0)}{y_2(x_0)}$$

at any point x_0, then

$$\frac{y_1'(x)}{y_1(x)} = \frac{y_2'(x)}{y_2(x)}$$

for all x. Show also that then $y_1 = cy_2$, i.e., y_1 and y_2 are linearly dependent.

5. If one applies the method of separation of variables (see Chapter 5) to the Helmholtz equation of three spatial dimensions in the spherical coordinates, one gets

$$\frac{1}{\rho^2}\frac{d}{d\rho}\left(\rho^2\frac{dy}{d\rho}\right) + \left(k^2 - \frac{l(l+1)}{\rho^2}\right)y = 0.$$

Find a change of the dependent variable so that the resulting equation has no term proportional to

$$\frac{dY}{d\rho},$$

where Y is the new dependent variable. Express the general solution of this equation by Bessel functions.

6. Consider the partial differential equation

$$i\frac{\partial}{\partial t}\Psi(x,t) = -\frac{\partial^2}{\partial x^2}\Psi(x,t) - \frac{1}{4}g(t)x^2\Psi(x,t).$$

a. Let the initial condition for Ψ be

$$\Psi(x,0) = \sqrt{\frac{1}{\pi\epsilon}}e^{-\frac{(x-x_0)^2}{\epsilon}};$$

find $\Psi(x,t)$.
Hint: Try $\Psi(x,t) = e^{a(t)x^2+b(t)x+c(t)}$.

b. Formulate the solution you obtain in (a) in the limit $\epsilon \to 0$. What is the meaning of this solution?

c. Find the solution in a closed form if $g(t) = e^{-At}$.

In a course, we assigned Problem 6 as an optional problem in lieu of the first exam. Three of the students out of a class of 24 handed in their solutions. All of them managed to solve parts (a) and (c) (for which they should be commended) but not part (b). The solution of this problem is below.

Solution for Problem 6

a. We shall try to solve this equation by an ansatz: We assume that the solution is of the form $\Psi(x,t) = e^{a(t)x^2+b(t)x+c(t)}$, and substitute into the PDE. We get, using (1.8),

$$i(\dot{a}x^2 + \dot{b}x + \dot{c}) = -\left(\frac{\partial}{\partial x} + 2ax + b\right)(2ax+b) - \frac{1}{4}gx^2,$$

or

$$i(\dot{a}x^2 + \dot{b}x + \dot{c}) = -2a - (2ax+b)^2 - \frac{1}{4}gx^2.$$

Equating the coefficients of x^2, x, and 1 of the two sides of the equation, we get

$$i\dot{a} = -4a^2 - \frac{1}{4}g, \tag{1.32}$$

$$i\dot{b} = -4ab, \tag{1.33}$$

and

$$i\dot{c} = -2a - b^2. \tag{1.34}$$

The initial conditions are

$$a(0) = -\frac{1}{\epsilon}, \tag{1.35}$$

$$b(0) = \frac{2x_0}{\epsilon}, \tag{1.36}$$

and

$$c(0) = -\frac{x_0^2}{\varepsilon} - \frac{1}{2}\ln(\pi\epsilon). \tag{1.37}$$

Equation (1.30) is a Riccati equation. By (1.25), we set

$$a = -\frac{\dot{u}}{4iu}; \tag{1.38}$$

then

$$\ddot{u}(t) = g(t)\, u(t), \tag{1.39}$$

which is the Newton equation for a particle with mass $\frac{1}{2}$ and coordinate u moving under the influence of the potential $-\frac{g}{4}u^2$.
Let the two independent solutions of (1.37) be $u_1(t)$ and $u_2(t)$, with $u_1(t)$ and $u_2(t)$ satisfying the initial conditions

$$u_1(0) = 0,\ \dot{u}_1(0) = 1 \tag{1.40}$$

and

$$u_2(0) = 1,\ \dot{u}_2(0) = 0. \tag{1.41}$$

We choose the linear superposition

$$u(t) = u_1(t) + \frac{\epsilon}{4i}u_2(t),$$

which we substitute into (1.36). We get

$$a(t) = -\frac{1}{4i}\frac{\dot{u}_1(t) + \frac{\epsilon}{4i}\dot{u}_2(t)}{u_1(t) + \frac{\epsilon}{4i}u_2(t)}. \tag{1.42}$$

The solution u has been chosen so that the initial condition (1.33) is satisfied.
Since

$$4i\int^t a(t')dt' = -\ln\left[u_1(t) + \frac{\epsilon}{4i}u_2(t)\right],$$

we have from (1.31) and (1.34) that

$$b(t) = e^{4i\int^t a(t')dt'} = \frac{x_0}{2i}\frac{1}{u_1(t) + \frac{\epsilon}{4i}u_2(t)}. \tag{1.43}$$

And we have from (1.32) that

$$c(t) = \ln \frac{1}{\sqrt{u_1(t) + \frac{\epsilon}{4i} u_2(t)}} - \int_0^t \frac{\frac{i x_0^2}{4} dt'}{\left[u_1(t') + \frac{\epsilon}{4i} u_2(t')\right]^2}$$
$$- \frac{x_0^2}{\epsilon} - \frac{1}{2} \ln(4\pi i), \tag{1.44}$$

which satisfies the initial condition (1.35).

Anticipating part (b) of this problem, we note that (1.44) is not convenient for obtaining $c(t)$ in the limit $\epsilon \to 0$. First of all, there is a term

$$-\frac{x_0^2}{\epsilon}$$

in (1.44), which blows up as $\epsilon \to 0$. Also, the integral in (1.44) is infinite if we set $\epsilon = 0$. We shall show that the infinities cancel, and that the function $c(t)$ is finite in the limit $\epsilon \to 0$.

First, it is easy to see that the integral in (1.44) is infinite at $\epsilon = 0$. This is because, at $\epsilon = 0$, this integral is equal to

$$-\int_0^t \frac{\frac{i x_0^2}{4} dt'}{u_1^2(t')}.$$

By (1.38), $u_1(t') \approx t'$ when t' is small. Thus the integral above diverges at the lower limit of integration.

To see that the term

$$-\frac{x_0^2}{\epsilon}$$

in $c(t)$ is canceled by a corresponding term in the integral in (1.44), we need to calculate the asymptotic form of this integral when ϵ is very small. As we have noted, the divergence of the integral comes from very small values of t'. Thus the asymptotic form of this integral can be calculated by making the approximation

$$u_1(t') + \frac{\epsilon}{4i} u_2(t') \approx t' + \frac{\epsilon}{4i},$$

which is valid at $t' \approx 0$. The integral in (1.44) is then approximately equal to

$$-\int_0^t \frac{\frac{ix_0^2}{4}dt'}{\left(t' + \frac{\epsilon}{4i}\right)^2} = \frac{x_0^2}{\epsilon} + \frac{\frac{ix_0^2}{4}}{t + \frac{\epsilon}{4i}} \approx \frac{x_0^2}{\epsilon},$$

which cancels the term

$$-\frac{x_0^2}{\epsilon}$$

in (1.44).

To facilitate the evaluation of $c(t)$ in limit $\epsilon \to 0$, we shall try to change the integral in (1.44) into something more manageable. We note that

$$\begin{aligned}\int_0^t \frac{dt'}{u^2(t')} &= -\int_0^t \frac{1}{\dot{u}(t')} d\frac{1}{u(t')} \\ &= -\frac{1}{\dot{u}(t)u(t)} + \frac{1}{\dot{u}(0)u(0)} - \int_0^t dt' \frac{\ddot{u}(t')}{\left[\dot{u}(t')\right]^2 u(t')}.\end{aligned}$$

If $u(t)$ satisfies (1.37), we get

$$\int_0^t \frac{dt'}{u^2(t')} = -\frac{1}{\dot{u}(t)u(t)} + \frac{1}{\dot{u}(0)u(0)} - \int_0^t dt' \frac{g(t')}{\left[\dot{u}(t')\right]^2}. \tag{1.45}$$

Replacing $u(t)$ with $u_1(t) + \frac{\epsilon}{4i}u_2(t)$, we find that the integrand of the integral in (1.45) does not blow up at $t' = 0$ if we set $\epsilon = 0$. This is because $\dot{u}(0) = 1$ at $\epsilon = 0$.

From (1.44) and (1.45) we get

$$\begin{aligned}c(t) = &-\frac{1}{2}\ln\left[u_1(t) + \frac{\epsilon}{4i}u_2(t)\right] \\ &+ \frac{\frac{ix_0^2}{4}}{u_1(t) + \frac{\epsilon}{4i}u_2(t)} \frac{1}{\dot{u}_1(t) + \frac{\epsilon}{4i}\dot{u}_2(t)} \\ &+ \int_0^t dt' \frac{\frac{ix_0^2}{4}g(t')}{\left[\dot{u}_1(t') + \frac{\epsilon}{4i}\dot{u}_2(t')\right]^2} - \frac{1}{2}\ln(4\pi i).\end{aligned} \tag{1.46}$$

b. One may have doubts about the generality of the solution obtained by using an ansatz. This is because it can give us only some, not all, of the solutions of the PDE. In the present case, the ansatz we use in (a) gives us only the solutions in the form of a Gaussian.

But it turns out that we may, with the result in (a), obtain the solution of the PDE satisfying any initial conditions.

The point is that the initial value $\Psi(x,0)$ treated in (a) is, in the limit $\epsilon \to 0$,

$$\lim_{\epsilon\to 0}\Psi(x,0)=\lim_{\epsilon\to 0}\sqrt{\frac{1}{\pi\epsilon}}e^{-\frac{(x-x_0)^2}{\epsilon}}=\delta(x-x_0),$$

the Dirac delta function. (See problem 3 in Chapter 4.) Thus the solution $\Psi(x,t)$ in the limit $\epsilon \to 0$ is the Green function for the PDE. We may obtain the solution of the PDE satisfying *any* initial condition with the use of this Green function. (See eq. 4.42 in Chapter 4.)

We have from (1.40), (1.41), and (1.44) that, in the limit $\epsilon \to 0$,

$$a(t)=-\frac{1}{4i}\frac{\dot{u}_1(t)}{u_1(t)}, \tag{1.47}$$

$$b(t)=\frac{x_0}{2i}\frac{1}{u_1(t)}, \tag{1.48}$$

and

$$c(t)=\ln\frac{1}{\sqrt{4\pi i u_1(t)}}+\frac{\frac{ix_0^2}{4}}{u_1(t)\dot{u}_1(t)}+\frac{ix_0^2}{4}\int_0^t\frac{dt'g(t')}{\dot{u}_1^2(t')}, \tag{1.49}$$

each of which is finite.

c. If $g(t)=e^{-At}$, then (1.39) becomes

$$\ddot{u}(t)=e^{-At}u(t).$$

Let

$$\frac{A}{2}\tau\equiv e^{-\frac{1}{2}At},\ t=-\frac{2}{A}\ln\frac{A\tau}{2};$$

then

$$\frac{d}{dt}=-\frac{A\tau}{2}\frac{d}{d\tau},$$

and the Newton equation is

$$\frac{du}{d\tau} + \frac{1}{\tau}\frac{du}{d\tau} = u.$$

The equation above is the modified Bessel equation of order zero. Thus

$$u_1(t) = \frac{K_0\left(\frac{2}{A}\right) I_0\left(\frac{2}{A}e^{-\frac{1}{2}At}\right) - I_0\left(\frac{2}{A}\right) K_0\left(\frac{2}{A}e^{-\frac{1}{2}At}\right)}{W\left(\frac{2}{A}\right)}, \tag{1.50}$$

and

$$u_2(t) = \frac{K_0'\left(\frac{2}{A}\right) I_0\left(\frac{2}{A}e^{-\frac{1}{2}At}\right) - I_0'\left(\frac{2}{A}\right) K_0\left(\frac{2}{A}e^{-\frac{1}{2}At}\right)}{W\left(\frac{2}{A}\right)}, \tag{1.51}$$

where

$$W\left(\frac{2}{A}\right) \equiv K_0'\left(\frac{2}{A}\right) I_0\left(\frac{2}{A}\right) - I_0'\left(\frac{2}{A}\right) K_0\left(\frac{2}{A}\right).$$

The Wronskians W can be explicitly evaluated. We have from (1.29) that

$$W(x) = \frac{c}{x}.$$

To evaluate c, we note that

$$c = xW(x).$$

Since c is a constant, independent of x, we may find its value by taking the limit $x \to \infty$ of the right side of the equation above. The asymptotic forms of the modified Bessel functions are, as $x \to \infty$,

$$K_0(x) \approx \sqrt{\frac{\pi}{2x}}e^{-x},$$

and

$$I_0(x) \approx \sqrt{\frac{1}{2\pi x}}e^{x}.$$

Thus

$$c = \lim_{x\to\infty} x\left[K_0'(x)I_0(x) - I_0'(x)K_0(x)\right] = -1.$$

Therefore,

$$W\left(\frac{2}{A}\right) = -\frac{A}{2}.$$

Thus we have

$$u_1(t) = \frac{2}{A}\left[I_0\left(\frac{2}{A}\right)K_0\left(\frac{2}{A}e^{-\frac{1}{2}At}\right) - K_0\left(\frac{2}{A}\right)I_0\left(\frac{2}{A}e^{-\frac{1}{2}At}\right)\right], \tag{1.52}$$

and

$$u_2(t) = \frac{2}{A}\left[I_0'\left(\frac{2}{A}\right)K_0\left(\frac{2}{A}e^{-\frac{1}{2}At}\right) - K_0'\left(\frac{2}{A}\right)I_0\left(\frac{2}{A}e^{-\frac{1}{2}At}\right)\right]. \tag{1.53}$$

The functions $a(t)$, $b(t)$, and $c(t)$ for $g(t) = e^{-At}$ are obtained by substituting (1.52) and (1.53) into the expressions for these functions. If t is very large, the arguments of the modified Bessel functions in (1.52) and (1.53) are very small. Since, when ρ is small,

$$I_0(\rho) \approx 1, \qquad K_0(\rho) \approx -\ln\frac{\rho}{2} - \gamma,$$

where γ is the Euler constant, we have, again when t is very large,

$$u_1(t) \approx I_0\left(\frac{2}{A}\right)t,$$

where a constant term has been neglected. Thus, in the limit $\epsilon \to 0$, we have from (1.45), (1.46), and (1.47) that when t is very large,

$$a(t) \approx -\frac{1}{4it},$$

$$b(t) \approx \frac{x_0}{2iI_0(\frac{2}{A})t},$$

and

$$c(t) \approx -\frac{1}{2}\ln t.$$

Hence the Green function at very large t is approximately equal to a constant times

$$\frac{1}{\sqrt{t}} \exp\left(i \frac{I(\frac{2}{A})x^2 - 2x_0 x}{4I(\frac{2}{A})t} \right).$$

Chapter 2

Complex Analysis

A. Complex Numbers and Complex Variables

In this chapter, I give a short discussion of complex numbers and the theory of a function of a complex variable.

Before we get to complex numbers, let me first say a few words about real numbers.

All real numbers have meanings in the real world. Ever since the beginning of civilization, people have found great use of real positive integers, say 2 and 30, which came up in conversations such as "my neighbor has two pigs, and I have thirty chickens." The concept of a negative real integer, say -5, is not quite as easy, but it became relevant when a person owed another person five copper coins. It was also natural to extend the concept of integers to rational numbers. For example, when six persons share equally a melon, the number describing the fraction of melon each of them has is not an integer but the rational number $1/6$. When we add, subtract, multiply, or divide integers or rational numbers, the result is always an integer or a rational number.

But the need for other real numbers came up as mathematicians pondered the length of the circumference of, say, a circular city wall. To express this length, the real number π must be introduced. This real number is neither an integer nor a rational number, and is called an irrational number. Another well-known irrational number found by mathematicians is the constant e.

Each of the real numbers, be it positive or negative, rational or irrational, can be geometrically represented by a point on a straight line. The converse is also true: a point on a straight line can always be represented by a real number.

When we add, subtract, multiply, or divide two real numbers, the outcome is always a real number. Thus the root of the linear equation

$$ax + b = c,$$

with a, b, and c real numbers, is always a real number. That is to say that if we make nothing but linear algebraic operations of real numbers, what comes out is invariably a real number. Thus the real numbers form a closed system under linear algebraic operations.

But as soon as we get to nonlinear operations, the system of real numbers alone becomes inadequate. As we all know, there are no real numbers that satisfy the quadratic equation

$$x^2 = -1.$$

Thus we use our imagination and denote i as a root of this equation. While we have gotten to be comfortable with the imaginary number i, the concept of the imaginary number was not always easy. Indeed, even Gauss once remarked that the "true metaphysics" of i was "hard."

The number

$$\alpha = a + ib,$$

where a and b are real numbers, is called a complex number. The numbers a and b are called the real part and the imaginary part of α, respectively.

While complex numbers might have once appeared to have no direct relevance in the real world, people have since found that the use of complex numbers enables them to handle more easily many physical problems in classical physics. For example, electrical engineers use the imaginary number i extensively, except that they call it j. And at the turn of the twentieth century, complex numbers became almost indispensable with the invention of quantum mechanics.

Let us enter the never-never land of the complex variable z denoted by

$$z = x + iy,$$

where x and y are real variables and

$$i^2 = -1.$$

The complex conjugate of z will be denoted as

$$z^* = x - iy.$$

The variable z can be represented geometrically by the point (x, y) in the Cartesian two-dimensional plane. In complex analysis, this same two-dimensional plane is called the complex plane. The x axis is called the real axis, and the y axis is called the imaginary axis. Let r and θ be the polar coordinates. Then we have

$$x = r\cos\theta, \quad y = r\sin\theta\ (r \geq 0).$$

The variable θ is defined modulo an integral of 2π. For many functions, the common way is to define θ to be either between 0 and 2π or between $-\pi$ and π. This will be further discussed in Section E of this chapter.

Expressed by the polar coordinates, z is

$$z = r(\cos\theta + i\sin\theta). \tag{2.1}$$

The Euler's formula says

$$e^{i\theta} = \cos\theta + i\sin\theta. \tag{2.2}$$

Thus we have

$$z = re^{i\theta}. \tag{2.3}$$

This is known as the polar form of z. The quantity $r = \sqrt{x^2 + y^2}$ is called the absolute value or the magnitude of z, which is also expressed as $|z|$. The quantity $\theta = \tan^{-1} y/x$ is called the argument or the phase of z.

Incidentally, (2.2) shows that $\cos\theta$ and $\sin\theta$ are respectively the real part and the imaginary part of $e^{i\theta}$, provided that θ is real. Note that the absolute value of $e^{i\theta}$ is

$$\sqrt{\cos^2\theta + \sin^2\theta} = 1.$$

If we set $\theta = 2n\pi$ in (2.2), where n is an integer, we get

$$e^{i2n\pi} = 1.$$

This result can be understood geometrically. The complex number $e^{i2n\pi}$ has the phase $2n\pi$, and is hence located on the positive real axis. This complex number has the magnitude unity, and is hence one unit distance away from the origin. Therefore it is equal to 1.

Setting θ in (2.2) to $(2n + 1)$, where n is an integer, we get

$$e^{i(2n+1)\pi} = -1.$$

Geometrically, $e^{i(2n+1)\pi}$ has the magnitude unity and the phase $(2n+1)\pi$, and is hence located at the negative real axis one unit distance away from the origin. Therefore it is equal to -1.

Setting $n = 1$, we have

$$e^{i\pi} = -1.$$

As we have mentioned, e and π are irrational numbers and there is no simple formula connecting these two numbers. Yet after we introduce the imaginary number i, which is a figment of our imagination, these three numbers are neatly joined together.

Since the argument of z is defined modulo an integral multiple of 2π, the polar form (2.3) can be written as

$$z = re^{i(\theta+2n\pi)}, \tag{2.4}$$

where n is an integer. Indeed, we have just shown that the factor $e^{i2n\pi}$ in (2.4) is equal to unity and hence (2.4) agrees with (2.3).

Problem for the Reader

Where is the complex number $(1 + i)$ in the complex plane? What are the phase and the magnitude of $(1 + i)$?

Solution

The Cartesian coordinates of the complex number $(1+i)$ are $(x, y) = (1, 1)$. Thus we put a dot on the point $(1, 1)$ in the xy-plane to represent this complex number geometrically.

Inspecting the location of this dot, we find that the phase and the magnitude of $1 + i$ are $\pi/4$ and $\sqrt{2}$, respectively.

The polar form is particularly convenient to use for carrying out the operations of multiplication or division of complex numbers. Let

$$z_1 = r_1 e^{i\theta_1}, \ z_2 = r_2 e^{i\theta_2},$$

then

$$z_1 z_2 = r_1 r_2 e^{i(\theta_1+\theta_2)}, \ z_1/z_2 = (r_1/r_2)e^{i(\theta_1-\theta_2)}.$$

The formulae above say that the absolute value of the product (ratio) of two complex numbers is equal to the product (ratio) of the absolute values of these complex numbers, and the phase of the product (ratio) of two complex numbers is equal to the sum (difference) of the phases of these complex numbers. These operations would have been a little more cumbersome to carry out if we had expressed the complex numbers with the Cartesian form.

Needless to say, using the polar form to do multiplication and division of more factors of complex numbers is even more laborsaving. In particular, we have

$$z^m = r^m e^{im\theta} = r^m(\cos m\theta + i \sin m\theta).$$

Setting $r = 1$, we obtain from the formula above that

$$(\cos\theta + i\sin\theta)^m = \cos m\theta + i \sin m\theta.$$

This single identity contains a number of identities we are familiar with. For example, choosing m to be 2, we get from this identity

$$\cos^2\theta - \sin^2\theta + 2i\cos\theta\sin\theta = \cos 2\theta + i\sin 2\theta.$$

By equating the real parts as well as the imaginary parts of the two sides of the equation above, we get the familiar identities

$$\cos 2\theta = \cos^2\theta - \sin^2\theta$$

and

$$\sin 2\theta = 2\cos\theta\sin\theta,$$

which express $\cos 2\theta$ and $\sin 2\theta$ as quadratic forms of $\cos\theta$ and $\sin\theta$.

Problem for the Reader

Find the roots of the equation

$$e^{e^z} = 1.$$

Solution

Let us call

$$\omega \equiv e^z;$$

then the equation under consideration is

$$e^{\omega} = 1.$$

This problem is easily solved if we write the equation as

$$e^{\omega} = e^{2n\pi i},$$

where n is any integer, and equate the exponents of the two sides of the equation. We get

$$\omega = 2n\pi i,$$

or

$$e^{z} = 2n\pi i,$$

where n is any integer. We shall solve the equation above for z. If $n = 0$, the equation above is

$$e^{z} = 0.$$

Since $|e^z| = e^x$, which is never equal to zero unless x is equal to minus infinity, $e^z = 0$ has no finite root.

If $n > 0$, the equation above is

$$e^{z} = e^{\ln(2n\pi)} e^{i\pi/2} e^{2m\pi i}.$$

Equating the exponents of the two sides, we get

$$z = \ln(2n\pi) + i\pi(1/2 + 2m),$$

where m is any integer.

Similarly, if $n < 0$, we get

$$z = \ln(2\,|n|\,\pi) + i\pi(-1/2 + 2m).$$

There are a doubly infinite number of solutions for equation $e^{e^z} = 1$.

Problem for the Reader

Find the phase and the magnitude of the power function z^a, where a is a number that is not an integer.

Solution

We again use the polar form (2.4) for z and get

$$z^a = \left[re^{i(\theta+2n\pi)}\right]^a = r^a e^{i(\theta+2n\pi)a}, \ (n = 0, \pm 1, \pm 2, \cdots).$$

The identity above shows that the magnitude of z^a is r^a, and the phase of z^a is equal to $(\theta + 2n\pi)a$, which has infinitely many values. While n is an integer, na is not necessarily an integer. Thus $e^{i2n\pi a}$ is not necessarily equal to unity and z^a generally has infinitely many values.

Exceptions occur when a is a rational number. Consider for example $z^{1/2}$. Setting $a = 1/2$ and $z = 1$ in the expression above, we get

$$\begin{aligned} 1^{1/2} = e^{in\pi} &= 1, \ n \text{ even}, \\ &= -1, n \text{ odd}, \end{aligned}$$

which is the familiar result that the equation $z^2 = 1$, or $z = 1^{1/2}$, has two roots: 1 and -1.

Similarly, setting $a = 1/N$ and $z = 1$, we find that $1^{1/N}$ is equal to

$$e^{2n\pi i/N}, n = 0, 1 \cdots (N-1),$$

with the other integral values of n giving no new roots. This corresponds to the familiar result that the equation

$$\omega^N = 1$$

has N roots.

B. Analytic Functions

Consider the limit

$$\lim_{\Delta z \to 0} \frac{\Delta f}{\Delta z} \tag{2.5}$$

for a complex-value function $f(z)$, where

$$\Delta f = f(z + \Delta z) - f(z).$$

The limit of the ratio above, if it exists, is called the derivative of $f(z)$, and is denoted as $f'(z)$.

The function $f(z)$ is said to be analytic at z_0 if $f'(z)$ exists in a neighborhood of z_0.The function $f(z)$ is said to be analytic in a region R in the complex z-plane if $f'(z)$ exists for every point z in R.

While (2.5) resembles the definition of the derivative of a function of a real variable x

$$f'(x) = \lim_{\Delta x \to 0} \frac{\Delta f}{\Delta x},$$

there is a substantive difference between them. The point is that Δz has both a real part and an imaginary part, i.e., $\Delta z = \Delta x + i\Delta y$. Therefore, if $f(z)$ is to have a derivative, the limit of (2.5) is required to exist for any Δx and Δy, as long as both of them go to zero. There is no restriction, for example, on the ratio of $\Delta y/\Delta x$, which may take any value. This is a strong condition on the function $f(z)$.

A strong condition has strong consequences. Let

$$f(z) = u(x, y) + iv(x, y),$$

where u and v are the real part and the imaginary part of $f(z)$, respectively. Then the expression in (2.5) is

$$\lim_{\Delta z \to 0} \frac{\Delta u + i\Delta v}{\Delta x + i\Delta y}, \tag{2.6}$$

where

$$\Delta u = u(x + \Delta x, y + \Delta y) - u(x, y),$$

and similarly for Δv.

We first consider the limit of (2.6) with Δz real, i.e., $\Delta z = \Delta x$. Then the limit of (2.6) is equal to

$$\lim_{\Delta x \to 0} \left. \frac{\Delta u + i\Delta v}{\Delta x} \right|_{y \text{ fixed}} = u_x + iv_x, \tag{2.7}$$

where u_x, for example, is the partial derivative of u with respect to x. Next we consider the limit (2.6) with Δz purely imaginary, i.e., $\Delta z = i\Delta y$. We have

$$\lim_{\Delta y \to 0} \left. \frac{\Delta u + i\Delta v}{i\Delta y} \right|_{x \text{ fixed}} = \frac{(u_y + iv_y)}{i}. \tag{2.8}$$

If $f(z)$ has a derivative, the expressions of (2.7) and (2.8) are, by definition, the same. This requires that

$$u_x = v_y, \; u_y = -v_x. \tag{2.9}$$

The equations in (2.9) are known as the Cauchy-Riemann equations. The real part and the imaginary part of an analytic function must satisfy these equations.

While we have only required that the limit of (2.5) is the same with Δz either purely real or purely imaginary, it is straightforward to prove that this limit is the same for any complex Δz when the Cauchy-Riemann equations are obeyed. (See homework problem 1 of this chapter.)

Differentiating the first Cauchy-Riemann equation with respect to x, we get

$$u_{xx} = v_{yx}.$$

Differentiating the second Cauchy-Riemann equation with respect to y, we get

$$u_{yy} = -v_{xy}.$$

Adding these two equations together, we get

$$u_{xx} + u_{yy} = 0.$$

The equation above is called the Laplace equation. We have shown that the real part of an analytic function must satisfy the Laplace equation.

We may similarly prove that v, the imaginary part of an analytic function, also satisfies the Laplace equation.

The Laplace equation is an important equation in physical sciences. From what we have just discussed, one may find the solution of a two-dimensional Laplace equation satisfying certain boundary conditions by looking for the analytic function the real part (or the imaginary part) of which satisfies these boundary conditions.

A function satisfying the Laplace equation is said to be harmonic. Thus the real part and the imaginary part of an analytic function are always harmonic. We call u and v the harmonic conjugate of each other.

Problem for the Reader

Find the real part and the imaginary part of e^z and show that they satisfy the Cauchy-Riemann equations.

Solution

We have

$$e^z = e^{x+iy} = e^x e^{iy} = u(x,y) + iv(x,y).$$

Using Euler's formula for e^{iy}, we get

$$u(x,y) = e^x \cos y, \ v(x,y) = e^x \sin y.$$

Thus

$$u_x = e^x \cos y, \ u_y = -e^x \sin y,$$
$$v_x = e^x \sin y, \ v_y = e^x \cos y.$$

We see that u and v satisfy the Cauchy-Riemann equations for all values of z. Thus the function e^z is analytic for all values of z. Incidentally, a function that is analytic at all points in the finite complex plane is called an entire function of z.

Problem for the Reader

Is the function $f(z) = zz^*$ analytic?

Solution

For the function zz^*,

$$u(x,y) = x^2 + y^2, \ v(x,y) = 0.$$

We have

$$u_x = 2x, \ u_y = 2y, \ v_x = v_y = 0.$$

Thus the Cauchy-Riemann equations are not satisfied except at the origin, which is but a point, not a region. Since the derivative of the function exists for no neighborhood of the origin, it is not analytic, even at the origin.

An intuitive way to understand why zz^* is not analytic is to think of this function as dependent on z^*.We shall show that an analytic function is independent of z^*.

To see this, let us first mention that we usually think of z^* to be dependent on z, and a function of z and z^* is just a function of z. Indeed,

$$\Delta z = \Delta x + i\Delta y, \; \Delta z^* = \Delta x - i\Delta y.$$

Since the magnitude of Δz is equal to that of Δz^*, we conclude that if Δz is nonzero, Δz^* is nonzero, which makes it impossible to vary z when keeping z^* fixed.

But this is true because we have implicitly accepted the premise that both Δx and Δy are real. As a matter of fact, that $\Delta z^* = 0$ implies

$$\Delta x = i\Delta y.$$

Therefore, if Δx and Δy are allowed to be complex, it is possible to have Δz be nonzero with Δz^* equal to zero. For example, when Δy is real, Δz^* is zero if Δx is equal to the imaginary number $i\Delta y$. With this provision we may regard z and z^* as independent variables.

We have

$$x = \frac{z + z^*}{2}, \; y = \frac{z - z^*}{2i}.$$

A function of the two variables x and y can now be considered as a function of z and z^*. The partial differentiation with respect to z can be defined with the chain rule of partial differentiation. We have

$$\frac{\partial}{\partial z} = \frac{\partial x}{\partial z}\frac{\partial}{\partial x} + \frac{\partial y}{\partial z}\frac{\partial}{\partial y} = \frac{1}{2}\left(\frac{\partial}{\partial x} - i\frac{\partial}{\partial y}\right).$$

Similarly,

$$\frac{\partial}{\partial z^*} = \frac{1}{2}\left(\frac{\partial}{\partial x} + i\frac{\partial}{\partial y}\right).$$

Let us now get back to the point that prompted this discussion. Let f be a complex-valued function of x and y. Let the real part and the imaginary part of f be denoted as u and v, respectively. We have

$$\frac{\partial}{\partial z^*} f = \frac{1}{2}\left(\frac{\partial}{\partial x} + i\frac{\partial}{\partial y}\right)(u + iv) = \frac{u_x - v_y + i(u_y + v_x)}{2}.$$

If f is an analytic function, then u and v satisfy the Cauchy-Riemann equations and hence

$$\frac{\partial}{\partial z^*} f = 0.$$

This says that an analytic function is independent of z^*.

Problem for the Reader

Show that

$$\frac{\partial^2}{\partial x^2} + \frac{\partial^2}{\partial y^2} = 4\frac{\partial}{\partial z}\frac{\partial}{\partial z^*}.$$

Solution

$$4\frac{\partial}{\partial z}\frac{\partial}{\partial z^*} = \left(\frac{\partial}{\partial x} - i\frac{\partial}{\partial y}\right)\left(\frac{\partial}{\partial x} + i\frac{\partial}{\partial y}\right) = \left(\frac{\partial^2}{\partial x^2} + \frac{\partial^2}{\partial y^2}\right).$$

It follows that the Laplace equation can be written as $\frac{\partial^z}{\partial z \partial z^*}u = 0$, and the solution of the Laplace equation is the sum of a function of z and a function of z^*.

Next, we give a few examples of functions that are analytic. The power function z^n with n an integer is analytic. While this result may very well be expected, I will give it a proof below. We have, by expressing $(z + \Delta z)^n$ with the binomial expansion,

$$\lim_{\Delta z \to 0} \frac{(z + \Delta z)^n - z^n}{\Delta z} = \lim_{\Delta z \to 0} \frac{nz^{n-1}\Delta z + \cdots}{\Delta z},$$

where the terms unexhibited are at least as small as the square of Δz. The limit above exists for all Δz, and we get

$$\frac{d}{dz}z^n = nz^{n-1},$$

which is the same formula we learned in calculus. Thus the derivative of the power function z^n exists for all values of z. Therefore, z^n is analytic for all values of z, or is an entire function of z.

Since the power function z^n is analytic, so is the linear superposition of a finite number of power functions. Indeed, so is the sum of an infinite number of power functions, as long as this sum is absolutely convergent. Conversely, we shall show in Section C that a function analytic at a point z_0 always has a convergent Taylor series expansion around z_0.

We have learned in calculus that the Maclaurin series for $\sin x$ is

$$\sin x = x - x^3/(3!) + x^5/(5!) + \cdots.$$

Let us replace x in the series above by z and define $\sin z$ to be this series, i.e.,

$$\sin z \equiv z - z^3/(3!) + z^5/(5!) + \cdots . \tag{2.10}$$

Similarly, we define

$$\cos z = 1 - z^2/2! + z^4/4! + \cdots . \tag{2.11}$$

We shall use the ratio test to prove that the series in (2.11) is absolutely convergent for all values of z. Let the nth term in the series of (2.11) be a_n. Then the ratio

$$\frac{a_n}{a_{n-1}} = -\frac{z^2}{(2n-2)(2n-3)}$$

vanishes in the limit $n \to \infty$ for all values of z. Since the ratio of the nth term and the $(n-1)^{\text{th}}$ term vanishes as $n \to \infty$ for all z, the ratio test asserts that the series of (2.11) converges absolutely for all values of z. Therefore, $\cos z$ defined by (2.11) is meaningful for all z. Being the same as $\cos x$ when $z = x$, $\cos z$ is called the analytic continuation of $\cos x$ into the complex plane. The analytic continuation of a function from the real line into the complex plane is unique. (For more general considerations of analytic continuation of a function, see homework problem 9.) In the case of $\cos z$, this means that the function defined in (2.11) is the only function possible that is analytic everywhere and agrees with $\cos x$ when $z = x$.

Similarly, the series in (2.10) is absolutely convergent for all z. Therefore, $\sin z$ defined by (2.10) is the unique analytic continuation of $\sin x$ into the complex plane.

C. The Cauchy Integral Theorem

The contour integral

$$I = \int_c f(z)dz,$$

where c is a contour in the complex plane, is defined to be

$$\int_c (u + iv)(dx + idy) = \int_c (udx - vdy) + i \int_c (udy + vdx). \tag{2.12}$$

We note that the two integrals on the right side of (2.12) are line integrals in the two-dimensional plane, which we already encountered in calculus.

An example of a line integral is the work done by a force. As we know, the work done by moving a particle from (x, y) to $(x + \Delta x, y + \Delta y)$ against a force

$$\overrightarrow{F} = M(x, y)\overrightarrow{i} + N(x, y)\overrightarrow{j}$$

is equal to the scalar product of the force with the displacement vector

$$\Delta x\overrightarrow{i} + \Delta y\overrightarrow{j}.$$

Thus the work done is

$$M(x, y)\Delta x + N(x, y)\Delta y.$$

Therefore, if A and B are two points in the xy-plane, the work done in moving a particle from A to B along a path c against the force is equal to the line integral

$$\int_c (Mdx + Ndy).$$

We also recall that $\overrightarrow{F}$ is known as a conservative force if the curl of $\overrightarrow{F}$ vanishes. Equivalently, $\overrightarrow{F}$ is a conservative force if there exists a potential V such that

$$\overrightarrow{F} = -\overrightarrow{\nabla}V.$$

If $\overrightarrow{F}$ is conservative, the work done in moving a particle from one point to another depends only on the difference of the values of the potential at these two points, and is independent of the path. To say this more precisely, let the potential V exist in a region R in the two-dimensional plane; then

$$\int_{c_1} (Mdx + Ndy) = \int_{c_2} (Mdx + Ndy),$$

provided that c_1 and c_2 are two curves with the same endpoints and both lie inside R.

If the potential V exists, we have

$$M = -V_x, \ N = -V_y,$$

and hence

$$M_y = N_x. \tag{2.13}$$

The converse is indeed also true: if (2.13) holds in a region R, then the force is the gradient of a potential.

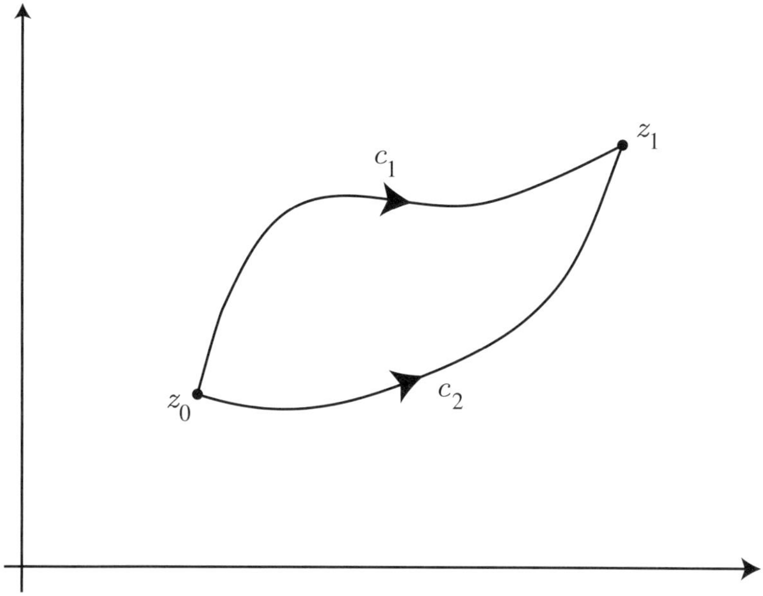

Figure 2.1

Now for the first line integral in (2.12), M is u and N is $-v$. Thus the condition (2.13) for this line integral is the second Cauchy-Riemann equation. For the second line integral in (2.12), M is v and N is u . Thus the condition (2.13) for this line integral is the first Cauchy-Riemann equation. The contour integral I in (2.12) is therefore path independent if $f(z)$ is analytic. More precisely, let c_1 and c_2 be two curves, both join the lower endpoint z_0 to the upper endpoint z_1 in the complex z-plane, and both lie inside the region R where $f(z)$ is analytic. Then we have

$$\int_{c_1} f(z)dz = \int_{c_2} f(z)dz. \tag{2.14}$$

Equation (2.14) tells us that we may deform the contour c_1 to the contour c_2, where c_1 and c_2 have the same endpoints, provided that $f(z)$ is analytic in the region lying between c_1 and c_2.

The contours c_1 and c_2 in (2.14) are open contours. We shall extend (2.14) to closed contours. Let c and c' be closed contours of the same sense of direction, i.e., either both counterclockwise or both clockwise, with no singularities of $f(z)$ lying between c and c'. We choose a point z_0 on c and think of the closed contour c as a contour joining the point z_0 to itself. Let us draw a line joining z_0 to a point z_0' on c', forming a bridge between c and c'. Then we may think of c' as another contour joining z_0 to itself. This is because c' is the contour that begins at

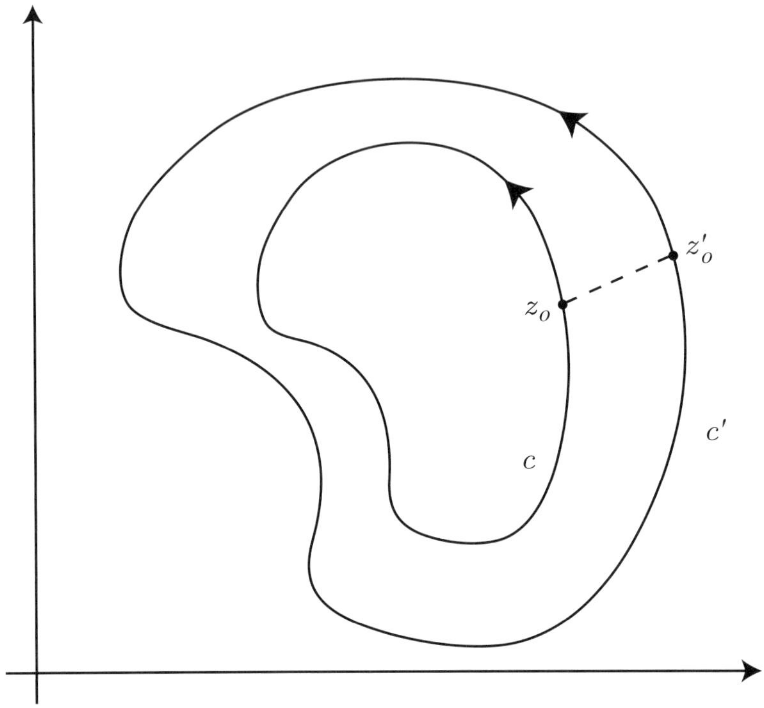

Figure 2.2

z_0, crosses the bridge to z'_0, and follows c' to return to z'_0, then crosses the bridge in the reverse direction to finally come back to z_0. As the bridge is crossed twice in opposite directions, the two contour integrals associated with the contour of the bridge cancel each other. Therefore, c' can also be considered as a closed contour joining z_0 to itself, and by (2.14) we have

$$\oint_c f(z)dz = \oint_{c'} f(z)dz, \tag{2.15}$$

where the symbol $\oint$ denotes an integration over a closed contour.

Equation (2.15) says that the contour c can be deformed into c' provided that $f(z)$ is analytic in the region lying between c and c'.

Let us go from z_0 to z_1 along contour c_1 in Figure 2.1, then go from z_1 back to z_0 along $-c_2$, which is c_2 in the reverse direction. The contour $c = c_1 - c_2$ is a closed contour. Thus (2.14) can be written as

$$\oint_c f(z)dz = 0 \tag{2.16}$$

provided that $f(z)$ is analytic in a region R and c is a closed contour c inside R. Equation (2.16) is the Cauchy integral theorem.

Next, we consider the integral

$$I_n = \oint_c \frac{dz}{(z-z_0)^n},$$

where c is a closed contour in the counterclockwise direction and n is a positive integer. The integrand blows up at $z = z_0$. We say that the integrand has a singularity at z_0. More generally, if a single-value function $f(z)$ is not analytic at point z_0, then we say that $f(z)$ has a singularity at z_0.

If c does not enclose z_0, I_n vanishes by Cauchy's integral theorem. But if c encloses z_0, as is illustrated in Figure 2.3, we may deform the contour into the circle C_R without crossing any singularity of the integrand, where C_R is the circle, the center of which is z_0, and the radius of which is R.

Now a point z on C_R satisfies

$$|z - z_0| = R,$$

and hence the polar form of $z - z_0$ is

$$z - z_0 = e^{i\theta} R.$$

Thus we get

$$dz = ie^{i\theta} R d\theta.$$

Therefore, we have

$$I_n = \frac{i}{R^{n-1}} \int_0^{2\pi} e^{i(1-n)\theta} d\theta.$$

The integral above is easily calculated. Indeed, we have

$$\begin{aligned} \int_0^{2\pi} e^{i(1-n)\theta} d\theta &= 2\pi,\ n = 1, \\ &= 0,\ n \neq 1. \end{aligned}$$

Thus we conclude that, if z_0 is inside the closed counterclockwise contour c, we have

$$\begin{aligned} I_n &= 2\pi i,\ n = 1, \\ &= 0,\ n \neq 1. \end{aligned} \tag{2.17}$$

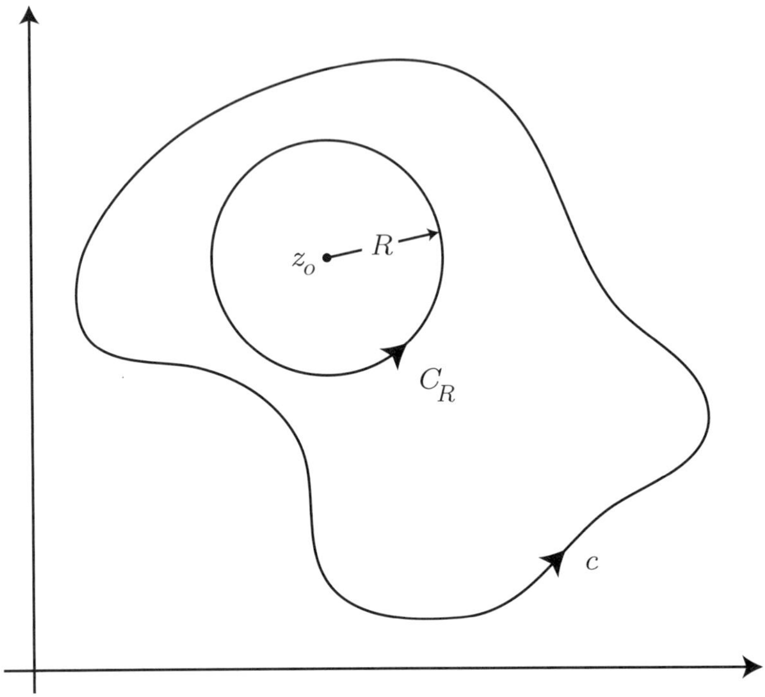

Figure 2.3

Equation (2.17) enables us to derive the Cauchy integral formula, which says that if $f(z)$ is analytic in the region R, and if z is an interior point of R, then we have

$$f(z) = \frac{1}{2\pi i} \oint_c \frac{f(z')}{z' - z} dz', \tag{2.18}$$

provided that c is a closed curve enclosing z once in the counterclockwise direction and lying inside R. To prove this, we deform the contour c into the circle c_ϵ with center at z and radius ϵ. This is allowed, as the integrand of (2.18) is analytic in the region lying between c and c_ϵ. As we make ϵ approach zero, z' approaches z and the integral approaches

$$\frac{f(z)}{2\pi i} \oint_{c_\epsilon} \frac{1}{z' - z} dz'.$$

By (2.17), Cauchy's integral formula is proved.

Differentiating the Cauchy integral formula with respect to z, we obtain the derivative of f as

$$f'(z) = \frac{1}{2\pi i} \oint_c \frac{f(z')}{(z' - z)^2} dz'.$$

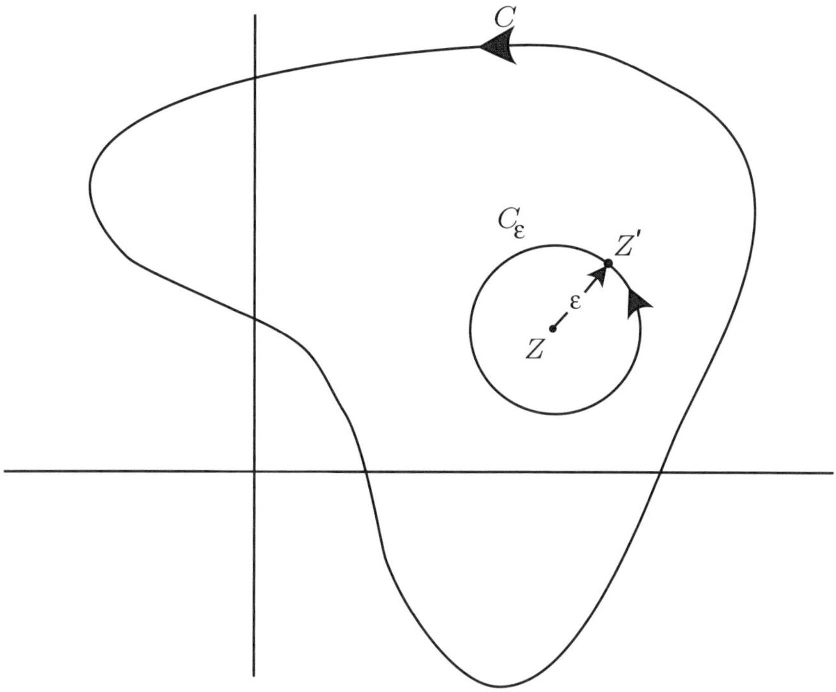

Figure 2.4

By differentiating the Cauchy integral formula n more times, we get

$$f^{(n)}(z) = \frac{n!}{2\pi i} \oint_c \frac{f(z')}{(z'-z)^{n+1}} dz'.$$

Thus if a function of a complex variable z is analytic, it has derivatives of all orders. But by definition, a function of a complex variable is analytic if it has the first derivative; thus a function of a complex variable has derivatives to all orders if it has the derivative of the first order.

This may seem surprising to readers who have learned in calculus that if a function of a real variable x has a first derivative, it does not necessarily have a second derivative, not to mention even higher-order derivatives. The seeming contradiction is resolved by the fact that the existence of the derivative of a function of a complex variable requires a stronger condition than that of a function of a real variable.

We may now prove that a function $f(z)$ analytic at z_0 has the Taylor series expansion

$$f(z) = \sum_{n=0}^{\infty} \frac{f^{(n)}(z_0)}{n!}(z - z_0)^n, \tag{2.19}$$

where z lies in a neighborhood of z_0 to be specified later. To prove (2.19), we choose c in (2.18) to be C_R, the circle with center at z_0 and radius R, where R is sufficiently large so that c encloses z. Now if z' is a point on C_R,

$$|z' - z_0| = R.$$

Also, since z is inside C_R,

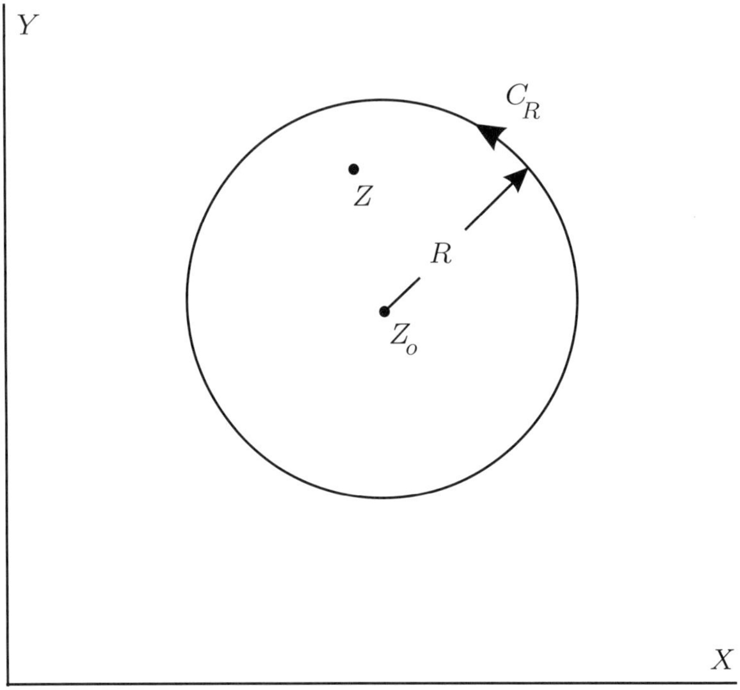

Figure 2.5

$$|z - z_0| < R.$$

Thus we have

$$|z - z_0|/|z' - z_0| < 1.$$

Now

$$(z' - z)^{-1} = \left[(z' - z_0) - (z - z_0)\right]^{-1} = (z' - z_0)^{-1}\left(1 - \frac{z - z_0}{z' - z_0}\right)^{-1}.$$

As we know, $(1 - \omega)^{-1}$ is equal to the convergent series

$$\sum_{n=0}^{\infty} \omega^n$$

provided that $|\omega| < 1$. Identifying ω with $(z - z_0)/(z' - z_0)$, we have,

$$(z' - z)^{-1} = (z' - z_0)^{-1} \sum_{n=0}^{\infty} \left(\frac{z - z_0}{z' - z_0}\right)^n.$$

Substituting the expression above into (2.18), we obtain the Taylor series expansion (2.19).

The contour C_R is a circle inside which $f(z)$ is analytic. Thus C_R is not allowed to enclose any singularity of $f(z)$. Let z_1 be the singularity of $f(z)$ closest to z_0; then the largest value of R possible is $|z_0 - z_1|$. Therefore, the radius of convergence of the Taylor series (2.19) is $|z_0 - z_1|$.

As an example, since $\sin n\pi$ vanishes for any integral value n, the function $z(\sin \pi z)^{-1}$ has singularities at $z = n$, where n is any integer not equal to zero. The point $z = 0$ is not a singularity of $z(\sin \pi z)^{-1}$ because the numerator of this function vanishes at $z = 0$. Indeed, by l'Hopital's rule, $z(\sin \pi z)^{-1}$ is equal to π^{-1} at $z = 0$. Let us consider the Maclaurin series for $z(\sin \pi z)^{-1}$. Since the singularities of $z(\sin \pi z)^{-1}$ closest to the origin are $z = \pm 1$, this series converges in a circle with center at the origin and radius unity.

As another example, we have shown that the function e^z is an entire function. Therefore, we know that its Maclaurin series

$$e^z = 1 + \frac{z}{1!} + \frac{z^2}{2!} + \cdots \tag{2.20}$$

converges for all finite values of z. This may also be directly verified with the use of the ratio test.

Problem for the Reader

Show that

$$\sin(iz) = i \sinh z, \ \cos(iz) = \cosh z.$$

Solution

From the Maclaurin series (2.10) for $\sin z$ we get

$$\sin(iz) = i\left[z + z^3/(3!) + z^5/(5!) + \cdots\right] = i \sinh z.$$

In a similar way, we get

$$\cos(iz) = 1 + z^2/2! + z^4/4! + \cdots = \cosh z.$$

From (2.10), (2.11), and (2.20), we find

$$e^{iz} = \cos z + i \sin z, \tag{2.21}$$

which is Euler's formula for a complex argument. Replacing z in (2.21) by $-z$, we have

$$e^{-iz} = \cos z - i \sin z.$$

Therefore,

$$\cos z = \frac{e^{iz} + e^{-iz}}{2}, \tag{2.22}$$

and

$$\sin z = \frac{e^{iz} - e^{-iz}}{2i}. \tag{2.23}$$

When $z = x$, (2.21), (2.22), and (2.23) are identities already established in calculus. We now see that they also hold when z is complex.

As we have mentioned, an equality between analytic functions valid for real $z = x$ is invariably valid when z is complex. This is a result of the uniqueness of analytic continuation. But a nonanalytic relation that holds when $z = x$ often does not hold when z is complex. For example, the inequality

$$-1 \leq \sin x \leq 1$$

does not hold when x is replaced by the complex variable z. Indeed, if z is complex, $\sin z$ is not even a real number. Note also that $\sin z$ is not the imaginary part of e^{iz} when z is complex, nor is $\cos z$ the real part of e^{iz} when z is complex.

If $f(z)$ is not analytic at z_0, it has no derivative at z_0 and naturally has no Taylor series expansion around z_0. But it may have another kind of series expansion around z_0. As an example, consider the function $f(z) = e^{1/z}$, which is not analytic at $z = 0$. Nevertheless, we have by (2.20) that

$$e^{1/z} = 1 + \frac{1}{z} + \frac{1}{2!z^2} + \cdots . \tag{2.24}$$

The series above is not a Taylor series, as it is not a sum of positive power functions. Instead, it is an example of a Laurent series defined by ((2.25)) below. Since the series for e^z converges for all finite z, the Laurent series for $e^{1/z}$ converges for all $z \neq 0$.

More generally, an analytic function has a Laurent series expansion around an isolated singularity, which we shall define as follows. Let z_0 be a singularity of $f(z)$, and z_1 be the singularity of $f(z)$ closest to z_0. If $|z_0 - z_1|$ is not equal to zero, then z_0 is called an isolated singularity of $f(z)$. Not all singularities of an analytic function are isolated. For example, the singularites of the function $1/\sin(\pi/z)$ are located at $z = 1/n$, $n = 0, \pm 1, \pm 2, \cdots$, where $\sin(\pi/z)$ vanishes. Since $1/n$ for n arbitrarily large is arbitrarily close to the origin, the point $z = 0$ is not an isolated singularity of $1/\sin(\pi/z)$.

If z_0 is an isolated singularity of $f(z)$, we may prove with the use of the Cauchy integral formula that $f(z)$ has the Laurent series expansion

$$f(z) = \sum_{n=-\infty}^{\infty} a_n (z - z_0)^n, \tag{2.25}$$

which contains not only positive powers of $(z - z_0)$, but also negative powers of $(z - z_0)$.

The series (2.25) is convergent at every point, with the exception of the point z_0, inside the circle with the center at z_0 and with the radius $|z_0 - z_1|$, where z_1 is the singularity of $f(z)$ closest to z_0. This can again be proven with the use of the Cauchy integral formula (2.18). (See homework problem 12.)

We note that the difference between a Taylor series and a Laurent series is that the latter has negative power functions $(z - z_0)^{-m}$, where $m > 0$. These power functions blow up at $z = z_0$. Indeed, the larger m is, the faster the power function blows up as z approaches z_0. If the term in the series that blows up the fastest is a $(z - z_0)^{-N}$ term, then we say that the function has a pole of order N at z_0.

If the Laurent series of $f(z)$ has nonvanishing $(z - z_0)^{-m}$ terms of arbitrarily large m, $f(z)$ is said to have an essential singularity at z_0. An example is the series of (2.24), which has an essential singularity at the origin.

Let us integrate the Laurent series over a closed contour c that encloses z_0 but no other singularities of $f(z)$. Let the direction of c be counterclockwise. By (2.15), all terms in this series except the term

$$a_{-1}/(z - z_0)$$

are integrated to zero. Thus we get

$$\oint_c f(z)dz = 2\pi i a_{-1}. \tag{2.26}$$

This is known as the Cauchy residue theorem. The coefficient a_{-1} is known to be the residue of $f(z)$ at z_0, which we shall denote as $\text{Res}(z_0)$. If the contour is clockwise, the integral will be equal to the negative of $2\pi i$ times the residue.

This formula is one of the most useful formulae in complex analysis. It tells us that the value of an integral over a closed contour can be obtained by simply evaluating the residue of its integrand.

If the contour c encloses more than one singularity of $f(z)$, we replace the right side of (2.26) by the sum of residues of $f(z)$ at these singularities. (Why?)

Before we close this section, let us show how to evaluate efficiently the residue of $f(z)$ at z_0 where the function has a pole of the first order, which is called a simple pole. If the singularity of $f(z)$ at z_0 is a simple pole,

$$f(z) = \frac{a_{-1}}{z - z_0} + a_0 + a_1(z - z_0) + \cdots .$$

Thus the residue of $f(z)$ at z_0 is equal to

$$\text{Res}(z_0) = \lim_{z \to z_0} (z - z_0) f(z). \tag{2.27a}$$

As an example, let us calculate the residue of $e^z/\sin z$ at $z = 0$. We have by (2.27a) that

$$\text{Res}(0) = \lim_{z \to 0} \frac{ze^z}{\sin z}.$$

We see that both the numerator and the denominator vanish as $z \to 0$; thus we apply l'Hopital's rule and get

$$\text{Res}(0) = \lim_{z \to -0} \frac{e^z + ze^z}{\cos z}.$$

Since ze^z vanishes as $z \to 0$, we have

$$\text{Res}(0) = \lim_{z\to 0} \frac{e^z}{\cos z} = 1.$$

Note that $e^z/\cos z$ is obtained from $e^z/\sin z$ by replacing the denominator of the latter function with the derivative of its denominator. Thus we may dispense with the formula (2.27a) and calculate the residue by differentiating the denominator.

More generally, let $f(z) = g(z)/h(z)$, where $g(z)$ and $h(z)$ are analytic at z_0. If $h(z)$ has a simple zero at z_0, then $f(z)$ has a simple pole at z_0. By (2.27a), we obtain with the use of l'Hopital's rule that the residue of $f(z)$ at z_0 is

$$\text{Res}(z_0) = \frac{g(z_0)}{h'(z_0)}. \tag{2.27b}$$

If $f(z)$ has a double pole at z_0, the Laurent series expansion of $f(z)$ in the neighborhood of z_0 is

$$f(z) = \frac{a_{-2}}{(z-z_0)^2} + \frac{a_{-1}}{z-z_0} + a_0 + a_1(z-z_0) + \cdots .$$

In this case, the limit on the right side of (2.27a) is equal to infinity, not the residue of $f(z)$ at z_0. To eliminate the singularity at z_0, we multiply $f(z)$ by $(z-z_0)^2$ and get

$$(z-z_0)^2 f(z) = a_{-2} + a_{-1}(z-z_0) + a_0(z-z_0)^2 + \cdots .$$

The expression above is finite in the limit $z \to z_0$, but this limit is equal to a_{-2}, not a_{-1}. To obtain a_{-1}, we differentiate the expression above and then set $z = z_0$. We get

$$a_{-1} = \lim_{z\to z_0} \frac{d}{dz}\left[(z-z_0)^2 f(z)\right].$$

D. Evaluation of Real Integrals

The Cauchy residue theorem provides us with a tool to evaluate a number of integrals in the real world, the integrands of which are functions of a real variable and the integration is over real values of the variable. Some of these integrations are difficult to carry out in closed form with the methods provided by calculus. We shall show that, by going into the never-never land of the complex plane, sometimes we can find the closed forms of these integrals.

As an example, let us consider the integral

$$I = \int_{-\infty}^{\infty} \frac{dx}{1+x^2}. \tag{2.28}$$

While this integral can be evaluated with calculus, we shall use it as an example to demonstrate how to do real integrals with contour integration.

We regard this integral as a contour integral over the real axis of the complex plane. But we cannot as yet apply the Cauchy residue theorem to it, as the real axis is not a closed contour. Let us think of the real axis as the contour from $-R$ to R along the real axis, in the limit as R approaches infinity. We add to this contour the counterclockwise semicircle in the upper half-plane with the origin as the center and R the radius. (See Figure 2.6.) Now we get a closed contour which we shall call c. As we shall see, the integral over the semicircle vanishes in the limit of $R \to \infty$. Thus the integral of (2.28) is equal to the integral over c. Since c is a closed contour we may apply the Cauchy residue theorem to the integral. The only singularity of the integrand enclosed by c is $z = i$. Thus we have by (2.27b),

$$I = 2\pi i \,\mathrm{Res}(i) = 2\pi i \frac{1}{2i} = \pi.$$

To finish the argument let us show that the contribution of the semicircle is zero in the limit $R \to \infty$. If z is a point on the semicircle,

$$z = e^{i\theta} R, \ 0 \leq \theta \leq \pi.$$

When R is very large, the integrand $1/(1+z^2)$ is approximately equal to $1/z^2$, the magnitude of which is $1/R^2$. We also have

$$dz = ie^{i\theta} R d\theta. \tag{2.29}$$

Thus we have

$$\int_{C_R} \frac{dz}{1+z^2} \approx \int_0^{\pi} \frac{ie\theta R d\theta}{R^2 e^{2i\theta}},$$

where C_R is the semicircle in the upper half-plane. In the limit $R \to \infty$, the integral above vanishes, as there are two factors of R in the denominator of the integrand and only one factor of R in the numerator of the integrand.

We may also close the contour of the integral in (2.28) by adding to it the semicircle in the lower half-plane in the clockwise direction. The only singularity enclosed by this contour is the one at $z = -i$. Thus we have

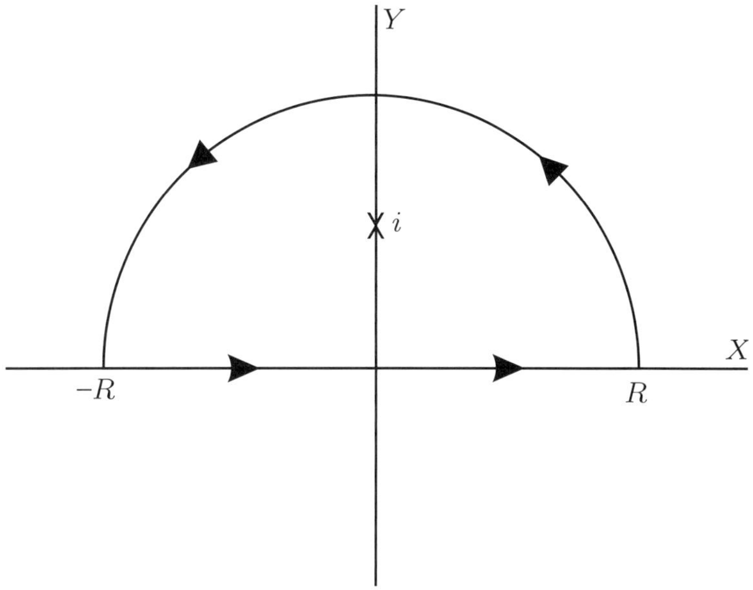

Figure 2.6

$$I = -2\pi i \,\mathrm{Res}(-i) = -2\pi i \frac{1}{-2i} = \pi,$$

which is the same answer. Note that the first minus sign above is due to the fact that the closed contour is clockwise.

One of the first things we do in applying the Cauchy residue theorem is to make sure that the contour is a closed one. If the contour is not closed, try to close it if possible. The second step is to locate the singularities of the integrand enclosed by the contour, and calculate the residues of the integrand at each of the singularities.

Problem for the Reader

Evaluate the integral

$$I = \int_{-\infty}^{\infty} \frac{dx}{(x-i)(x-2i)(x-3i)(x-4i)}.$$

Solution

We may close the contour upstairs by adding to the contour of integration the semicircle in the upper half-plane. This is justified, as the contribution of the semicircle of radius R is of the order of

$$R/R^4,$$

where the numerator factor R comes from dz given by (2.29) and the factor R^4 is the order of the denominator of the integrand when R is large. Thus the contribution of the semicircle vanishes as $R \to \infty$. With the semicircle included, the contour is closed and encloses the four singularities of the integrand located at i, $2i$, $3i$, and $4i$. Therefore, we may obtain the value of I by adding up the residues at these singularities.

It is far simpler, however, to evaluate this integral by closing the contour downstairs. The contribution from the semicircle in the lower half-plane vanishes as before. Since the integrand has no singularities in the lower half-plane, we get

$$I = 0.$$

For the integral

$$I = \int_{-\infty}^{\infty} \frac{dx}{(x+i)(x-2i)(x-3i)(x-4i)},$$

the integrand of which has three poles in the upper half-plane and one pole in the lower half-plane, it is easier if we close the contour downstairs. This is because if we do so, we need to evaluate only the residue at $z = -i$.

For other integrals, there is no way to close the contour before we make some changes. Consider

$$I = \int_{-\infty}^{\infty} \frac{\cos x}{1+x^2} dx. \tag{2.30}$$

We emphasize that it is impossible to close the contour for the integral of (2.30) either upstairs or downstairs. The culprit is the factor $\cos x$ in the numerator of the integrand. While $\cos x$ is of finite values between -1 and 1, no matter how large x is, $\cos z$ is very large when z is complex and large. To see this, we have from (2.22) that

$$\cos z = \frac{e^{ix-y} + e^{-ix+y}}{2}.$$

The first term on the right side of this equation blows up if y goes to $-\infty$, and the second term on the right side of this equation blows up if y goes to ∞. Thus $\cos z$ blows up exponentially when either $y \to \infty$ or $y \to -\infty$. As a result, the contribution from the infinite semicircles, either the one in the upper half-plane or the one in the lower half-plane, is not zero.

Now we note that

$$\cos x = \text{Re } e^{ix}.$$

Since $(1+x^2)$ is real, we have

$$I = \text{Re } J,$$

where

$$J = \int_{-\infty}^{\infty} \frac{e^{ix}}{1+x^2} dx.$$

It is possible to close the contour upstairs for the integral J, as $e^{iz} = e^{ix-y}$ is exponentially small when $y \to \infty$. Therefore, we are allowed to close the contour upstairs. A more detailed discussion of this can be found in Appendix B of Chapter 8. The only singularity of the integrand enclosed by this contour is at $z = i$. We easily get, using (2.27b),

$$J = 2\pi i \, \text{Res}(i) = 2\pi i \frac{e^{iz}}{2z}|_{z=i} = \frac{\pi}{e}.$$

Hence we have

$$I = \int_{-\infty}^{\infty} \frac{\cos x}{1+x^2} dx = \frac{\pi}{e}. \tag{2.31}$$

Alternately, we may make use of the relation

$$\cos x = \text{Re } e^{-ix}$$

and get

$$I = \text{Re } K,$$

where

$$K = \int_{-\infty}^{\infty} \frac{e^{-ix}}{1+x^2} dx.$$

Since $e^{-iz} = e^{-ix+y}$ vanishes as y goes to minus infinity, we may close the contour downstairs for the integral K and get the same answer for I.

Problem for the Reader

Evaluate the integral

$$I = \int_{-\infty}^{\infty} \frac{x \sin x}{1 + x^2} dx. \tag{2.32}$$

Solution

Again, the first step is to make the contour a closed contour. If this is done, we may evaluate the integral by the use of Cauchy's residue theorem.

But it is not possible to close the contour either upstairs or downstairs with the factor $\sin x$ in the numerator. This is because, just like the magnitude of $\cos z$, the magnitude of $\sin z$ is exponentially large if y, the imaginary part of z, is large, regardless of whether y is positive or negative.

Let us try another way. We have

$$\sin x = \mathrm{Im} e^{ix},$$

and since the other factors of the integrand are real, we get

$$I = \mathrm{Im} J,$$

where

$$J = \int_{-\infty}^{\infty} \frac{x e^{ix}}{1 + x^2} dx.$$

We close the contour upstairs and get, using (2.27b),

$$J = 2\pi i \frac{i e^{-1}}{2i} = \frac{\pi i}{e}.$$

Thus

$$I = \int_{-\infty}^{\infty} \frac{x \sin x}{1 + x^2} dx = \frac{\pi}{e}. \tag{2.33}$$

Next we consider the integral

$$I = \int_{-\infty}^{\infty} \frac{\sin x}{x} dx. \tag{2.34}$$

Note that, although its denominator vanishes at $x = 0$, $\sin x/x$ is finite at $x = 0$, as its numerator vanishes at $x = 0$ as well. Indeed, $\sin z/z$ is an entire function of z.

Again, it is not possible to close the contour of integration with $\sin x$ as the numerator. So let us try the trick of replacing I by the imaginary part of J, where

$$J = \int_{-\infty}^{\infty} \frac{e^{ix}}{x} dx.$$

But the integrand of J blows up at $x = 0$. This is because the denominator of the integrand vanishes at $x = 0$, while the numerator of the integrand does not. Therefore, the integral J is divergent and meaningless, and the trick of replacing $\sin x$ with e^{ix} fails.

Since the origin is a troublesome point, let us deform the contour away from the origin. This is possible as the integrand $\sin z/z$ is analytic at $z = 0$. It does not matter what precisely the contour is. We may, for example, deform the contour into c, where c goes from $-\infty$ to -1 along the real axis, from -1 to 1 along a curve lying in the upper half-plane, and from 1 to ∞ along the real axis. Thus we have

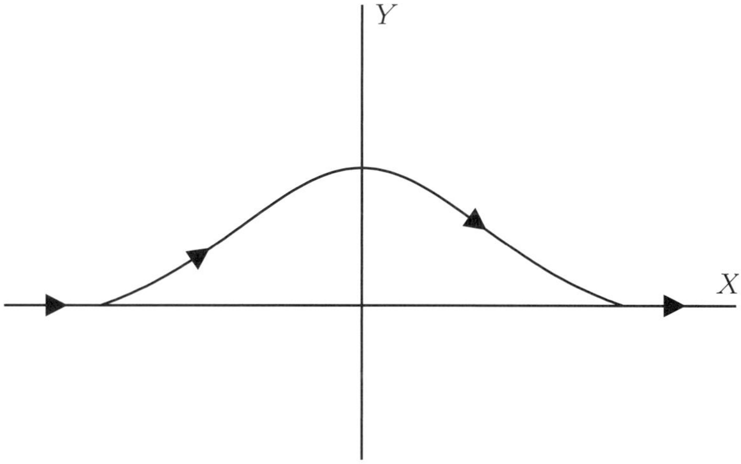

Figure 2.7

$$I = \int_c \frac{\sin z}{z} dz. \tag{2.35}$$

By (2.23) we have

$$I = I_1 + I_2,$$

where

$$I_1 = \int_c \frac{e^{iz}}{2iz} dz, \tag{2.36}$$

and

$$I_2 = -\int_c \frac{e^{-iz}}{2iz} dz. \tag{2.37}$$

The numerator of the integrand of I_1 is e^{iz}, which allows us to close the contour upstairs. Since the integrand of I_1 is analytic in the upper half-plane, we have

$$I_1 = 0.$$

For the integral I_2 we close the contour downstairs. Since the contour is clockwise and since the only singularity of the integrand enclosed by the contour is located at $z = 0$, we get, using (2.27b),

$$I_2 = (-2\pi i)\left(-\frac{1}{2i}\right) = \pi.$$

Thus we find

$$I = \int_{-\infty}^{\infty} \frac{\sin x}{x} dx = \pi. \tag{2.38}$$

Had we not deformed the contour away from the origin, both I_1 and I_2 would have been divergent integrals, as the integrands of these integrals blow up at the origin.

The contour of integration for an integral is not always the entire real axis, and it is not always possible to close the contour either upstairs or downstairs. Nevertheless, sometimes we are still able to do so after making some minor changes. As an example, consider the integral

$$\int_0^{\infty} \frac{dx}{1 + x^4}, \tag{2.39}$$

the contour of integration of which is the positive real axis only. Now the integrand of this integral is an even function of x. This is to say that the integrand at $x = -r$

on the negative real axis is equal to the integrand at $x = r$ on the positive real axis. Thus we may integrate over both the positive real axis and the negative real axis and divide the result by two. Therefore, we have

$$\int_0^\infty \frac{dx}{1+x^4} = \frac{1}{2}\int_{-\infty}^\infty \frac{dx}{1+x^4}. \tag{2.40}$$

Now that the contour of integration is the entire real axis, we may evaluate the integral by closing the contour either upstairs or downstairs. Let us close the contour in the upper half-plane. The singularities of the integrand enclosed are $e^{i\pi/4}$ and $e^{3i\pi/4}$. We get, after adding up the residues of these two points,

$$\int_0^\infty \frac{dx}{1+x^4} = \frac{\sqrt{2}\pi}{4}. \tag{2.41}$$

Consider next the integral

$$I = \int_0^\infty \frac{dx}{1+x^5}. \tag{2.42}$$

Since the integrand is not an even function of x, it will not help if we add the negative real axis to the contour of integration and divide by two. Instead, let us consider the integrand on the ray of argument $2\pi/5$ in the complex plane, i.e.,

$$z = re^{2\pi i/5}. \tag{2.43}$$

On this ray,

$$z^5 = r^5.$$

Therefore, the integrand at $z = re^{2\pi i/5}$ is $1/(1+r^5)$, which is the same as the integrand at $z = r$ on the positive real axis.

Let us therefore consider the integral

$$J = \oint_c \frac{dz}{1+z^5}, \tag{2.44}$$

where c is the closed contour consisting of the ray that is the positive real axis going from zero to R, the ray that given by (2.43) going from $r = R$ to $r = 0$, and the arc that is of the distance R from the origin and joins these two rays. We shall take the limit $R \to \infty$ at the end.

On the arc, the integrand vanishes like R^{-5} as $R \to \infty$. On the other hand, the length of the arc is $2\pi R/5$, which is merely linearly proportional to R. Thus

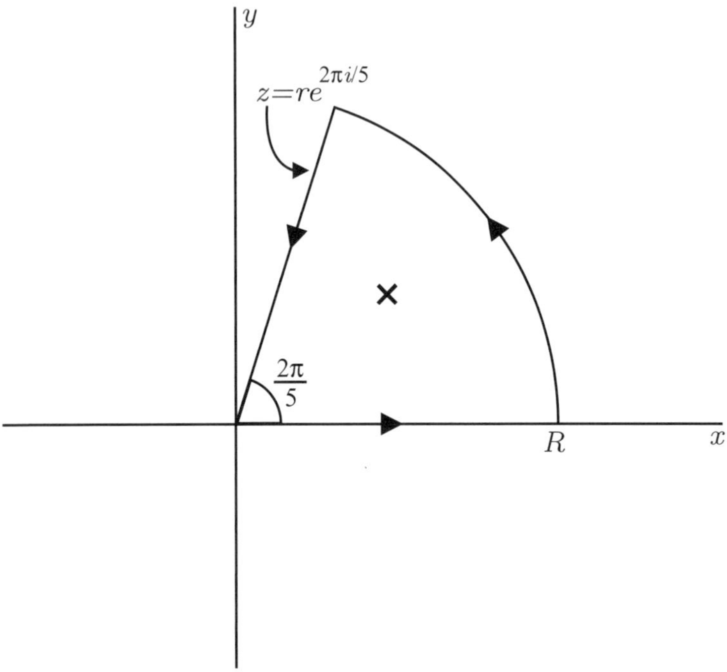

Figure 2.8

the integrand times the length of the arc vanishes as $R \to \infty$. Therefore, the contribution from the arc is zero as $R \to \infty$.

The integral over the ray of (2.43) in the limit $R \to \infty$ is related to I of (2.42) by

$$\int_{\infty}^{0} \frac{e^{2\pi i/5} dr}{1 + r^5} = -e^{2\pi i/5} I.$$

Therefore, we have

$$J = (1 - e^{2\pi i/5}) I. \tag{2.45}$$

We may evaluate J with the Cauchy residue theorem. The singularities of the integrand are located at

$$e^{i(2n+1)\pi/5}. \tag{2.46}$$

The only singularity enclosed by c is $e^{i\pi/5}$. Thus we have, by (2.27b),

$$J = \frac{2\pi i}{5e^{4i\pi/5}}. \tag{2.47}$$

Hence we get

$$\int_0^\infty \frac{dx}{1+x^5} = \frac{J}{1-e^{2\pi i/5}} = \frac{\pi}{5\sin\pi/5}. \tag{2.48}$$

The exact value of $\sin\pi/5$ can be deduced from the result in homework problem 2.

As the final example of contour integration, we evaluate the integral

$$I = \int_0^{2\pi} \frac{d\theta}{a+b\cos\theta}, \tag{2.49}$$

where a and b are positive. We require that $a > b$ so that the denominator of the integrand does not vanish for any θ.

While the contour of integration is the finite interval $[0, 2\pi]$, which is not closed, we may transform it into a closed contour by making a change of variable. We put

$$z \equiv e^{i\theta}.$$

As θ varies from 0 to 2π, z traverses in the counterclockwise direction the unit circle with the center at the origin. This circle is a closed contour. We have

$$dz = e^{i\theta} i d\theta,$$

or

$$d\theta = \frac{dz}{iz}.$$

We also have

$$\cos\theta = \frac{1}{2}(z+z^{-1}).$$

Thus

$$I = \oint_{c_1} \frac{2dz}{ib(z^2+2az/b+1)},$$

where c_1 is the unit circle with center at the origin. The singularities of the integrand are located at the zeroes of the denominator of the integrand, which are

$$z = -a/b \pm \sqrt{a^2/b^2-1}.$$

The singularity enclosed by c_1 is the one above with the plus sign. Applying the Cauchy residue theorem, we get, by (2.27b),

$$\int_0^{2\pi} \frac{d\theta}{a + b\cos\theta} = \frac{2\pi}{\sqrt{a^2 - b^2}}. \tag{2.50}$$

E. Branch Points and Branch Cuts

Consider the function

$$\log z = \log(re^{i\theta}) = \ln r + i\theta. \tag{2.51}$$

Let us start out at the point A in the figure below, follow the closed path c_1 in the figure in the counterclockwise sense, and come back to point A.

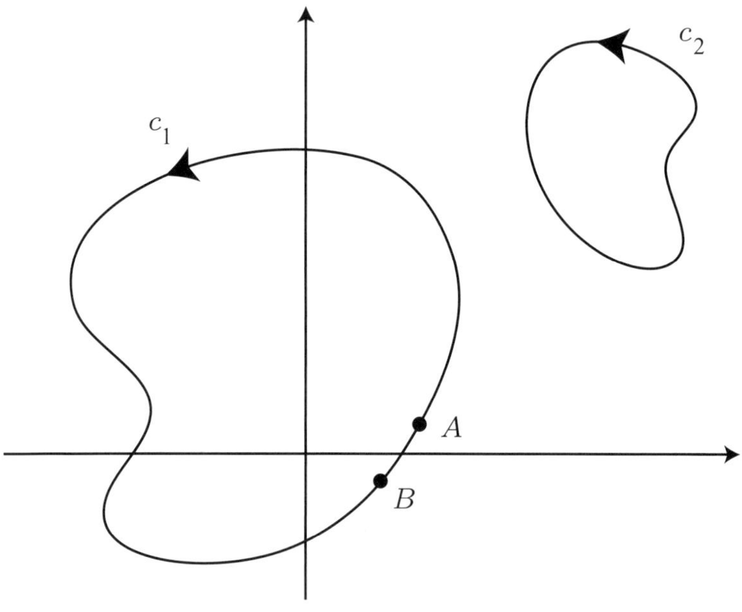

Figure 2.9

It is clear that while we return to the same location, the value of θ is not the same anymore. Indeed it changes to $\theta + 2\pi$. Therefore, upon traversing the closed path, we have

$$\log z \to \log z + 2\pi i,$$

which means that $\log z$ does not come back to its original value. This is true as long as one defines it to be a continuous function along the closed path.

Note that if we follow the same closed path around the origin in the clockwise sense and return to the same point, the value of θ changes to $\theta - 2\pi$. Thus the function $\log z$ changes to $(\log z - 2\pi i)$. And if the closed path goes around the origin in the counterclockwise sense n times, $\log z$ changes to $\log z + 2n\pi i$.

Thus we have an identity crisis: Which value should we choose to be the value of $\log z$ at A?

It may help us to find the answer to this question if we repeat the considerations above with the curve c_2 in Figure 2.9. We note that the value of θ does not change as one starts at a point on c_2 in the figure, traverses the closed path and comes back to the starting point. As a consequence, the value of $\log z$ does not change after the closed path c_2 is traversed.

What is the difference between the paths c_1 and c_2? The answer is that c_1 encloses the origin, while c_2 does not. Since the polar angle θ is the angle of the position vector joining the origin to z, θ increases by 2π as one goes once around c_1, and does not change as one goes around c_2. As a consequence, the function $\log z$ changes its value if the traversed path is c_1, which encloses the origin, but does not change its value if the traversed path is c_2, which does not enclose the origin. The origin is a special point with respect to the function $\log z$; it is said to be a branch point of the function $\log z$.

More generally, we define the point z_0 as a branch point of the function $f(z)$ if $f(z)$ changes its value as one traces a closed path enclosing z_0.

If we restrict ourselves to an open region that does not include the origin as an interior point, the function $\log z$ can be uniquely defined in this region. This is because there is no closed path inside the region that encloses the origin. Therefore, we may choose a point inside this region and define the value of θ at this point to be between 0 and 2π, say. Then the value of θ for any point inside this region is uniquely defined, and so is the function $\log z$. The designation of the value of θ at the chosen point is not unique. For example, we may define θ at the chosen point to be between 2π and 4π, or between $-\pi$ and π. The values of $\log z$ in this region are different with different definitions, but the function $\log z$ is single-valued with each designation of the value of θ at the chosen point. There is, therefore, more than one consistent definition of $\log z$ in a region. They are called different branches of $\log z$.

We may verify that the real part and the imaginary part of $\log z$ satisfy the Cauchy-Riemann equations everywhere except at the origin. Then $\log z$ is analytic in a region in which a branch of $\log z$ is chosen and the function is uniquely defined.

Let us cut up the complex z-plane by drawing a curve joining the origin to infinity. To be specific, we shall choose this curve to be the positive real axis. In the cut z-plane there is no closed contour that encloses the origin. We call the positive real axis a branch cut of $\log z$.

Let us start at point A, which is above and infinitesimally close to the positive real axis, and go along a circle in the counterclockwise sense, arriving at point B, which is below and infinitesimally close to the positive real axis. Then the value of $\log z$ at B differs from that at A by $2\pi i$. Thus $\log z$ is discontinuous across the positive x axis. Nevertheless, if we restrict ourselves in the cut z-plane, the function is single-valued. This is because the points A and B, separated by the branch cut, are not regarded as the same point.

This is like thinking of $f(z)$ as a function not on a plane but on a parking garage that has many levels. Let the point A be the entry point of the garage. Let us choose the value of θ at this point to be zero. As we go around a full circle in the counterclockwise sense, we arrive not at the entry point of the garage but at the point one level above it. The value of $\log z$ is taken to be dependent on which level we are at. Therefore, while the value of $\log z$ changes by $2\pi i$ as we go up one level, the function $\log z$ is uniquely defined at each point of the parking garage.

If we go around the garage in the counterclockwise sense n times, we arrive at the $(n+1)^{\text{th}}$ level of the garage. This level is called the $(n+1)^{\text{th}}$ Riemann sheet of the function. Since n can be any positive or negative integer, $\log z$ has infinitely many Riemann sheets.

There is no reason to restrict the branch cut to be on the positive real axis. We may choose the branch cut to be on the negative real axis. If we start at the entry point A with $\theta = 0$ as before, then we are choosing the branch of the function on the garage that is half a level below ground and half a level above ground. The values of θ in this branch are between $-\pi$ and π. We may, indeed, choose the branch cut to be any curve joining the origin to infinity in any way.

While the point $z = 0$ is a singularity of $\log z$ in the finite z-plane, the point infinity is also a branch point of $\log z$. To see this, we put

$$z = \frac{1}{\omega}.$$

Then the point $\omega = 0$ corresponds to the point $z = \infty$. Since

$$\log z = -\log \omega,$$

the function $\log z$ has a branch point at $\omega = 0$, or $z = \infty$. We also realize that a branch cut of $\log z$, chosen in any way we just described, is always a curve joining the only two branch points of the function.

It is straightforward to find the branch points for the function $\log(z - z_0)$. Let $\omega = z - z_0$; then

$$\log(z - z_0) = \log \omega.$$

Since $\omega = 0$ and $\omega = \infty$ are the branch points of the function $\log \omega$, $z = z_0$ and $z = \infty$ are the branch points of $\log(z - z_0)$.

Problem for the Reader

Find the branch points of $\log(z^2 - 1)$. Draw some possible sets of branch cuts.

Solution

We have

$$\log(z^2 - 1) = \log(z - 1) + \log(z + 1).$$

Thus the points -1, 1, and ∞ are the branch points of the function $\log(z^2 - 1)$.

We shall draw branch cuts to ensure the function is single-valued in the cut plane. Some possible sets of branch cuts for the function $\log(z^2 - 1)$ are drawn below.

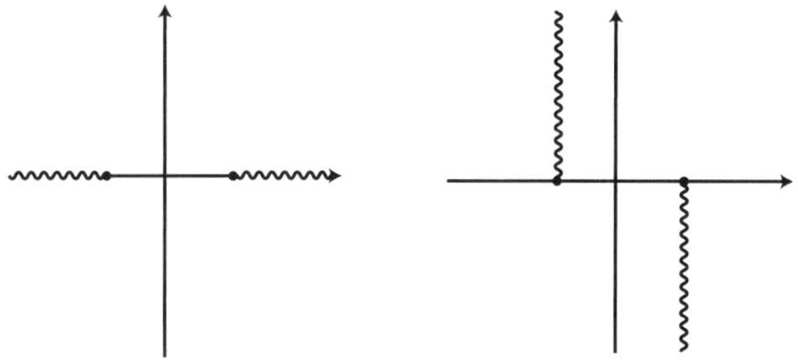

Figure 2.10

Problem for the Reader

Find the branch points of the function

$$\log\left(\frac{z - i}{z + i}\right).$$

Draw two sets of branch cuts for this function.

Solution

Since

$$\log\left(\frac{z-i}{z+i}\right) = \log(z-i) - \log(z+i),$$

the branch points of the function in the finite plane are i and $-i$.

We draw two possible sets of branch cuts in the figure below. While the set of branch cuts in the right-hand figure is self-explanatory, we shall say a few words about the left-hand figure. Let us traverse in the counterclockwise sense a closed path enclosing both branch points. Since this closed path encloses the point i, we have, as we return to the starting point,

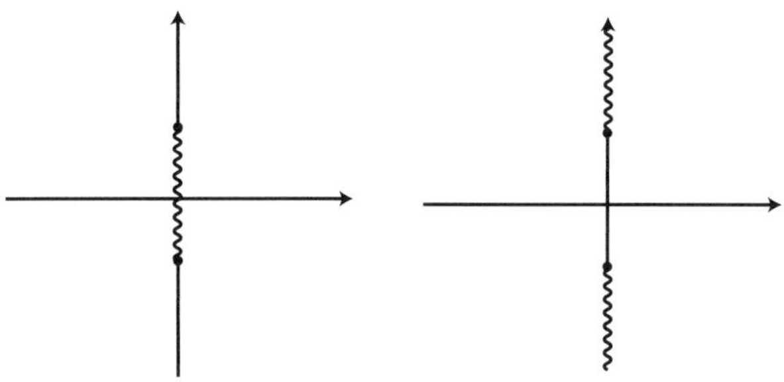

Figure 2.11

$$\log(z-i) \to \log(z-i) + 2\pi i.$$

And since this closed path also encloses the point $-i$, we have

$$\log(z+i) \to \log(z+i) + 2\pi i.$$

Since

$$\log\left(\frac{z-i}{z+i}\right)$$

is equal to the difference of $\log(z-i)$ and $\log(z+i)$, its value does not change as the closed path is traversed. Therefore, it suffices to draw just one finite branch cut between i and $-i$ to make the function

$$\log\left(\frac{z-i}{z+i}\right)$$

single-valued in the cut plane. This is because there is no closed path in this cut plane, which encloses just one of the branch points.

Note that this branch cut is a finite branch cut. That this is possible means that infinity is not a branch point. This is easily verified directly. Let $z = \omega^{-1}$; then

$$\log\left(\frac{z-i}{z+i}\right) = \log\left(\frac{1-i\omega}{1+i\omega}\right).$$

We see from the right side of the equation above that $\omega = 0$ is not a branch point of the function.

Since infinity is not a branch point, it is not necessary to draw a branch cut joining infinity with a finite point. We may therefore think of the two branch cuts in the right-hand figure above as joining with each other at infinity, forming just one continuous branch cut. We may draw this branch cut by starting from $-i$ and moving downward along the negative imaginary axis, passing through infinity and following the positive imaginary axis to the point i. This is somewhat like the way Columbus tried to get to India.

Next we discuss the function

$$z^a = r^a e^{ia\theta},$$

where a is a complex number. As we traverse a closed path enclosing the origin once in the counterclockwise sense, we have

$$z^a \to e^{i2\pi a} z^a.$$

Since the value of z^a changes after such a trip, the origin is a branch point of z^a. Let $z = \omega^{-1}$. Then we have $z^a = \omega^{-a}$, and we conclude that the point $z = \infty$ is also a branch point of z^a. Indeed, the points 0 and ∞ are the only two branch points of the function z^a. Thus this function is singled-valued in the plane with a cut joining the origin with infinity.

We call attention to the fact if $a = n$, where n is an integer of either sign, then $e^{i2\pi a}$ is equal to $e^{i2\pi n}$, which is unity. Thus the value of z^n does not change after a closed path enclosing the origin is traversed. Therefore, $z = 0$ and $z = \infty$ are not branch points for the function z^n, where n is either a positive integer or a negative integer.

Next we consider the case when a is equal to the rational number m/n, where m and n are integers having no common factors. The value of the function $z^{m/n}$ changes by a multiple of $e^{i2\pi m/n}$ after a path enclosing the origin is traversed once in the counterclockwise sense. Thus the origin is a branch point of the function $z^{m/n}$. But after we go around the origin in the counterclockwise sense n times, the function changes by a multiple of $e^{i2\pi m}$, which is equal to unity. This says that the function $z^{m/n}$ has only n Riemann sheets.

Problem for the Reader

Find the branch points of the function $z^a(1-z)^b$.

Solution

If neither a nor b is an integer, the function has branch points at $z = 0$ and $z = 1$ in the finite plane.

Let $z = \omega^{-1}$; then

$$z^a(1-z)^b = \omega^{-a-b}(\omega - 1)^b.$$

The function has a branch point at $\omega = 0$ unless $(a+b)$ is an integer. Thus $z^a(1-z)^b$ has a branch point at $z = \infty$ unless $(a+b)$ is an integer.

As an example, consider the function $(z-i)^{-1/2}(1-z)^{3/2}$. The points i and 1 are the branch points of this function, but infinity is not. Therefore, this function is single-valued in the plane with a finite branch cut connecting i to 1.

We give a few pointers below to help the reader find the branch points of a function.

a. If z_0 is a branch point of $f(z)$, it is also a branch point of $\log f(z)$. This is because as the value of $f(z)$ changes, so does the value of $\log f(z)$.

Similarly, a branch point of $f(z)$ is a possible branch point of $[f(z)]^a$. We qualify with the word "possible," as there are exceptions. An example is $f(z) = \sqrt{z}$ and $a = 2$. In this case, the origin is a branch point for $f(z)$, but not that of $[f(z)]^2$.

b. If z_0 is a zero of $f(z)$, it is a branch point of $\log f(z)$. It is also a possible zero of $[f(z)]^a$. To prove this, let z_0 be an nth-order zero of $f(z)$, i.e.,

$$f(z) = (z - z_0)^n F(z),$$

where $F(z)$ is analytic at z_0 and $F(z_0) \neq 0$. Then

$$\log f(z) = \log F(z) + n \log(z - z_0).$$

We see from the formula above that z_0 is a branch point of $\log f(z)$. Also, we have

$$[f(z)]^a = (z - z_0)^{na} [F(z)]^a .$$

From the formula above, we see that z_0 is a branch point of $[f(z)]^a$ unless na is an integer.

c. If z_0 is a pole of $f(z)$, it is a branch point of $\log f(z)$. It is also a possible branch point of $[f(z)]^a$. To prove this, let z_0 be an nth-order pole of $f(z)$, i.e.,

$$f(z) = \frac{F(z)}{(z - z_0)^n},$$

where $F(z)$ is analytic at z_0 and $F(z_0) \neq 0$. Thus

$$\log f(z) = \log F(z) - n \log(z - z_0).$$

We see from the formula above that z_0 is a branch point of $\log f(z)$.

Also, we have

$$[f(z)]^a = (z - z_0)^{-na} [F(z)]^a .$$

From the formula above, we see that z_0 is a branch point of $[f(z)]^a$ unless na is an integer.

Therefore, to locate the branch points of $\log f(z)$ or $[f(z)]^a$, we look for the zeroes, the poles, and the branch points of $f(z)$.

Problem for the Reader

Find the branch points for $\left[1 - (1 - z^2)^{1/2}\right]^{1/3}$.

Solution

Let $f(z) = 1 - (1 - z^2)^{1/2}$. Since $z = \pm 1$ are the branch points of $f(z)$, by point (a) they are also branch points of the function $[f(z)]^{1/3}$. To find the zeroes of $1 - (1 - z^2)^{1/2}$, or the roots of the equation

$$1 = (1 - z^2)^{1/2},$$

we square the equation above and get

$$z = 0.$$

But squaring an equation may produce roots that are not the roots of the equation. So let us see if $z = 0$ satisfies $\sqrt{1 - z^2} = 1$. We have

$$\sqrt{(1 - z^2)}|_{z=0} = \pm 1.$$

Thus the equation is not satisfied unless we choose the value of $\sqrt{1 - z^2}|_{z=0}$ to be 1. To say this in another way, whether $z = 0$ is a branch point for the function $[f(z)]^{1/3}$ depends on the branch we choose for the function $\sqrt{1 - z^2}$. And if we choose the branch so that the function $\sqrt{1 - z^2}$ at $z = 0$ is unity, then for z very small, we have

$$\sqrt{1 - z^2} \simeq 1 - z^2/2.$$

Thus

$$[f(z)]^{1/3} \simeq z^{2/3}/2^{1/3},$$

and $z = 0$ is a cubic-root branch point of the function.

To see whether infinity is a branch point for $[f(z)]^{1/3}$, we set $z = \omega^{-1}$, and we get

$$[f(z)]^{1/3} = \left[\omega - \sqrt{\omega^2 - 1}\right]^{1/3} \omega^{-1/3}.$$

Therefore, $z = \infty$ is a cubic-root branch point for the function $\left[1 - (1 - z^2)^{1/2}\right]^{1/3}$.

In summary, the branch points of the function in the z-plane are $\pm 1, \infty$. In addition, the origin is a branch point of the function provided that we choose the branch in which $\sqrt{1 - z^2}$ is equal to unity at $z = 0$.

Problem for the Reader

Find the branch points of the function $\log\left(z+\sqrt{1-z^2}\right)$.

Solution

Let $f(z) = \left(z+\sqrt{1-z^2}\right)$. Since the points ± 1 are the branch points of $f(z)$, they are the branch points of the function $\log f(z)$.

Next, the zeroes of $f(z)$ are given by

$$\sqrt{1-z^2} = -z.$$

There is no root for this equation.

Let $z = \omega^{-1}$, and we get

$$\log\left(z+\sqrt{1-z^2}\right) = \log\left(\frac{1+\sqrt{\omega^2-1}}{\omega}\right).$$

We see from the formula above that $\omega = 0$ or $z = \infty$ is a logarithmic branch point of the function.

To summarize, the branch points for the function in the z-plane are ± 1 and ∞.

We'll give an example to illustrate how to use the concept of branch cuts to calculate a real integral.

Example:

Evaluate the integral

$$I = \int_0^\infty \frac{\ln x}{4+x^2}dx$$

with contour integration.

The first thing to do is to relate this integral with another integral the contour of integration of which is a closed contour. We apply Cauchy's residue theorem to evaluate the latter, and obtain the value of the former after the value of the latter is found.

We draw a branch cut joining the origin to $-i\infty$, and choose θ, the argument of z, to be zero on the positive real axis. The function $\log z$ on this Riemann sheet is then uniquely defined. In particular, $\log z$ is equal to $\ln x$ on the positive real axis. Therefore, we may express I as $\log z/(4+z^2)$ integrated over the positive real axis. This contour of I is not closed.

Let us define another integral

$$J \equiv \int_{-\infty}^{\infty} \frac{\log z}{4+z^2} dz.$$

The contour of J can be closed upstairs, as the contribution of the infinite semicircle in the upper half-plane is of the order of the limit of $(R \ln R)/R^2$ as $R \to \infty$, where the factor $\ln R$ in the numerator of this quantity comes from the numerator of the integrand, the factor R^2 in the denominator of this quantity comes from the denominator of the integrand, and the factor R in the numerator of this quantity is the order of magnitude of the length of the arc. This quantity is equal to zero in the limit $R \to \infty$.

The only singularity enclosed by this closed contour is at $z = 2i$. Note that, starting at the point 2, we may reach the point $2i$ by traversing the counterclockwise circular arc of radius 2 with angular width $\pi/2$. Thus the argument of $2i$ is $\pi/2$, and we have

$$2i = 2e^{i\pi/2}.$$

As a result,

$$J = 2\pi i \frac{\log(2e^{i\pi/2})}{4i} = \pi \frac{\ln 2 + i\pi/2}{2}.$$

To obtain I from J, we express J as

$$J = J_1 + J_2,$$

where

$$J_1 = \int_0^{\infty} \frac{\log z}{4+z^2} dz,$$

and

$$J_2 = \int_{-\infty}^{0} \frac{\log z}{4+z^2} dz.$$

First of all, J_1 is simply I. Next we shall show that J_2 is related to I as well. We have, on the negative real axis,

$$\log z = \log(re^{i\pi}) = \ln r + i\pi.$$

Thus

$$J_2 = \int_0^\infty \frac{i\pi + \ln r}{4 + r^2} dr = \frac{i\pi^2}{4} + I.$$

Therefore,

$$J = 2I + \frac{i\pi^2}{4}.$$

With the value of J obtained earlier with the help of the Cauchy residue theorem, we get

$$I = \int_0^\infty \frac{\ln x}{4 + x^2} dx = \pi \frac{\ln 2}{4}. \tag{2.52}$$

We close this section with two topics: (a) a discussion of the principal value of an integral, and (b) a discussion of the Plemelj formulae.

Back in the high school days when we were first introduced to integrations over a real variable, some of us might have puzzled over the value of real integrals such as

$$\int_{-2}^{3} \frac{dx}{x}.$$

For, if we carry out the integration in a straightforward way, we get

$$\int_{-2}^{3} \frac{dx}{x} = \ln x \,|_{-2}^{3} = \ln 3 - \ln(-2).$$

But what is the value of $\ln(-2)$?

This ambiguity is due to a difficulty with this integral. The fact is that the integrand $\frac{1}{x}$ has a simple pole at the origin. Since the contour of integration passes through the origin, the integral

$$\int_{-2}^{3} \frac{dx}{x}$$

is actually undefined as it is.

Let us try to define this integral by identifying it with

$$\int_{-2}^{-\epsilon_1} \frac{dx}{x} + \int_{\epsilon_2}^{3} \frac{dx}{x},$$

where ϵ_1 and ϵ_2 are positive and infinitesimally small constants. This means that we first integrate from $x = -2$ to the point $-\epsilon_1$, which is to the left of the origin, jump over the origin, and continue to integrate from the point ϵ_2, which is to the right of the origin, to the point $x = 3$. In this way we make sure that the contour of integration does not pass through the origin. We define

$$\int_{-2}^{3} \frac{dx}{x}$$

to be this integral in the limit ϵ_1 and ϵ_2 go to zero.

There is something encouraging about this defintion of the divergent integral

$$\int_{-2}^{3} \frac{dx}{x}.$$

For, while both integrals are infinite as ϵ_1 and ϵ_2 go to zero, the first integral is negative and the second integral is positive. Indeed, the integrands of both integrals are the same function $1/x$, which is an odd function of x. Thus there is cancellation between these two integrals, and we hope that the sum of them is finite as ϵ_1 and ϵ_2 go to zero.

To see if this true, we calculate these integrals explicitly. We have

$$\int_{-2}^{-\epsilon_1} \frac{dx}{x} = \int_{\epsilon_1}^{2} \frac{dx}{x} = \ln \frac{\epsilon_1}{2},$$

and

$$\int_{\epsilon_2}^{3} \frac{dx}{x} = \ln \frac{3}{\epsilon_2}.$$

Thus the sum of the two integrals above is

$$\ln \frac{3\epsilon_1}{2\epsilon_2}.$$

This value depends on the ratio

$$\frac{\epsilon_1}{\epsilon_2}$$

and is not unique.

Let us define the principal value of the divergent integral

$$\int_{-2}^{3} \frac{dx}{x}$$

to be the one with $\epsilon_1 = \epsilon_2$:

$$P\int_{-2}^{3} \frac{dx}{x} = \int_{-2}^{-\epsilon} \frac{dx}{x} + \int_{\epsilon}^{3} \frac{dx}{x}$$

with ϵ infinitesimal. We then have

$$P\int_{-2}^{3} \frac{dx}{x} = \ln 3/2,$$

which is unique.

For some integrals, the singularity of the integrand may be located at x_0 rather than at the origin. Let $f(x)$ blow up like a simple pole at x_0, an interior point of the interval of integration. We define the principal value of the divergent integral $\int_a^b f(x)dx$ to be

$$P\int_{a}^{b} f(x)dx = \int_{a}^{x_0-\epsilon} f(x)dx + \int_{x_0+\epsilon}^{b} f(x)dx,$$

where ϵ is positive and infinitesimal.

We mention that not all divergent integrals can be made finite and unique as we take their principal values. For example, the principal value of the divergent integral

$$\int_{-1}^{2} \frac{dx}{x^2}$$

remains divergent, as $1/x^2$ is positive at the two sides of the origin and there is no cancellation between the integral over positive values of x and that over negative values of x. On the other hand, the principal value of a convergent integral such as

$$\int_{-1}^{2} \frac{\sin x}{x} dx$$

is equal to the integral itself. This is because omitting an infinitesimally small interval of integration does not change the value of a convergent integral.

We also mention that there are other ways to define divergent integrals such as

$$\int_{-2}^{2} \frac{dx}{x}.$$

As an example, the contour of this integral may be chosen to be the half-circle of radius 2 which has the origin as its center and lies in the upper half-plane. We may also make other choices of the contour of integration, but as long as the curve does not pass through the origin, the integral is defined. We may then carry out the integration explicitly and get

$$\int_{-2}^{2} \frac{dz}{z} = \ln \frac{2}{-2}.$$

The value of the integral depends on the difference of the phase of the upper endpoint $z = 2$ and that of the lower endpoint $z = -2$ as the contour is traversed. Let us choose this contour to lie entirely in the upper half-plane, the endpoints being excluded. While there are many such contours, by Cauchy's integral theorem the value of the integral is independent of these contours we choose. Now as we go from -2 to 2 on such a contour, we go clockwise and the phase of the point $z = 2$ is smaller than that of the point $z = -2$ by π. Thus we have, if the entire contour of integration lies above the origin,

$$\int_{-2}^{2} \frac{dz}{z} = \ln \left(\frac{2}{-2} \right) = -i\pi.$$

By choosing the contour to be the one that is infinitesimally above the real axis, we may express this result as

$$\int_{-2}^{2} \frac{dx}{x + i\epsilon} - P \int_{-2}^{2} \frac{dx}{x} = -i\pi,$$

the second integral above being zero as its integrand is an odd function of x.

Problem for the Reader

If we define the contour for the integral

$$\int_{-2}^{3} \frac{dz}{z}$$

to be a curve joining -2 to 3 and lying entirely in the lower half-plane, the endpoints being excluded, find the value of this integral.

Solution

If we follow such a curve from -2 to 3, we go counterclockwise, and the phase of the point $z = 3$ is larger than that of the point $z = -2$ by π. Thus for all such contours, we have

$$\int_{-2}^{3} \frac{dz}{z} = \ln(3/2) + i\pi,$$

which differs from the principal value of this integral by $i\pi$.

By choosing the contour to be infinitesimally below the real axis, we may express this result as

$$\int_{-2}^{3} \frac{dx}{x - i\epsilon} - P\int_{-2}^{3} \frac{dx}{x} = i\pi.$$

We are now ready to prove the Plemelj formulae, which say that

$$\int_{a}^{b} \frac{r(x')}{x' - (x \pm i\epsilon)} dx = P\int_{a}^{b} \frac{r(x')}{x' - x} dx \pm i\pi r(x)$$

where $a < x < b$. We assume that the function $r(x')$ is such that the integral above is convergent. This does not exclude the possibility for $r(x')$ to be infinite at some point inside the interval of integration. It just means that $r(x')$ must not blow up too badly at any such points for the integral to diverge. For example, it must not blow up as fast as a simple pole at the endpoints a and b.

We also point out that considered as a function of x', the integrand above has a pole at $x \pm i\epsilon$. Since the contour of integration is on the real axis, and since ϵ is infinitesimal, the plus and minus signs in front of ϵ in the integral merely signify whether the pole is above or below the contour of integration.

We write

$$\int_{a}^{b} \frac{r(x')}{x' - x - i\epsilon} dx' = \int_{a}^{b} \frac{r(x') - r(x)}{x' - x - i\epsilon} dx' + \int_{a}^{b} \frac{r(x)}{x' - x - i\epsilon} dx'.$$

Note that the factor $r(x)$ in the last integral above , being independent of x^1, may be placed at the left of the integral sign. We also note that the integrand of the first integral on the right side of the equation above is finite in the limit as ϵ goes to zero. This is because the numerator of the integrand vanishes at $x' = x$. Thus we may ignore the term $-i\epsilon$ in the denominator of this integral. Since the principal value of such an integral is equal to the integral itself, we have

$$\int_a^b \frac{r(x') - r(x)}{x' - x} dx' = P\int_a^b \frac{r(x') - r(x)}{x' - x - i\epsilon} dx' = P\int_a^b \frac{r(x')}{x' - x} dx' - P\int_a^b \frac{r(x)}{x' - x} dx'.$$

Again, $r(x)$ in the last integral above may be taken to the left of the integral sign. Generalizing slightly what we have discussed, it is possible to prove that

$$\int_a^b \frac{dx'}{x' - x - i\epsilon} - P\int_a^b \frac{dx'}{x' - x} = i\pi.$$

With all these considerations, the first of the Plemelj formulae is obtained.

The second Plemelj formula is obtained from the first Plemelj formula by taking complex conjugation.

We mention that if we replace, in the integral of the Plemelj formula, the real variable x by the complex variable z, we get the function

$$f(z) = \int_a^b \frac{r(x')}{x' - z} dx',$$

which is analytic provided that z is not a point on the interval $[a, b]$. This is because the values of x', the variable of integration, are restricted to the real values between a and b. Therefore, the denominator $(x' - z)$ never vanishes as long as z is not equal to some real value between a and b. The factor $1/(x' - z)$ in the integrand is thus an analytic function of z provided that z is not a point in the interval $[a, b]$, and hence so is $f(z)$.

The function $f(z)$ is discontinuous across the interval $[a, b]$, as we shall presently see.

Problem for the Reader

Find the values of $f(x + i\epsilon)$ and $f(x - i\epsilon)$ when $a < x < b$. Find also the discontinuity of $f(z)$ across the branch cut $a < x < b$.

Solution

By the first of the Plemelj formulae, we have

$$f(x + i\epsilon) = P\int_a^b \frac{r(x')}{x' - x} dx' + i\pi r(x).$$

By the second Plemelj formula, we have

$$f(x - i\epsilon) = P \int_a^b \frac{r(x')}{x' - x} dx' - i\pi r(x).$$

Therefore,

$$f(x + i\epsilon) - f(x - i\epsilon) = 2i\pi r(x)$$

is the discontinuity of $f(z)$ across the branch cut $a < x < b$.

Thus the function $f(z)$ defined by the integral above has a branch cut from a to b, and is analytic everywhere else. We also note that, since $1/(x' - z)$ vanishes as z goes to infinity, $f(z)$ vanishes as z goes to infinity.

The converse is also true: If $f(z)$ is analytic in the complex plane with the exception of a branch cut from a to b on the real axis, does not blow up as fast as a pole either at a or at b, and vanishes at infinity, then $f(z)$ is given by the integral above, with $2\pi i r(x)$ the discontinuity of $f(z)$ across the branch cut. (See homework problem 15.)

This means that we can construct the function $f(z)$ by merely knowing its discontinuity across the branch cut, provided that all the conditions mentioned above are satisified.

These results are easily generalized to the case in which the branch cut is not a straight line on the real axis but a curve in the complex plane.

F. Fourier Integrals and Fourier Series

In this section we shall discuss the Fourier integral and the Fourier series, which are necessary tools for solving many physical problems. We shall encounter some of these problems in Chapters 4 and 5.

Let $F(x)$ be a function defined for all values of x from $-\infty$ to ∞. We define the Fourier transform of $F(x)$ as

$$\tilde{F}(k) \equiv \int_{-\infty}^{\infty} e^{-ikx} F(x) dx. \tag{2.53}$$

The Fourier transform $\tilde{F}(k)$ is uniquely determined once $F(x)$ is given, provided that the integral above is convergent.

The Fourier integral theorem says that the converse is also true, that once $\tilde{F}(k)$ is given, one is able to determine $F(x)$ from $\tilde{F}(k)$. Indeed, the inversion formula of Fourier transform is

$$F(x) = \int_{-\infty}^{\infty} e^{ikx}\tilde{F}(k)\frac{dk}{2\pi} \ (-\infty < x < \infty), \tag{2.54}$$

which has a striking resemblance to (2.53).

With some partial differential equations, it is easier to find $\tilde{F}(k)$ than to find $F(x)$. One of the best ways to solve such equations is to find $\tilde{F}(k)$ first and then use Fourier's inversion theorem to determine $F(x)$.

To prove Fourier's inversion formula, we define

$$I_\lambda(x) \equiv \int_{-\lambda}^{\lambda} e^{ikx}\tilde{F}(k)\frac{dk}{2\pi}. \tag{2.55}$$

Then (2.54) is exactly

$$F(x) = \lim_{\lambda\to\infty} I_\lambda(x).$$

We substitute (2.53) into (2.55) and get

$$I_\lambda(x) \equiv \int_{-\lambda}^{\lambda} e^{ikx}\frac{dk}{2\pi}\left[\int_{-\infty}^{\infty} e^{-ikx'}F(x')dx'\right].$$

In the double integral above, we are supposed to integrate over x' first before we integrate over k. We shall reverse the order of the integration, integrating over k before integrating over x'. There are conditions on $F(x)$ under which this change of the order of integration is justified, but we will not elaborate on it here.

It is easy to carry out the integration over k and get

$$\int_{-\lambda}^{\lambda} e^{ik(x-x')}\frac{dk}{2\pi} = \frac{\sin\left[\lambda(x'-x)\right]}{\pi(x'-x)}. \tag{2.56}$$

Thus $I_\lambda(x)$ is given by

$$I_\lambda(x) \equiv \int_{-\infty}^{\infty} \frac{\sin\left[\lambda(x'-x)\right]}{\pi(x'-x)}F(x')dx'. \tag{2.57}$$

For the sake of getting the point across quickly, we shall first give a heuristic proof of the Fourier integral theorem without regard to rigor. We put

$$(x'-x)\lambda \equiv X; \tag{2.58}$$

then (2.57) becomes

$$I_\lambda(x) \equiv \int_{-\infty}^{\infty} \frac{\sin X}{\pi X} F\left(x + \frac{X}{\lambda}\right) dX. \tag{2.59}$$

Since

$$\lim_{\lambda\to\infty} F\left(x + \frac{X}{\lambda}\right) = F(x), \tag{2.60}$$

we have

$$\lim_{\lambda\to\infty} I_\lambda(x) = F(x) \int_{-\infty}^{\infty} \frac{\sin X}{\pi X} dX = F(x), \tag{2.61}$$

where we have made use of (2.38). This is the Fourier integral theorem (2.54).

As I have warned the reader, there is a lack of mathematical rigor with the arguments given above. This is because (2.61) is obtained after we replace $F(x + X/\lambda)$ in (2.59) with $F(x)$. Such a replacement is valid only if X/λ can be regarded as very small when λ is very large. But X, the integration variable of (2.59), ranges from $-\infty$ to ∞, and can surely be larger than λ. A rigorous proof must therefore include the argument that, as λ is very large, the region of integration that contributes to the integral (2.59) is restricted to

$$|X| \ll \lambda,$$

or, by (2.58),

$$|x' - x| \ll 1.$$

To prove this, we return to the integral in (2.57). The integration variable of this integral is x'. At $x' = x$, we have, by l'Hopital's rule,

$$\lim_{x'\to x} \frac{\sin\left[\lambda(x'-x)\right]}{\pi(x'-x)} = \frac{\lambda}{\pi}.$$

Therefore, the integrand of (2.57) at $x' = x$ is equal to $\lambda F(x)/\pi$, a very large number when λ is very large. It is clear that the integrand of (2.57) continues to be of the order of $\lambda F(x)$ in a sufficiently small interval around the point $x' = x$. To be specific, if $(x'-x)$ is as small as $1/\lambda$, then $(x'-x)^{-1}$ is as large as λ. Furthermore, since $(x'-x)$ is small, $F(x')$ is approximately $F(x)$. Thus the integrand of (2.57) is of the order of $\lambda F(x)$ in the small interval in which $(x'-x)$ is of the order of $1/\lambda$. The value of the integral over this small interval is of the order of the magnitude of the integrand times the width of the interval, or

$$[\lambda F(x)] \cdot (1/\lambda) = F(x).$$

This order of magnitude estimate is in agreement with (2.61).

But, as we have mentioned, we need to prove that we may replace $F(x')$ by $F(x)$ for the entire integral of (2.57). The problem is that when x' is considerably different from x, the integrand of (2.57) does not vanish as $\lambda \to \infty$. For such values of x', it is not justified to replace $F(x')$ by $F(x)$.

We note that when λ is very large, the factor $\sin[\lambda(x'-x)]$ in the integrand is a rapidly oscillatory function of x'. As we shall discuss in more detail in Chapter 9, integrating a rapidly oscillatory function over an interval gives a very small number. For example,

$$\int_0^1 \cos(\lambda x')dx' = \frac{\sin\lambda}{\lambda},$$

which vanishes as λ goes to infinity. Therefore, with the exception of the small interval around $x' = x$ we just discussed, the contributions to the integral of (2.57) from any other region are very small. And as $\lambda \to \infty$, the contributions from any region other than an infinitesimal interval around $x' = x$ vanish. Thus, in the limit $\lambda \to \infty$, the contributions to the integral (2.57) come exclusively from an infinitesimal neighborhood of the point $x' = x$. For this reason, we may replace $F(x')$ in the integrand of (2.57) by $F(x)$ and get

$$\lim_{\lambda\to\infty} I_\lambda(x) = F(x) \lim_{\lambda\to\infty} \int_{-\infty}^{\infty} \frac{\sin[\lambda(x'-x)]}{\pi(x'-x)} dx'.$$

By (2.38), we get precisely (2.61), and the inversion formula for Fourier transform has been proven.

As a side remark,

$$\lim_{\lambda\to\infty} \frac{\sin[\lambda(x'-x)]}{\pi(x'-x)} \equiv \delta(x'-x) \tag{2.62}$$

is known to be the Dirac delta function. The Dirac delta function is very useful with many problems in science and engineering, and I shall say a few words about its properties here. First of all, by (2.38) we have

$$\int_{-\infty}^{\infty} \delta(x'-x)dx' = 1.$$

Furthermore, since there is no contribution from any region outside an infinitesimal neighborhood of $x' = x$, we have, for any function $g(x')$,

$$\int_{-\infty}^{\infty} \delta(x'-x) g(x') dx' = g(x) \int_{-\infty}^{\infty} \delta(x'-x) dx' = g(x).$$

In particular,

$$\int_{-\infty}^{\infty} \delta(x') e^{-ikx'} dx' = 1,$$

which says that the Fourier transform of $\delta(x)$ is unity. By the inversion formula of Fourier transform, we find

$$\delta(x) = \int_{-\infty}^{\infty} e^{ikx} \frac{dk}{2\pi}, \tag{2.63}$$

which is the integral representation for the Dirac delta function. (The Dirac delta function will be discussed more fully in Chapter 4.)

Problem for the Reader

Find the Fourier transform of $F(x) = \dfrac{1}{4+x^2}$, $-\infty < x < \infty$.

Solution

We have

$$\tilde{F}(k) = \int_{-\infty}^{\infty} \frac{e^{-ikx}}{4+x^2} dx.$$

If $k > 0$, we close the contour downstairs and get

$$\tilde{F}(k) = -2\pi i \frac{e^{-2k}}{-4i} = \frac{\pi e^{-2k}}{2}.$$

If $k < 0$, we close the contour upstairs and get

$$\tilde{F}(k) = 2\pi i \frac{e^{2k}}{4i} = \frac{\pi e^{2k}}{2}.$$

Thus

$$\tilde{F}(k) = \frac{\pi e^{-2|k|}}{2}.$$

In the proof of the Fourier integral theorem, we have implicitly assumed that $F(x)$ is continuous. Consider now the case when $F(x)$ is discontinuous at x_0. We shall denote $F(x_0^+)$ as the value of $F(x)$ as x approaches x_0 from the right, and $F(x_0^-)$ as the value of $F(x)$ as x approaches x_0 from the left. We write (2.59) as

$$I_\lambda(x_0) \equiv \int_{-\infty}^{0} \frac{\sin X}{\pi X} F\left(x_0 + \frac{X}{\lambda}\right) dX + \int_{0}^{\infty} \frac{\sin X}{\pi X} F\left(x_0 + \frac{X}{\lambda}\right) dX. \quad (2.64)$$

In the limit $\lambda \to \infty$, $F(x_0 + \frac{X}{\lambda})$ approaches $F(x_0^-)$ if X is negative, and approaches $F(x_0^+)$ if X is positive. Therefore, we have

$$\lim_{\lambda\to\infty} I_\lambda(x_0) = \frac{F(x_0^-) + F(x_0^+)}{2},$$

which is, explicitly,

$$\int_{-\infty}^{\infty} e^{ikx_0} \tilde{F}(k) \frac{dk}{2\pi} = \frac{F(x_0^-) + F(x_0^+)}{2}. \quad (2.65)$$

This says that the Fourier integral of a function $F(x)$ is equal to the mean of the two values of $F(x)$ at x_0, at which $F(x)$ is discontinuous.

Problem for the Reader

Let

$$F(x) = e^{-x}, \; x > 0,$$
$$= 0, \quad x < 0.$$

This function is continuous at $x = 0$. Find $\tilde{F}(k)$ and verify if (2.65) is valid.

Solution

It is easy to find that

$$\tilde{F}(k) = \int_0^\infty e^{-ikx} e^{-x} dx = \frac{1}{1+ik}.$$

Verifying (2.65) is less trivial. The left side of (2.65) with $x_0 = 0$ is

$$\int_{-\infty}^{\infty} \frac{dk}{2\pi} \frac{1}{1+ik},$$

which is divergent.

We note that by (2.55), the lower limit and the upper limit of the Fourier inversion integral are $-\lambda$ and λ, respectively, with the limit $\lambda \to \infty$ taken after the integration has been carried out. Thus the divergent integral above should be regarded as

$$\begin{aligned}\lim_{\lambda\to\infty} \int_{-\lambda}^{\lambda} \frac{dk}{2\pi} \frac{1}{1+ik} &= \lim_{\lambda\to\infty} \int_{-\lambda}^{\lambda} \frac{dk}{4\pi} \left(\frac{1}{1+ik} + \frac{1}{1-ik} \right) \\ &= \lim_{\lambda\to\infty} \int_{-\lambda}^{\lambda} \frac{dk}{2\pi} \frac{1}{1+k^2} = \frac{1}{2},\end{aligned}$$

a finite result which agrees with the right side of (2.65).

Next we write, in (2.53),

$$e^{-ikx} = \cos kx - i \sin kx.$$

Then (2.53) becomes

$$\tilde{F}(k) = A(k) - iB(k),$$

where

$$A(k) = \int_{-\infty}^{\infty} \cos kx F(x) dx, \; B(k) = \int_{-\infty}^{\infty} \sin kx F(x) dx. \tag{2.66}$$

We find from (2.66) that

$$A(k) = A(-k), \quad B(k) = -B(-k).$$

Replacing $\tilde{F}(k)$ in the inversion formula (2.54) by $A(k) - iB(k)$, we get, after making use of the facts that $A(k)$ is an even functon of k and $B(k)$ is an odd function of k,

$$F(x) = \int_{-\infty}^{\infty} [A(k) \cos kx + B(k) \sin kx] \frac{dk}{2\pi}. \tag{2.67}$$

If $F(x)$ is an even function of x, (2.66) shows that $B(k) = 0$. In this case, $F(x)$ has the integral representation

$$F(x) = \int_{-\infty}^{\infty} A(k) \cos kx \frac{dk}{2\pi}. \tag{2.68}$$

If $F(x)$ is an odd function of x, (2.66) shows that $A(k) = 0$. Therefore, $F(x)$ has the integral representation

$$F(x) = \int_{-\infty}^{\infty} B(k) \sin kx \frac{dk}{2\pi}. \tag{2.69}$$

We shall next derive the Fourier integral representations of a function $F(x)$, which is given only in the semi-infinite region $0 < x < \infty$.

We may, as a mathematical artifact, define the function $F(x)$ for x negative by

$$F(-x) = F(x).$$

The function $F(x)$ is now defined for all values of x, and is an even function of x. As we have shown, an even function of x has the Fourier cosine integral representation. Thus a function originally defined only in the semi-infinite region $0 < x < \infty$ has a Fourier cosine integral representation given by (2.68). By (2.66) and the fact that $F(x)$ is even, we may express the coefficient $A(k)$ as

$$A(k) = 2 \int_{0}^{\infty} \cos kx F(x) dx. \tag{2.70}$$

In this expression for $A(k)$, the region of integration is the positive x axis, on which the value of $F(x)$ is originally given.

Alternately, we may also define $F(x)$ for negative values of x by

$$F(-x) = -F(x).$$

The function is now defined for all values of x, and is an odd function of x. An odd function of x has the Fourier sine integral representation. Thus a function $F(x)$ given only for $x > 0$ has the Fourier sine integral representation (2.69), where

$$B(k) = 2 \int_{0}^{\infty} \sin kx F(x) dx. \tag{2.71}$$

Note that the Fourier sine integral (2.69) vanishes as we set $x = 0$. If $F(0^+)$ is not zero, the Fourier sine integral at $x = 0$ is not equal to $F(0^+)$. Instead, the Fourier sine integral at $x = 0$ is equal to half of $F(0^+) + F(0^-)$, which is zero as $F(x)$ is by construction an odd function of x.

Next we turn to representations of functions given in a finite region. As we will see, such a function is represented by an infinite sum, not an integral. Let $f(\theta)$ be a function given in the interval $[-\pi, \pi]$, and let

$$a_n \equiv \frac{1}{2\pi} \int_{-\pi}^{\pi} e^{-in\theta} f(\theta) d\theta. \tag{2.72}$$

The coefficient a_n, known as the Fourier coefficient of $f(\theta)$, is determined once $f(\theta)$ is given. We shall prove that

$$f(\theta) = \sum_{n=-\infty}^{\infty} a_n e^{in\theta}. \tag{2.73}$$

The series in (2.73) is known as the Fourier series of $f(\theta)$. It shows that the function $f(\theta)$ is determined once the Fourier coefficient a_n is given.

Since $e^{in\theta}$ is a periodic function of θ with the period 2π, so is the Fourier series in (2.73). Therefore, if only implicitly so, the function $f(\theta)$ outside the interval $[-\pi, \pi]$ is assumed to be a periodic function of θ satisfying

$$f(\theta + 2\pi) = f(\theta). \tag{2.74}$$

Before we prove the inversion formula (2.73), we first say a few words about why such a formula is expected. We multiply (2.73) by $e^{-im\theta}/(2\pi)$ and integrate the equation from $-\pi$ to π. We get

$$\frac{1}{2\pi} \int_{0}^{2\pi} e^{-im\theta} f(\theta) d\theta = \frac{1}{2\pi} \int_{0}^{2\pi} e^{-im\theta} \sum_{n=-\infty}^{\infty} a_n e^{in\theta} d\theta. \tag{2.75}$$

The integrand on the right side of (2.75) is an infinite series, and there are conditions under which it is justified to carry out the integration term by term. We shall assume such conditions are satisfied. Since

$$\begin{aligned} \frac{1}{2\pi} \int_{-\pi}^{\pi} e^{-i(m-n)\theta} d\theta &= 1, \ (m = n), \\ &= 0, \ (m \neq n), \end{aligned} \tag{2.76}$$

the only term on the right side of (2.75) that does not vanish after the integration has been performed is the term with $m = n$. Thus the right side of (2.75) is equal to a_m and we get (2.72). We have therefore shown that, if the function $f(\theta)$ given in the region $-\pi < \theta < \pi$ has a Fourier series expansion, its Fourier coefficient must be the one given by (2.72).

The Fourier coefficient a_n of $f(\theta)$ can be compared with the component A_n of a vector $\vec{v}$ in space:

$$\vec{v} = A_1\vec{e_1} + A_2\vec{e_2} + A_3\vec{e_3}, \tag{2.77}$$

where $\vec{e_1}$, $\vec{e_2}$, and $\vec{e_3}$ are the unit vectors in the directions of the x, y,and z axes, respectively. Since the unit vectors $\vec{e_1}$, $\vec{e_2}$, and $\vec{e_3}$ are mutually orthogonal, the scalar product between them is given by

$$\begin{aligned} \vec{e_m} \cdot \vec{e_n} &= 1, \ m = n, \\ &= 0, \ m \neq n. \end{aligned} \tag{2.78}$$

From (2.77) and (2.78) we have

$$A_n = \vec{e_n} \cdot \vec{v}. \tag{2.79}$$

We note that (2.77) is the analogue of (2.73), with $\vec{e_n}$ the counterpart of $e^{in\theta}$. Also, (2.78) is the analogue of (2.76), and (2.79) is the analogue of (2.72). Note also that it is not possible to represent a vector $\vec{v}$ in the three-dimensional space by only two of its components, say A_1 and A_2, for

$$\vec{v} \neq A_1\vec{e_1} + A_2\vec{e_2}.$$

Rather, we need all three of the basis vectors $\vec{e_1}$, $\vec{e_2}$, and $\vec{e_3}$ to represent a three-dimensional vector $\vec{v}$. Such a set of basis vectors is said to be a complete set of basis vectors in the three-dimensional space. Returning to the issue of the Fourier series, (2.73) says that the set of functions $e^{in\theta}, n = 0, \pm 1, \pm 2, \cdots$ is complete. In other words, it is possible to use this set of functions to represent a function $f(\theta)$ in the form (2.73) for $-\pi \leq \theta \leq \pi$.

To prove (2.73), we define

$$S_N(\theta) \equiv \sum_{n=-N}^{N} a_n e^{in\theta}. \tag{2.80}$$

The remaining task is to show that

$$f(\theta) = \lim_{N\to\infty} S_N(\theta).$$

Substituting (2.72) into (2.80), we get

$$S_N(\theta) \equiv \sum_{n=-N}^{N} \frac{1}{2\pi} \int_{-\pi}^{\pi} e^{in(\theta-\theta')} f(\theta')d\theta'. \tag{2.81}$$

Next we carry out the summation over n in (2.81). Since

$$\sum_{n=-N}^{N} \omega^n = \omega^{-N}(1+\omega+\cdots+\omega^{2N}) = \omega^{-N}\frac{1-\omega^{2N+1}}{1-\omega}$$
$$= \frac{\omega^{-N}-\omega^{N+1}}{1-\omega},$$

we have, by multiplying both the numerator and the denominator of the expression above by $\omega^{-1/2}$,

$$\sum_{n=-N}^{N} \omega^n = \frac{\omega^{-(N+1/2)}-\omega^{N+1/2}}{\omega^{-1/2}-\omega^{1/2}}.$$

Identifying $e^{i\theta}$ with ω, we get

$$\sum_{n=-N}^{N} e^{in\theta} = \frac{e^{-i(N+1/2)\theta}-e^{i(N+1/2)\theta}}{e^{-i\theta/2}-e^{i\theta/2}} = \frac{\sin(N+1/2)\theta}{\sin(\theta/2)}. \tag{2.82}$$

Thus (2.81) is

$$S_N(\theta) = \frac{1}{2\pi}\int_{-\pi}^{\pi} \frac{\sin\left[(N+1/2)(\theta'-\theta)\right]}{\sin\left[(\theta'-\theta)/2\right]} f(\theta')d\theta'. \tag{2.83}$$

We shall consider S_n when N is very large. At a point $\theta' \neq \theta$, the integrand above is a rapidly varying function of θ'. And at $\theta' = \theta$, the integrand is as large as $(2N+1)f(\theta)$. As we have just explained, this means that the contributions to the integral come from a small neighborhood of $\theta' = \theta$. We may therefore make the approximation $\sin\left[(\theta'-\theta)/2\right] \approx (\theta'-\theta)/2$, and (2.83) becomes

$$S_N(\theta) \approx \frac{1}{\pi}\int_{-\pi}^{\pi} \frac{\sin\left[(N+1/2)(\theta'-\theta)\right]}{(\theta'-\theta)} f(\theta')d\theta'. \tag{2.84}$$

By (2.62) and (2.84), we get

$$\lim_{N\to\infty} S_n(\theta) = f(\theta), \tag{2.85}$$

which is (2.73).

In the argument above, we have implicitly assumed that $f(\theta)$ is continuous. Let $f(\theta)$ be discontinuous at θ_0, with $f(\theta_0^+)$ the value of $f(\theta)$ as θ approaches θ_0 from the right, and $f(\theta_0^-)$ the value of $f(\theta)$ as θ approaches θ_0 from the left. It is then straightforward to prove that

$$\lim_{N\to\infty} S_N = \frac{1}{2}\left[f(\theta_0^+) + f(\theta_0^-)\right]. \tag{2.86}$$

Thus at a point of discontinuity, the Fourier series of a function is equal to the mean of the left and right limiting values of the function.

Problem for the Reader

Find the Fourier coefficient a_n for the function

$$\begin{aligned} f(\theta) &= 1, \ (0 < \theta < \pi) \\ &= -1, (-\pi < \theta < 0). \end{aligned} \tag{2.87}$$

What are the values of the series at $\theta = 0, \pi/2$, and π? Can you explain why the series takes such values at these points?

Solution

We have

$$a_n = -\frac{1}{2\pi}\int_{-\pi}^{0} e^{-in\theta}d\theta + \frac{1}{2\pi}\int_{0}^{\pi} e^{-in\theta}d\theta = \frac{1-(-1)^n}{\pi in},$$

or

$$\begin{aligned} a_n &= \frac{2}{\pi in}, (n \text{ odd}) \\ &= 0, (n \text{ even}). \end{aligned}$$

By (2.73), the Fourier series for $f(\theta)$ is

$$\frac{2}{\pi i}\left[\frac{e^{i\theta}}{1} - \frac{e^{-i\theta}}{1} + \frac{e^{3i\theta}}{3} - \frac{e^{3i\theta}}{3} + \cdots\right],$$

or

$$\frac{4}{\pi}\left[\frac{\sin\theta}{1} + \frac{\sin 3\theta}{3} + \cdots\right]. \tag{2.88}$$

At the point $\theta = \pi/2$, $f(\theta)$ is continuous. Therefore, the value of the Fourier series at $\theta = \pi/2$ is equal to the value of $f(\theta)$ at $\theta = \pi/2$. This gives us the identity

$$\frac{4}{\pi}\left[1 - \frac{1}{3} + \frac{1}{5} - \cdots\right] = 1.$$

The Fourier series (2.88) vanishes at $\theta = 0$. To understand why, we note that the function $f(\theta)$ is discontinuous at $\theta = 0$, with

$$f(0^-) = -1, f(0^+) = 1.$$

By (2.86), the Fourier series of $f(\theta)$ must vanish at $\theta = 0$. The Fourier series (2.88) also vanishes at $\theta = \pi$. To understand why, we note that $f(\theta)$ is equal to unity for $0 < \theta < \pi$. Thus we have

$$f(\pi^-) = 1.$$

The value $f(\pi^+)$ is not explicitly given. But the function $f(\theta)$ is regarded as periodic with the period 2π. By (2.74), we have

$$f(\theta) = -1, (\pi < \theta < 2\pi),$$

and hence

$$f(\pi^+) = -1.$$

By (2.86), the Fourier series of $f(\theta)$ must vanish at $\theta = \pi$.

Next we derive another series known as the complete Fourier series for a function $f(\theta)$. We substitute

$$e^{in\theta} = \cos n\theta + i \sin n\theta$$

into (2.73) and get

$$f(\theta) = a_0 + \sum_{n=1}^{\infty} \left[A_n \cos n\theta + B_n \sin n\theta\right], \tag{2.89}$$

where

$$A_n = a_n + a_{-n}, B_n = i(a_n - a_{-n}), (n > 0). \tag{2.90}$$

Equation (2.89) is the complete Fourier series for $f(\theta)$. From (2.72) and (2.90), we have

$$A_n = \frac{1}{\pi} \int_{-\pi}^{\pi} f(\theta) \cos n\theta d\theta, B_n = \frac{1}{\pi} \int_{-\pi}^{\pi} f(\theta) \sin n\theta d\theta. \tag{2.91}$$

If $f(\theta)$ is an even function of θ, we find from (2.91) that

$$B_n = 0,$$

and the complete Fourier series (2.89) is reduced to

$$f(\theta) = a_0 + \sum_{n=1}^{\infty} A_n \cos n\theta, \tag{2.92}$$

which is called a Fourier cosine series. The coefficients A_n and a_0 are given by

$$A_n = \frac{2}{\pi}\int_0^{\pi} f(\theta)\cos n\theta d\theta, \tag{2.93}$$

and

$$a_0 = \frac{1}{\pi}\int_0^{\pi} f(\theta) d\theta, \tag{2.94}$$

which are integrals over positive values of θ only.

Similarly, if the function $f(\theta)$ is an odd function of θ, we have from (2.72) and (2.91) that

$$a_0 = A_n = 0.$$

Therefore, the complete Fourier series (2.89) is reduced to the Fourier sine series

$$f(\theta) = \sum_{n=1}^{\infty} B_n \sin n\theta, \tag{2.95}$$

where

$$B_n = \frac{2}{\pi}\int_0^{\pi} f(\theta)\sin n\theta d\theta. \tag{2.96}$$

An example of an odd function of θ is the function of (2.87). As we have found, the series (2.88) for this function is indeed a Fourier sine series.

If a function $f(\theta)$ is given only in the interval $0 < \theta < \pi$, we may, as a mathematical artifact, define this function for θ negative to be

$$f(-\theta) = -f(\theta).$$

In this way, we have given meaning to the function $f(\theta)$ in the region $-\pi < \theta < 0$; the function so defined is an odd function of θ. As we have seen, an odd function can be represented by the Fourier sine series (2.95). This means that a function $f(\theta)$ given for $0 < \theta < \pi$ can be represented by the Fourier sine series (2.95) with the coefficient B_n given by (2.96).

We may also extend the domain of a function $f(\theta)$ given for $0 < \theta < \pi$ by choosing it to satisfy

$$f(-\theta) = f(\theta).$$

The function $f(\theta)$ is then an even function of θ with the domain extended to $[-\pi, \pi]$. This function can be represented by the Fourier cosine series (2.92) with the coefficients A_n and a_0 given by (2.93) and (2.94).

We may also extend the results above to a function of x, which has the domain of $[-L, L]$, where L does not have to be equal to π. We define

$$\theta \equiv \frac{\pi}{L}x; \tag{2.97}$$

then the domain of θ is $[-\pi, \pi]$. By replacing θ in (2.73) with $\pi x/L$, we find that

$$F(x) = \sum_{n=-\infty}^{\infty} a_n e^{in\pi x/L}, (-L < x < L). \tag{2.98}$$

Making the same replacement in (2.72), we find that

$$a_n = \frac{1}{2L}\int_{-L}^{L} e^{-in\pi x/L} F(x)dx. \tag{2.99}$$

If the domain of a function $F(x)$ is $0 < x < L$, we may express this function as a Fourier cosine series

$$F(x) = a_0 + \sum_{n=1}^{\infty} A_n \cos(n\pi x/L), (0 < x < L). \tag{2.100}$$

The coefficients in the series above are given by

$$a_0 = \frac{1}{L}\int_0^L F(x)dx \tag{2.101}$$

and

$$A_n = \frac{2}{L}\int_0^L F(x)\cos(n\pi x/L)dx. \tag{2.102}$$

This function can also be expressed by the Fourier sine series

$$F(x) = \sum_{n=1}^{\infty} B_n \sin(n\pi x/L), (0 < x < L) \tag{2.103}$$

where

$$B_n = \frac{2}{L}\int_0^L F(x)\sin(nx/L)dx. \tag{2.104}$$

We close this section with two remarks:

a. Let us compare the Fourier series $\sum_{n=-\infty}^{\infty} a_n e^{in\theta}$ with the Maclaurin series, with which the readers are perhaps more familiar. The convergence of the Maclaurin series depends on the magnitude of z. Specifically, the series converges faster if the magnitude of z is smaller. The situation is different for a Fourier series. This is because the magnitude of $e^{in\theta}$ is unity for all real values of θ. Therefore, there is no a priori reason why a Fourier series should always converge faster for $\theta = 0$ than for $\theta = \pi/2$, say. As a matter of fact, if we put $z \equiv e^{i\theta}$, the Fourier series (2.73) becomes $\sum_{n=-\infty}^{\infty} a_n z^n$, which is a Laurent series. Since the magnitude of $z = e^{i\theta}$ is equal to unity for all real values of θ, there is no a priori reason why the convergence of the series favors any particular value of θ.

In Chapters 4 and 5, we shall show how to express a solution of a partial differential equation by a Fourier series. We shall find that each term of the series represents a mode of the solution of the equation. The sum over the first few modes often approximates the solution equally well for all values of the independent variable.

b. Next we discuss an interesting phenomenon of the Fourier series known as the Gibbs phenomenon. As we know, $S_N(\theta)$ of (2.80) is the sum of a finite number of terms each of which is a continuous function of θ. Thus $S_N(\theta)$ is a continuous function of θ. Let $f(\theta)$ be discontinuous at θ_0. When N is very large, we expect S_N to approximate $f(\theta)$ very well. But how does the continuous function $S_N(\theta)$ approximate the discontinuous function $f(\theta)$ in the neighborhood of θ_0?

To be more specific, take the example of the function defined in (2.87). When N is large, we expext that $S_N(\theta)$ is approximately 1 when $0 < \theta < \pi$, and is approximately -1 when $-\pi < \theta < 0$. How does this continuous function $S_N(\theta)$ transit from roughly -1 to roughly 1 within a very short interval of θ around $\theta = 0$?

One may expect that the function $S_N(\theta)$ increases monotonically from approximately -1 to approximately 1 over a small neighborhood of the origin. But as it turns out, this is not exactly the case.

If $f(\theta)$ is the function of (2.87), we have from (2.84) that

$$\begin{aligned} S_N(\theta) \approx & -\frac{1}{\pi}\int_{-\pi}^{0} \frac{\sin\left[(N+1/2)(\theta'-\theta)\right]}{(\theta'-\theta)} d\theta' \\ & + \frac{1}{\pi}\int_{0}^{\pi} \frac{\sin\left[(N+1/2)(\theta'-\theta)\right]}{(\theta'-\theta)} d\theta'. \end{aligned}$$

Let

$$(N + 1/2)(\theta' - \theta) = t;$$

then

$$S_N(\theta) \approx -\frac{1}{\pi}\int_{-(N+1/2)(\pi+\theta)}^{-(N+1/2)\theta} \frac{\sin t}{t} dt + \frac{1}{\pi}\int_{-(N+1/2)\theta}^{(N+1/2)(\pi-\theta)} \frac{\sin t}{t} dt.$$

In the limit of N approaching ∞, $(N + 1/2)(\pi + \theta)$ and $(N + 1/2)(\pi - \theta)$ both approach ∞, provided that θ is not near the endpoint $-\pi$ or the endpoint π. Thus we have

$$S_N(\theta) \approx -\frac{1}{\pi}\int_{-\infty}^{-(N+1/2)\theta} \frac{\sin t}{t} dt + \frac{1}{\pi}\int_{-(N+1/2)\theta}^{\infty} \frac{\sin t}{t} dt. \tag{2.105}$$

Since its integrand is an even function of t, the first integral in (2.105) is equal to

$$-\frac{1}{\pi}\int_{(N+1/2)\theta}^{\infty} \frac{\sin t}{t} dt.$$

The two integrals in (2.105) can be combined to give

$$S_N(\theta) \approx \frac{1}{\pi}\int_{-\Theta}^{\Theta} \frac{\sin t}{t} dt = \frac{2}{\pi}\int_{0}^{\Theta} \frac{\sin t}{t} dt, \tag{2.106}$$

where

$$\Theta = (N + 1/2)\theta. \tag{2.107}$$

The integral of (2.106) is an odd function of Θ. Thus it suffices to discuss the behavior of this integral for nonnegative values of Θ only.

First, we consider $S_N(\theta)$ at $\theta = 0$. We have from (2.106) that

$$\lim_{N\to\infty} S_N(0) = 0,$$

which agrees with the average value of $f(0^+)$ and $f(0^-)$.

Next we consider $S_N(\theta)$ when θ takes a fixed and nonzero value. As $N \to \infty$, Θ goes to infinity. Thus we have

$$\lim_{N\to\infty} S_N(\theta) \approx \frac{2}{\pi}\int_0^\infty \frac{\sin t}{t}dt = 1,$$

which agrees with the value of $f(\theta)$ for $0 < \theta < \pi$.

Thus, when N is very large, the value of $S_N(\theta)$ changes from zero when θ is equal to zero to approximately unity when θ is equal to a fixed value, as expected.

But what are the values of $S_N(\theta)$ when the value of θ is in between? Let Θ, instead of θ, take a fixed, nonzero and finite value. This means that θ is equal to a fixed, nonzero and finite value divided by $(N + 1/2)$.

When Θ is finite, we have by (2.106) that $S_N(\theta)$ is approximately a function of Θ. Now $\sin t$ is positive for $0 < t < \pi$, is negative for $\pi < t < 2\pi$, and keeps changing its sign after a period of π. Thus the integrand $\sin t/t$ is positive for $0 < t < \pi$, is negative for $\pi < t < 2\pi$, and keeps changing its sign after a period of π. Since $1/t$ monotonically decreases as t increases, the maximum value of the integral

$$\frac{2}{\pi}\int_0^\Theta \frac{\sin t}{t}dt$$

is reached at $\Theta = \pi$, as the interval $0 < t < \pi$ is the maximum interval possible for $\sin t/t$ to be positive throughout. The value of $S_N(\theta)$ at $\Theta = \pi$ is

$$S_N\left(\frac{\pi}{N + 1/2}\right) = \frac{2}{\pi}\int_0^\pi \frac{\sin t}{t}dt.$$

Numerically, this value is approximately 1.179, almost eighteen percent higher than the value of unity, the value of $f(\theta)$ for θ positive. Note that the value of θ at this point is $\pi/(N + 1/2)$, which goes to zero as $N \to \infty$. This shows that the value of the Fourier partial sum $S_N(\theta)$ does not change monotonically from 0 to 1. Rather, it starts at the value of zero at $\theta = 0$ and overshoots its target value by almost eighteen percent at $\theta = \pi/(N + 1/2)$.

Neither does $S_N(\theta)$ move down from its peak value to its target value monotonically. As Θ becomes larger than π, $S_N(\theta)$ decreases as the integrand of (2.106) becomes negative in the region of integration $t > \pi$. The function $S_N(\theta)$ reaches a minimum at $\theta = 2\pi/(N + 1/2)$ and then starts increasing again. In fact, it oscillates around the value of unity many times before it approaches the asymptotic value of unity. If we plot the function $S_N(\theta)$ as a function of θ, the distances between peaks and valleys shrink to zero as N increases to infinity, while the heights of the peaks and valleys stay constant.

This oscillatory behavior of $S_n(\theta)$ is called the Gibbs phenomenon.

G. The Laplace Transform

Let a function $f(x)$ be given for $x \geq 0$. The Laplace transform for such a function is defined to be

$$L(s) = \int_0^\infty e^{-sx} f(x) dx, \tag{2.108}$$

provided that the integral converges. In the above, s is independent of x.

Problem for the Reader

Find the Laplace transforms of the following functions:

a. $x^n/n!$, $n \geq 0$,

b. e^{-ax},

c. $\cos kx$ and $\sin kx$,

d. e^{x^2}.

What is the analytic property of each of these Laplace transforms as a function of the complex variable s?

Solution

a. $x^n/n!$

It is known that

$$\int_0^\infty e^{-t} t^n dt = n!.$$

Therefore,

$$\int_0^\infty dx e^{-sx} x^n / n! = \frac{1}{s^{n+1}}. \tag{2.109}$$

The integral above is convergent only if $\mathrm{Re}\, s > 0$. If $\mathrm{Re}\, s \leq 0$, the Laplace integral above is divergent. Therefore, the Laplace transform of x^n is defined by the Laplace integral above only in the right half-plane $\mathrm{Re}\, s > 0$. We see that in this half-plane, s^{-n-1} is an analytic function of s.

When the integral above is not convergent, we call the right side of (2.109) the Laplace transform of $x^n/n!$. In other words, we simply define the Laplace transform of $x^n/n!$ in the left half-plane $\mathrm{Re}\, s \leq 0$ to be s^{-n-1}. The Laplace transform of $x^n/n!$ so defined is an analytic function of s with the exception of an nth-order pole at $s = 0$. Thus we have analytically continued the Laplace transform of $x^n/n!$, defined by the Laplace integral in the right half-plane $\operatorname{Re} s > 0$, to the left half-plane $\operatorname{Re} s < 0$.

b. e^{-ax}

The Laplace transform of e^{-ax} is

$$\int_0^\infty dx e^{-sx} e^{-ax} = \frac{1}{s+a}. \tag{2.110}$$

The integral above is convergent only if $\operatorname{Re}(s+a) > 0$. We analytically continue this Laplace transform by defining it as $(s+a)^{-1}$ for all s. It is analytic everywhere with the exception of a simple pole at $s = a$.

c. $\cos kx$ and $\sin kx$

Replacing a in (2.110) by ik, we have

$$\int_0^\infty dx e^{-sx} e^{-ikx} = \frac{1}{s+ik}.$$

Restricting to real values of s and k, we obtain from the real part and the imaginary part of the equation above

$$\int_0^\infty dx e^{-sx} \cos kx = \frac{s}{s^2+k^2}, \tag{2.111}$$

and

$$\int_0^\infty dx e^{-sx} \sin kx = \frac{k}{s^2+k^2}. \tag{2.112}$$

The integrals in (2.111) and (2.112) converge if s is greater than zero. For complex values of s and k, we simply define the Laplace transforms of $\cos kx$ and $\sin kx$ to be the right sides of (2.111) and (2.112), respectively. The Laplace transforms so defined are analytic functions of s with simple poles at $s = \pm ik$.

d. e^{x^2}

As $x \to \infty$, the function e^{x^2} blows up so rapidly that $e^{-sx}e^{x^2}$ for any value of s always blows up as $x \to \infty$. As a result, the integral $\int_0^\infty e^{-sx}e^{x^2}dx$ is divergent no matter what s is. Thus the function e^{x^2} has no Laplace transform.

We note that the Laplace transforms of 1, e^{-ax}, and $\cos kx$ all approach $1/s$ as $|s| \to \infty$. As a general rule, the Laplace integral in (2.108) approaches $f(0)/s$ as $|s| \to \infty$ provided that $f(0)$ is finite and does not vanish. This can be shown by the methods we shall give in Chapter 8.

We observe that each $L(s)$ in the examples given is an analytic function of s in the region where its Laplace integral converges. This is true in a more general context. The integrand in (2.108) is an analytic function of s. Consequently, the integral in (2.108) is an analytic function of s in the region of s where it is convergent. Let

$$s = s_1 + is_2,$$

where s_1 and s_2 are the real part and the imaginary part of s, respectively. Now we have

$$\left|e^{-sx}\right| = e^{-s_1 x}.$$

As the value of x in the Laplace integral is always positive, $e^{-s_1 x}$ becomes smaller as s_1 becomes larger. Thus, if the integral of (2.108) converges at $s = \xi_0$, then the integral of (2.108) converges for all s satisfying $s_1 > \operatorname{Re}\xi_0$. Hence $L(s)$ is an analytic function of s in the right half-plane $\operatorname{Re} s > \operatorname{Re}\xi_0$. In particular, if $L(s)$ exists when s is purely imaginary, $L(s)$ is analytic in the right half-plane $s_1 \geq 0$. If such is the case, all of the possible singularities of $L(s)$ lie to the left of the imaginary axis.

The Laplace transform is a special case of the Fourier transform. Let

$$\begin{aligned} F(x) &= f(x), \ x > 0, \\ &= 0, \qquad x < 0, \end{aligned} \tag{2.113}$$

and let the Fourier integral of $F(x)$ converge. Then the Fourier integral of (2.53) is the same as the Laplace integral of (2.108) with s identified with ik. Thus we have

$$L(ik) = \tilde{F}(k). \tag{2.114}$$

This shows that if $\tilde{F}(k)$ exists when k is real, $L(s)$ exists when s is purely imaginary. If such is the case, (2.54) gives

$$f(x) = \int_{-\infty}^{\infty} e^{ikx} L(ik) \frac{dk}{2\pi}, \quad x > 0. \tag{2.115}$$

Replacing k by $-is$, we obtain

$$f(x) = \int_{-i\infty}^{i\infty} e^{sx} L(s) \frac{ds}{2\pi i}, \quad x > 0, \tag{2.116}$$

where the integration is over the imaginary s-axis. Equation (2.116) is an inversion formula that enables us to determine the values of $f(x)$ once the values of $L(s)$ on the imaginary axis of the s-plane are obtained.

We have implicitly assumed in the above that $f(x)$ is a continuous function of x. At the point x_0 where $f(x)$ is discontinuous, the left side of (2.116) should be replaced by

$$\frac{f(x_0^+) + f(x_0^-)}{2}.$$

In particular, if $f(0) \neq 0$, the function $F(x)$ of (2.113) is discontinuous at $x = 0$, and the inversion integral of (2.116) at $x = 0$ is equal to $f(0)/2$.

We may think of (2.116) as the contour integral

$$f(x) = \int_c e^{sx} L(s) \frac{ds}{2\pi i}, \quad x > 0, \tag{2.117}$$

where c is the imaginary axis of the complex s-plane. Since $L(s)$ is analytic in the right half-plane $\mathrm{Re}\, s \geq 0$ by assumption, all possible singularities of the integrand lie to the left of c.

The contour c does not have to be a straight line. We are allowed to deform c into another contour in conformity to the Cauchy integral theorem.

If we differentiate (2.108) with respect to s, we get

$$\frac{dL(s)}{ds} = -\int_0^\infty e^{-sx} x f(x)\, dx.$$

The integral above is convergent for $s_1 > \mathrm{Re}\, \xi_0$ if the Laplace integral converges at $s = \xi_0$. This is because $e^{-s_1 x}$ is smaller if s_1 is larger, as we have mentioned above. Therefore, the Laplace transform of $xf(x)$ is $-dL(s)/ds$.

On the other hand, if we differentiate (2.117) with respect to x, we get

$$\frac{df(x)}{dx} = \int_c e^{sx} sL(s) \frac{ds}{2\pi i}. \tag{2.118}$$

However, (2.118) is not always valid, as the integral in (2.118) is not always convergent. Indeed, if $f(0)$ is finite and is not equal to zero, $sL(s)$ approaches the constant $f(0)$ as $s \to \pm i\infty$, while e^{sx} is oscillatory and does not vanish as $s \to \pm i\infty$. Thus the integral in (2.118) does not converge if $f(0)$ does not vanish, and (2.118) is not meaningful. Therefore, $sL(s)$ is not always the Laplace transform of $df(x)/dx$.

Let us find out what the Laplace transform of df/dx is. We have

$$\int_0^\infty dx e^{-sx} df(x)/dx = -f(0) + \int_0^\infty dx\, s e^{-sx} f(x),$$

where we have performed an integration by parts. Thus the Laplace transform of df/dx is

$$-f(0) + sL(s), \tag{2.119}$$

which is equal to $sL(s)$ only if $f(0) = 0$.

Problem for the Reader

Express the Laplace transform of d^2f/dx^2 with $L(s)$.

Solution

After performing two integrations by parts successively, we find that

$$\int_0^\infty dx e^{-sx} d^2 f(x)/dx^2 = -f'(0) - sf(0) + s^2 L(s). \tag{2.120}$$

We mention that the integral in (2.117) is equal to zero if $x < 0$. This is because if $x < 0$, $e^{s_1 x}$ approaches zero as s_1 approaches $+\infty$. Therefore, the integrand in (2.117) vanishes at the infinity of the right-half s-plane and we may close the contour to the right. Since the integrand has no singularities in the right half-plane, the integral is zero by Cauchy's integral theorem. We may add that this is consistent with the definition of $F(x)$ for $x < 0$ as given by (2.113).

We have shown that the contour integral (2.117) is valid if $L(s)$ is analytic in the region $\text{Re}\, s \geq 0$. If this condition is not met, i.e., if $L(s)$ has singularities in the region $\text{Re}\, s \geq 0$, the only modification needed is to make c in (2.117) a vertical line that lies to the right of all of the singularities of $L(s)$. To see this, let us multiply $f(x)$ by the exponentially vanishing function e^{-ax} and call

$$g(x) = e^{-ax} f(x). \tag{2.121}$$

We shall assume that it is possible to choose a so that $g(x)$ vanishes sufficiently rapidly as $x \to \infty$ and the Laplace transform of $g(x)$, given by

$$G(s) = \int_0^\infty e^{-sx} g(x) dx, \tag{2.122}$$

is analytic in the right half-plane Re $s \geq 0$. By the inversion formula of Laplace transform, we get

$$g(x) = \int_{-i\infty}^{i\infty} e^{sx} G(s) \frac{ds}{2\pi i}. \tag{2.123}$$

Replacing $g(x)$ with $e^{-ax}f(x)$, we obtain from (2.122) that

$$G(s) = \int_0^{\infty} e^{-sx} e^{-ax} f(x) dx = L(s+a).$$

Since $G(s)$ is by assumption analytic in the region Re $s \geq 0$, $L(s) = G(s-a)$ is analytic in the region Re $s \geq a$.

Replacing $g(x)$ with $e^{-ax}f(x)$, we obtain from (2.123) that

$$f(x) = \int_{-i\infty}^{i\infty} e^{(s+a)x} L(s+a) \frac{ds}{2\pi i}.$$

This formula can be written as

$$f(x) = \int_c e^{sx} L(s) \frac{ds}{2\pi i}, \tag{2.124}$$

where c is the straight line $s = a + is_2$. Since $L(s)$ is analytic in the region Re $s \geq a$, all of the possible singularities of $L(s)$ are at the left of the contour c.

Laplace transform is useful in solving differential equations with initial conditions. As an example, let us consider the solution of the differential equation

$$\frac{d^2y}{dx^2} + y = 1, x > 0,$$

satisfying the initial conditions $y(0) = 1$, $y'(0) = 2$.

Problem for the Reader

Solve the initial-value problem above by Laplace transform.

Solution

By (2.120), the Laplace transform of the left side of the differential equation is

$$-2 - s + (s^2 + 1)L(s).$$

Note that the first two terms in the expression above are given by the initial conditions. The Laplace transform of the right side is equal to s^{-1}. Equating these two Laplace transforms gives

$$L(s) = \frac{s^{-1} + 2 + s}{1 + s^2}.$$

By (2.124), we have

$$y(x) = \int_c e^{sx} \frac{s^{-1} + 2 + s}{1 + s^2} \frac{ds}{2\pi i},$$

where c is a vertical line lying at the right of the imaginary s axis. For $x > 0$, e^{sx} vanishes as $s_1 \to -\infty$. Therefore, we may evaluate the integral above by closing the contour to the left. By Cauchy's residue theorem the integral is equal to the sum of the residues of the integrand at $s = 0, i$, and $-i$. Thus we get

$$y(x) = 1 + 2\sin x, \quad x \geq 0.$$

Note that the initial conditions have already been incorporated. Therefore, unlike under the method used in Chapter 1, there are no arbitrary constants in the solution obtained here, and we are spared the chore of determining the arbitrary constants with the use of the initial conditions. It is easy to verify that the initial conditions are indeed satisfied by this solution.

Homework Problems for This Chapter

Solutions to the Homework Problems can be found at www.lubanpress.com.

1. Prove that the limit of (2.5) is the same for any Δz if the Cauchy-Riemann equations are satisfied by the real part and the imaginary part of $f(z)$.

2. Prove that $\cos\dfrac{2\pi}{5} = \dfrac{\sqrt{5} - 1}{4}$.

Hint: Let $e^{\pi i/5} \equiv c + is$; then

$$(c + is)^5 = -1.$$

Equate the imaginary parts of the two sides of this equation.

3. A function that is analytic everywhere in the finite complex plane is called an entire function. Prove the Liouville theorem that an entire function $f(z)$ is a constant if it is bounded at infinity.

Hint: Show that $f'(z) = 0$ with the formula

$$f'(z) = \frac{1}{2\pi i} \lim_{R\to\infty} \oint_{c_R} \frac{f(z')}{(z'-z)^2} dz',$$

where c_R is a circle with its center at the origin.

4. Evaluate the following integrals with contour integration:

a. $\displaystyle\int_{-\infty}^{\infty} \frac{dx}{(x^2+1)(x-2i)(x-3i)(x-4i)}.$

Ans. $-\dfrac{i\pi}{60}.$

b. $\displaystyle\int_{-\infty}^{\infty} \frac{1-\cos 2x}{x^2} dx.$

Ans. $2\pi.$

c. $\displaystyle\int_{-\infty}^{\infty} \frac{\sin^3 x}{x^3} dx.$

Ans. $3\pi/4.$

d. $\displaystyle\int_{0}^{2\pi} \frac{1}{(a+b\cos\theta)^2} d\theta \ (a > b > 0).$

Ans. $2\pi a/(a^2-b^2)^{3/2}.$

e. $\displaystyle\int_{-\infty}^{\infty} \frac{x \sin x}{x^2 - 4\pi^2} dx.$

Ans. $\pi.$

5. Explain why the integral of (2.35) is not equal to the imaginary part of

$$\int_c \frac{e^{iz}}{z} dz$$

if c is the contour in Figure 2.7

6. Evaluate the integral

$$\int_{-\infty}^{\infty} \frac{\sin x}{x+i} dx.$$

Ans. $\dfrac{\pi}{e}.$

Explain why it is not fruitful to evaluate the integral

$$J = \int_{-\infty}^{\infty} \frac{e^{ix}}{x+i} dx.$$

7. Consider $f(z) = \log \dfrac{1+\sqrt{1+z^2}}{2}$.

a. Find all possible branch points of this function.
Ans. $0, \pm i, \infty$.

b. If we define $\sqrt{1+z^2}\ |_{z=0} = 1$, show that the origin is not a branch point of this function. Draw a set of branch cuts to make the function single-valued.

8. Show that the Taylor series (2.19) is convergent inside the circle with center at z_0 and with the radius equal to $|z_0 - z_1|$, where z_1 is the singularity of $f(z)$ nearest to z_0.

Hint: Estimate the magnitude of $f^{(n)}(z_0)$ with the use of the Cauchy integral formula.

9. Let $f(z)$ and $g(z)$ be analytic in a region R, and let z_0 be an interior point of R. If $f(z) = g(z)$ has at least one root in any neighborhood of z_0, no matter how small this neighborhood is, prove that $f(z) = g(z)$ in R.

Hint: Let $G(z) \equiv f(z) - g(z)$ and consider the Taylor series expansion of $G(z)$ around z_0. Show that unless this series vanishes identically, it cannot vanish at z if z is sufficiently close to z_0 but not equal to z_0.

10. Let $I_n = \displaystyle\int_0^\infty \frac{dx}{1+x^n}$.

a. Prove that $I_n = \dfrac{\pi}{n\sin(\pi/n)}$. What is I_n in the limit $n \to \infty$?

b. Show that as $n \to \infty$, the limit of the integral I_n is equal to unity integrated over $[0, 1]$.

11. Evaluate the following integrals making use of branch cuts:

a. $\displaystyle\int_0^\infty \frac{\ln^2 x}{1+x^2} dx$. Ans. $\pi^3/8$.

b. $\displaystyle\int_0^1 \frac{1}{1+x^2}\sqrt{\frac{x^3}{1-x}}\, dx$.

Hint: The integrand has a square-root branch cut from 0 to 1. One considers the sum of the following five contours: 1. The straight line joining $i\epsilon$ to $1 + i\epsilon$; 2. The straight line joining $1 - i\epsilon$ to $-i\epsilon$; 3. The right half-circle in the clockwise direction with center at 1 and with the radius ϵ; 4. The left half-circle in the clockwise direction with center at

0 and with the radius ϵ; 5. The infinite circle in the counterclockwise direction. You may prove that as ϵ approaches zero, the contributions from contour 3 and contour 4 are zero, and the contribution from contour 2 is equal to that from contour 1. The contribution from contour 5 can be calculated. The sum of the contributions from these five contours is equal to, by Cauchy's residue theorem, $2\pi i$ times the sum of the residues of the integrand in the cut plane.

Ans. $\pi - \dfrac{\pi\cos(\pi/8)}{2^{1/4}}$.

c. $\displaystyle\int_0^\infty \frac{\ln x}{1+x^5}dx.$

Hint: Add to the contour of the positive real axis the ray joining zero to infinity with the argument $2\pi/5$. On this ray, $z^5 = r^5$, and $\ln z = \ln r + 2\pi/5$.

12. Let z_0 be an isolated singularity of $f(z)$, and let z be a point in the neighborhood of z_0. Show that

$$f(z) = \frac{1}{2\pi i}\int_{C_R} \frac{f(z')}{z'-z}dz' - \frac{1}{2\pi i}\int_{C_\epsilon} \frac{f(z')}{z'-z}dz',$$

where C_R and C_ϵ are counterclockwise circles, the centers of which are at z_0 and the radii of which are R and ϵ, respectively. Also, R is sufficiently large so that z is inside C_R, and ϵ is sufficiently small so that z is outside C_ϵ. Derive the Laurent series expansion of $f(z)$ from the equation above and discuss the region where the series is convergent.

13. Find the Fourier coefficient a_n for the following functions. What is the value of the Fourier series at $\theta = \pi$?

a. e^θ.

Ans. $a_n = \dfrac{(-1)^n}{2\pi}\dfrac{e^\pi - e^{-\pi}}{1-in}$. The value of the series at $\theta = \pi$ is $\dfrac{1}{2}(e^\pi + e^{-\pi})$.

b. $\dfrac{1}{a+b\cos\theta}$.

Ans. $a_n = (-1)^n \left(a - \sqrt{a^2-b^2}\right)^n / \left(b^n\sqrt{a^2-b^2}\right)$ $(n>0)$, $a_{-n} = a_n$. The value of the Fourier series at $\theta = \pi$ is $(a-b)^{-1}$.

14. Let u and v be functions of x and y. Prove that

$$(\vec{\nabla} u)\cdot(\vec{\nabla} v) \equiv u_x v_x + u_y v_y = 2(u_z v_{z^*} + u_{z^*} v_z).$$

15. Show that if $f(z)$ is analytic everywhere in the complex plane with the exception of a branch cut from a to b on the real axis, does not blow up as fast as a pole at the endpoints a and b, and vanishes at infinity, then

$$f(z) = \int_a^b \frac{r(x')}{x' - z} dx',$$

with $2\pi i r(x)$ the discontinuity of $f(z)$ across the branch cut.

Hint: Consider the integral $\dfrac{1}{2\pi i} \displaystyle\int \frac{f(z')dz'}{z' - z}$ with the contour of integration a closed contour wrapping around the interval $[a, b]$ in the clockwise direction. Show that this integral is equal to $f(z)$ provided that z is outside this contour. Now make this contour to be infinitesimally close to the line from a to b on the real axis.

16. Find the Fourier transforms of the functions $e^{-|x|}$ and $(1 + x^2)^{-2}$.
Ans. $2(1 + k^2)^{-1}$ and $\pi(1 + |k|)e^{-|k|}/2$.

17. The convolution of $F(x)$ and $G(x)$ is defined to be

$$H(x) = \int_{-\infty}^{\infty} F(x - x')G(x')\, dx'.$$

a. Calculate the Fourier transform of $H(x)$ and show that it is equal to $\tilde{F}(k)\tilde{G}(k)$, where $\tilde{F}(k)$ and $\tilde{G}(k)$ are the Fourier transforms of $F(x)$ and $G(x)$, respectively.

b. Use the result above and show that $\int_{-\infty}^{\infty} F(-x)G(x)\, dx$ $= \int_{-\infty}^{\infty} \tilde{F}(k)\tilde{G}(k)\dfrac{dk}{2\pi}$.

18. The convolution of $F_m(x)$, $m = 1, 2, \cdots, n$, is

$$H(x) = \int_{-\infty}^{\infty} dx_1 \int_{-\infty}^{\infty} dx_2 \cdots \int_{-\infty}^{\infty} dx_n\, \delta\left(x - \sum_{m=1}^{n} x_m\right) \prod_{n=1}^{n} F_m(x_m).$$

Show that the Fourier transform of $H(x)$ is $\Pi_{m=1}^{n}\tilde{F}_m(k)$, where $\tilde{F}_m(k)$ is the Fourier transform of $F_m(x)$.

19. Let $f(x)$ and $g(x)$ be defined for $x > 0$. The convolution of $f(x)$ and $g(x)$ is defined to be $\int_0^x f(x - x')g(x')\, dx'$. Show that the Laplace transform of the convolution of $f(x)$ and $g(x)$ is equal to the product of the Laplace transform of $f(x)$ and that of $g(x)$.

20. Solve with the method of Laplace transform the initial-value problem

$$\frac{d^2y}{dx^2} + y = \sqrt{x}, \; x > 0,$$

with the initial condition $y(0) = 1$, $y'(0) = 0$.

Hint: the Laplace inversion integral for $y(x)$ cannot be evaluated with Cauchy's residue theorem, as the integrand of this integral has a branch point. Make use of the result of Problem 19 and express $y(x)$ as a convolution of elementary functions.

Chapter 3

First-Order Partial Differential Equations

In this chapter and the next two chapters, we'll briefly review partial differential equations (PDEs). As PDEs are much more difficult to solve than ODEs, we shall start with the simplest of PDEs, those of the first order.

The good thing about a first-order PDE is this: It can always be "solved" in a closed form. This is true whether the PDE is linear or nonlinear, and in the former case, whether it is homogeneous or inhomogeneous.

Having made such a strong statement about the first-order PDE, I must add that the solutions of the first-order PDE are expressed by the solutions of a system of ODEs, which may not be exactly solvable. But since an ODE is so much simpler to solve than a PDE, we consider a PDE "solved" if its solution can be expressed as those of an ODE.

Here I would like to relate a story, the like of which must have happened innumerous times. I was asked, in 1965, to help out with a mathematical problem encountered by a team doing research on the laser. The problem that impeded their research progress at that point was a partial differential equation. While being very capable physicists, the researchers did not recognize that the PDE they had was of the first order and hence had a closed-form solution. After this was pointed out to them, they were able to solve the equation very quickly. This story illustrates how some useful and elementary mathematics often have been left out in a graduate science education.

We shall start by making a comparison of the solution of a first-order ODE with those of a first-order PDE.

A. Trivial Example

Let us consider the ODE

$$\frac{du(x)}{dx} = 0.$$

The answer is, trivially,

$$u(x) = c,$$

an arbitrary constant. The solution of the ODE is unique if an initial condition is imposed. For example, if we require

$$u(0) = 1,$$

then we have

$$u(x) = 1.$$

In contrast, let us consider the PDE

$$u_x(x, y) = 0, \tag{3.1}$$

where

$$u_x \equiv \frac{\partial u}{\partial x}.$$

The solution of (3.1) is, again trivially,

$$u(x, y) = f(y). \tag{3.2}$$

Let us describe what this obvious result means in some detail. Equation (3.2) says that $u(x, y)$ is a function of y only, and is independent of x. Therefore, if the value of $u(x, y)$ at an initial point on the y axis, say $(0, a)$, is equal to b, i.e.,

$$u(0, a) = b,$$

then $u(x, y)$ is equal to b on the entire line $y = a$, i.e.,

$$u(x, a) = b.$$

We call the line $y = a$ a characteristic curve of the PDE (3.1). It is remarkable that u on this characteristic curve is independent of the value of u anywhere else. To find the value of u on this characteristic curve, we only need the value of u at any one point on this characteristic curve, just as with the case of the first-order ODE discussed above. This initial point does not have to be on the y axis.

If the values of $u(x, y)$ on the y axis between $a_1 \leq y \leq a_2$ are given, then the values of $u(x, y)$ are known in the strip of the xy-plane with $a_1 \leq y \leq a_2$.

If the values of $u(x, y)$ are given on a curve, then the values of $u(x, y)$ are known in a strip obtained by sweeping the curve horizontally.

But if the curve intercepts a horizontal line at two points, then there is a problem. For there can be no continuous solution unless the values assigned for u at these two points happen to be equal.

The situation gets worse if the curve on which the initial data is assigned is a horizontal line. In such a case, the initial data is in conflict with the PDE unless all initial values on this horizontal line are equal to the same constant. In such a case, the value of the solution at any point not on the horizontal line remains unknown.

While this example is admittedly trivial, the solution of many first-order PDEs can be obtained by slightly generalizing what we've just done.

B. Linear Homogeneous PDEs

Consider the PDE

$$a(x, y)u_x(x, y) + b(x, y)u_y(x, y) = 0, \tag{3.3}$$

which is linear and homogeneous.

Let

$$\Delta u(x, y) \equiv u(x + \Delta x, y + \Delta y) - u(x, y)$$

be the variation of u as the coordinates are varied by $(\Delta x, \Delta y)$. By the Taylor expansion,

$$\Delta u = u_x \Delta x + u_y \Delta y,$$

where the higher order terms, negligible as Δx and Δy become infinitesimal, are ignored.

Question for the Reader

What values should we choose for Δx and Δy in order for Δu to vanish?

Solution

If u is a solution of (3.3), we have, for $b \neq 0$,

$$u_y = -\frac{a}{b} u_x.$$

Substituting this relationship into the expression for Δu, we find that

$$\Delta u = (\Delta x - \frac{a}{b} \Delta y) u_x.$$

In particular, if

$$\Delta x = \frac{a(x, y)}{b(x, y)} \Delta y, \tag{3.4a}$$

or, in the limit when the variations are infinitesimal,

$$\frac{dx}{a(x, y)} = \frac{dy}{b(x, y)}, \tag{3.4b}$$

the variation of u vanishes. This means that $u(x, y)$ is constant on a curve the slope of which is dictated by (3.4b). We may rewrite (3.4b) as

$$\frac{dy}{dx} = \frac{b(x, y)}{a(x, y)}.$$

This equation does not depend on u. It is self-contained and can be solved on its own, regardless of what u is. We call the solutions of (3.4b) the characteristic curve of the PDE (3.3) on each of which u is constant.

If b vanishes at some points in the region of interest, there is the difficulty of $1/b$ blowing up at these points. But if a does not vanish at these points, we may express (3.3) as $u_x = -bu_y/a$. Thus u is constant on the curve determined by

$$\Delta y = \frac{b}{a} \Delta x,$$

which is the same as (3.4a) if b does not vanish.

If both a and b vanish at a point, the slope of the characteristic curve may be ill defined at this point, and there may be more than one characteristic curve passing through this point. An example is given by homework problem 13.

Problem for the Reader

Find the characteristic curves of the PDE

$$u_t - u_x = 0.$$

In particular, find the characteristic curve that passes through the point (x_0, t_0).

Solution

The equation for the characteristic curve is

$$\frac{dt}{1} = -\frac{dx}{1},$$

the solutions of which are

$$a = x + t, \tag{3.5}$$

where a is a constant that differs from one characteristic curve to another. For example, the characteristic curve with $a = 0$ is the straight line of the slope -1 passing through the origin, and the characteristic curve with $a = 1$ is the straight line of the slope -1 intersecting the x axis at $x = 1$. By varying the value of a, we get a family of parallel straight lines with different intercepts of the x axis. Thus each characteristic curve is designated by a value of a, and the characteristic curves are a family of straight lines of one parameter.

Conversely, we may easily find the parameter a associated with the characteristic curve passing through a point (x, t). For example, the parameter associated with the characteristic curve passing through $(2, 3)$ is, by (3.5),

$$a = 2 + 3 = 5.$$

More generally, the parameter a for the characteristic curve passing through (x_0, t_0) is given by

$$a = x_0 + t_0.$$

Problem for the Reader

Find the analytic form of the general solution of the PDE

$$u_t - u_x = 0.$$

Solution

Since the values of u are the same on a characteristic curve, its value at (x, t) depends only on the value a associated with the characteristic on which the point (x, t) lies. The value a at the point (x, t) is equal to $(x + t)$. Thus we have

$$u(x, t) = F(x + t), \tag{3.6}$$

where F is an arbitrary function.

Problem for the Reader

If the initial values of u(x, t) are given by $u(x, 0) = f(x)$, find $u(x, t)$.

Solution

The solution can be expressed in two forms. We may express it in the analytic form by setting $t = 0$ in (3.6). We get

$$f(x) = F(x). \tag{3.7}$$

Thus

$$u(x, t) = f(x + t),$$

which is a wave travelling to the left with the uniform velocity unity.

The solution may also be described graphically by assigning a value of u for each characteristic line in the figure on the right. For example, the value of u on the characteristic passing through the origin is $f(0)$.

The analytic solution is simpler to use for the wave equation given above. Indeed, it is simpler to use for equations in the form of (3.3). But the graphical solutions are preferable for some of the more complicated first-order PDEs not in the form of (3.3).

Problem for the Reader

Find the characteristic curves of the PDE

$$u_x + (x + y)u_y = 0, \tag{3.8}$$

and draw some of these characteristic curves in the xy-plane.

Solution

The characteristic curves of the PDE (3.8) satisfy

$$\frac{dx}{1} = \frac{dy}{x+y}$$

or

$$\frac{dy}{dx} = x + y.$$

This is a linear and inhomogeneous ODE of the first order with the complementary solution e^x. A particular solution is

$$y_P(x) = \frac{1}{D-1}x = -(1 + D + D^2 + \cdots)x = -x - 1.$$

Thus the characteristic curves are given by

$$y = ce^x - x - 1, \tag{3.9}$$

or

$$e^{-x}(x + y + 1) = c. \tag{3.10}$$

The characteristic curves for $c = 0, 1, -1$ are drawn in Figure 3.1.

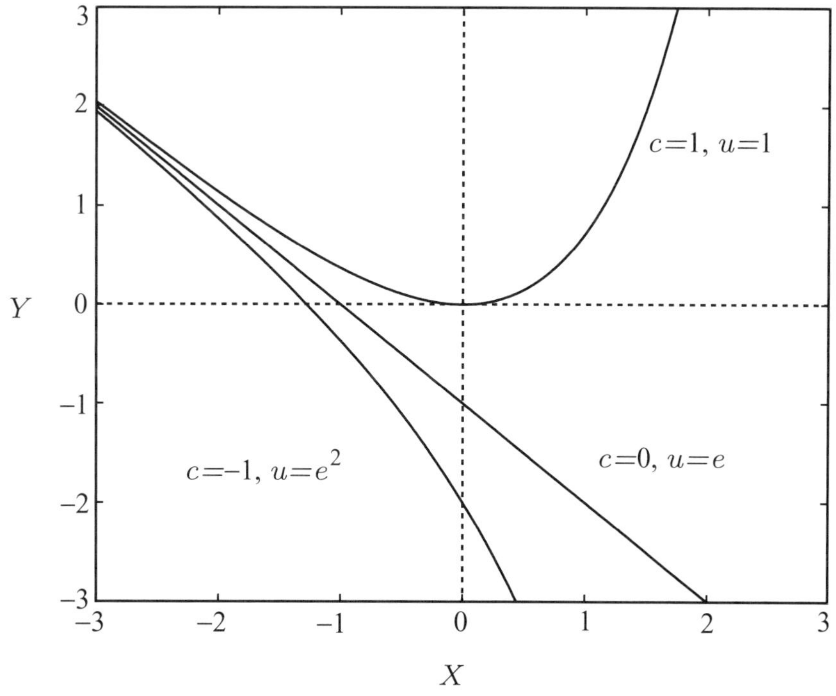

Figure 3.1

Problem for the Reader

If $u(x, y)$ of (3.8) satisfies the initial condition

$$u(0, y) = e^{-y}, \tag{3.11}$$

state the values of $u(x, y)$ on the characteristic curves drawn in Figure 3.1.

Solution

The coordinates of the intercepts of the characteristic curves (3.10) of $c = 0, 1, -1$ at the y axis are $y = -1, 0, -2$, respectively. According to (3.11), the values of u at these three points are e, 1, e^2, as indicated in the figure.

We may obtain $u(x, y)$ in an analytic form. As the value of u depends only on the characteristic curve it is on, and as each characteristic curve is defined by the parameter c, u depends only on c; we have

$$u(x, y) = F(c) = F(e^{-x}(x + y + 1)).$$

The function F is determined from the initial condition (3.11). We have

$$e^{-y} = F(y + 1),$$

or

$$F(y) = e^{1-y}.$$

Thus we have

$$u(x, y) = \exp\left[1 - e^{-x}(x + y + 1)\right]. \tag{3.12}$$

The solution (3.12) can be viewed geometrically as a surface in the three-dimensional xyu space.

What we've done in the examples above is easily extended to the PDE in the general form (3.3). In particular, let the initial data of the solution of (3.3) be given on a curve in the xy-plane. After we solve the equation (3.4) and obtain the characteristic curves of (3.3), the solution satisfying the given initial condition can be obtained in two equivalent ways:

1. The graphical way. We draw the characteristic curves in the xy-plane and indicate the value of u on each of the characteristic curves. These values are obtained from the initial condition. Alternately, we may plot each of the characteristic curves in the three-dimensional xyu-space, with the coordinate u the value of u on this characteristic curve. We note that the aggregate of these curves spans a two-dimensional surface.

2. The analytic way. The analytic solution can be obtained in three steps:

Step a

Obtain the characteristic curves of the PDE and express them in the form

$$f(x, y) = a,$$

where a is an arbitrary constant that varies from one characteristic curve to another.

Step b

The general solution of (3.3) is

$$u(x, y) = F(f(x, y)), \tag{3.13}$$

where F is an arbitrary function.

Step c

Let the initial data be $u(0, y) = g(y)$; then we get

$$g(y) = F(f(0, y)),$$

which determines the function F.

To prove that these two kinds of solutions are equivalent, we note that the form (3.13) says that u is a constant on a characteristic curve, on which $f(x, y)$ is a constant. This means that (3.13) satisfies (3.3). The PDE (3.3) does not dictate how u varies from one characteristic curve to the other. Hence the general solution (3.13) contains an arbitrary function F that is determined only after an initial condition is given.

If the curve on which the initial data is given intercepts a characteristic curve in the xy-plane at two points, and if the values of u specified at these two points are different, there is no solution satisfying both the PDE and the initial data unless u is allowed to be discontinuous on a characteristic curve. It is even worse if the curve on which we assign the initial values is itself a characteristic curve. In such a case, the initial values assigned must be, to be consistent, constant throughout the curve, and such an initial condition yields no information on what u is elsewhere.

Problem for the Reader

Show how to determine the characteristic curves for

$$A(x, y, z)u_x + B(x, y, z)u_y + C(x, y, z)u_z = 0,$$

which is a PDE with three independent variables. Give an example of an initial datum.

Solution

The equations for the characteristic curves are

$$\frac{dx}{A} = \frac{dy}{B} = \frac{dz}{C},$$

on which $u(x, y, z)$ is constant. An example of an initial datum for this PDE is

$$u(0, y, z) = h(y, z).$$

We may express the equations for the characteristic curves given above by

$$\frac{dy}{dx} = \frac{B(x, y, z)}{A(x, y, z)},$$

and

$$\frac{dz}{dx} = \frac{C(x, y, z)}{A(x, y, z)}.$$

This is a system of two coupled first-order ordinary differential equations, the dependent variables being y and z and the independent variable being x. Thus the solutions have two arbitrary constants a and b. Let the solution of this system of equations be expressed by

$$a = f(x, y, z)$$

and

$$b = g(x, y, z).$$

To understand the geometric meaning of this solution, each of these equations with a and b taking specific values generally describes a two-dimensional surface in the xyz-space. (If you need an example, just think of the equation $z = 1$, which describes a two-dimensional plane in the xyz-space.) The interesection of two surfaces is a curve in the three-dimensional space. Thus the two equations above with assigned values for a and b describe a curve in the xyz-space. The value of u on each of these curves is a constant.

The analytic form for the general solution of the PDE is given by

$$u(x, y, z) = F(f(x, y, z), g(x, y, z)),$$

where F is an arbitrary function of two variables. To satisfy the initial data, we require

$$h(y,z) = F(f(0,y,z), g(0,y,z)),$$

which determines F.

C. Quasi-Linear PDEs

Let us plot a characteristic curve for the linear and homogeneous PDE (3.3) in the xyu-space. We see that since u is a constant on a characteristic curve of (3.3), such a curve is parallel to the xy-plane. But this is not necessarily so with more general PDEs.

Let us first consider the linear and inhomogeneous PDE

$$a(x,y)u_x(x,y) + b(x,y)u_y(x,y) = c(x,y),$$

which is obtained by adding an inhomogeneous term to the right side of (3.3). Putting

$$\Delta x = a(x,y)\varepsilon, \Delta y = b(x,y)\varepsilon,$$

we have

$$\Delta u = u_x \Delta x + u_y \Delta y = [u_x a + u_y b]\,\epsilon = c(x,y)\epsilon, \tag{3.14}$$

which gives us the variation of u along a curve described by (3.4b). Since (3.4b) is independent of u, we may solve it without knowing what u is. Curves determined by (3.4b) are called characteristic curves in the xy-plane. The variation of u on a characteristic curve is given by (3.14). Thus we have

$$du = \frac{c(x,y)}{a(x,y)}dx.$$

On a characteristic curve in the xy-plane, y is a function of x. We may, if we want, substitute this functional dependence of y into the equation above, and obtain u by a simple integration. Thus u along a characteristic curve can be calculated precisely if the initial value of u is prescribed. The value of u at a point of this characteristic depends on this initial value only, and does not depend on initial values on other characteristic curves. If we plot this characteristic curve in the xyu-space, the curve is not necessarily parallel to the xy-plane, as u is not necessarily a constant.

We shall be able to treat the more general PDE:

$$a(x,y,u)u_x + b(x,y,u)u_y = c(x,y,u). \tag{3.15}$$

Equation (3.15) is linear in u_x and u_y but not necessarily so in u. We call such a PDE quasi-linear.

Similar to what we have done in (3.4a), we choose Δx and Δy to be

$$\Delta x = a(x, y, u)\varepsilon, \ \Delta y = b(x, y, u)\varepsilon. \tag{3.16}$$

The difference here is that the ratio $\Delta x/\Delta y$ depends not only on x and y but also on u. In the limit of Δx and Δy being infinitesimal, we have

$$\frac{dy}{dx} = \frac{b(x, y, u)}{a(x, y, u)}.$$

Since the right side of the equation above depends on u, it is not self-contained. In other words, in order to solve the equation above, we need u as a function of x and y, which is yet unknown. Indeed, it is the answer we seek. Now the variation of u in the direction of (3.16) is given by (3.14). Thus we have

$$\frac{dx}{a(x, y, u)} = \frac{dy}{b(x, y, u)} = \frac{du}{c(x, y, u)} \equiv dt. \tag{3.17}$$

Introducing the variable t makes it easier to visualize the equations we just derived. Indeed, we get from (3.17) that

$$\frac{dx}{dt} = a(x, y, u),$$
$$\frac{dy}{dt} = b(x, y, u),$$

and

$$\frac{du}{dt} = c(x, y, u). \tag{3.18}$$

In the equations above, we have included u as a dependent variable, enlarging the dimension of the system of equations. The enlarged system of equations is self-contained. If we think of (x, y, u) as the coordinate of a particle in the xyu-space, and think of the introduced variable t as the time variable, then (3.18) specifies the velocity of this particle at time t. Therefore, if the initial position (x_0, y_0, u_0) of the particle is given, (3.18) enables us to determine the motion of the particle at subsequent times. Thus (3.18) forms a closed system of three ordinary differential equations.

The motion of the particle traces a curve in the xyu-space. We are interested mostly in what this curve looks like in space, and not where this particle is at a particular time t. The variable t just offers a way to parametrize the curve.

When $c \neq 0$, u is not a constant on the curve. We'll demonstrate this using an example.

Problem for the Reader

Write down the equations of characteristics for the PDE

$$xu_x + yu_y = 1 + y^2 \tag{3.19}$$

and solve these equations. If u satisfies the initial condition

$$u(x, 1) = 1 + x, \tag{3.20}$$

determine $u(x, y)$ uniquely.

Solution

The equations of characteristics for the PDE in question are

$$\frac{dx}{x} = \frac{dy}{y} = \frac{du}{1 + y^2}. \tag{3.21}$$

Equating the first and the second term in (3.20), we get

$$y = x/c, \tag{3.22}$$

where c is a constant.

Equating the second and the third term in (3.21), we get

$$u = \ln y + \frac{y^2}{2} + a. \tag{3.23}$$

We note that there are two parameters, c and a, in these characteristic solutions. With the value of c given, eq. (3.22) describes a two-dimensional plane in the three-dimensional space of xyu. Similarly, with the value of a given, eq. (3.23) describes a two-dimensional surface in the xyu-space. The intersection of these two surfaces is a curve. Indeed, it is a characteristic curve of the PDE (3.19). By varying all possible values of a and c, one obtains all possible characteristic curves of the PDE.

How do we pick out the solution that satisfies, say, $u(x, 1) = g(x)$? Graphically, we choose the characteristic curves that pass through the points $(x, 1, g(x))$ in the space of xyu. The aggregate of these characteristic curves form a surface in the xyu plane.

Alternately, we may find an analytic solution of the PDE in the following way. Let us restrict ourselves to values of a and c that satisfy the relation

$$a = F(c),$$

where F is a function to be determined later. All of the characteristic curves associated with such a special relation between a and c satisfy

$$u - \ln y - \frac{y^2}{2} = F\left(\frac{x}{y}\right). \tag{3.24}$$

Equation (3.24) is the analytic form for the general solution of the PDE, with F an arbitrary function. The function F is determined from the requirement that the initial condition be satisfied:

$$g(x) - \frac{1}{2} = F(x).$$

If $g(x)$ is given by (3.20), we have

$$1 + x - \frac{1}{2} = F(x),$$

or

$$F(x) = \frac{1}{2} + x.$$

Thus we have

$$u(x, y) = \ln y + \frac{y^2}{2} + \frac{1}{2} + \frac{x}{y}. \tag{3.25}$$

We note from (3.22) that there are an infinite number of characteristic curves intersecting at the point $x = y = 0$. Thus the solution of the PDE in this example is not continuous at the origin. This behavior is also exhibited by the solution (3.25).

The procedure for obtaining an analytic form for the general solution of the quasi-linear PDE (3.15) is:

Step 1

Obtain the characteristic curves of (3.15) and express them in the form

$$f(x, y, u) = a$$

and

$$g(x, y, u) = c.$$

Step 2

The general solution of (3.15) is

$$g(x, y, u) = F(f(x, y, u)),$$

where F is an arbitrary function determined from the initial data. For example, if the initial data is given by

$$u(0, y) = h(y),$$

then we have

$$g(0, y, h(y)) = F(f(0, y, h(y))).$$

Thus F is determined from the equation above.

Let us give an example in which two characteristic curves associated with different values of u happen to cross. Consider the problem of traffic in a straight, one-dimensional highway. We shall ignore the fact that the cars on the highway are discrete objects and assume that they can be described by the density function $\rho(x, t)$, a continuous function of the position x and time t. Let the velocity of the cars be $v(x, t)$. Then the flux of cars is given by

$$q(x, t) = \rho(x, t)v(x, t).$$

Let us consider the case in which there are no exits on the highway, so the equation of continuity (see Chapter 4 for the derivation of this equation) holds and we have

$$\rho_t + q_x = 0.$$

We shall model the velocity of the traffic by

$$v = 1 - \rho. \tag{3.26}$$

Equation (3.26) says that when there are no other cars around ($\rho = 0$), a car would go at the speed limit, which we choose to be unity, and when the traffic is bumper to bumper ($\rho = 1$), a car would be stationary. Since the car flux is equal to the product of the car density times the car speed, we have

$$q = \rho(1 - \rho).$$

In particular, the flux is maximum at $\rho = 1/2$. The equation governing the traffic pattern is then given by

$$\rho_t + (1 - 2\rho)\rho_x = 0, \tag{3.27}$$

a first-order, nonlinear PDE.

Problem for the Reader

Find the characteristics for the PDE (3.27).

Solution

The equation of characteristics for (3.27) is

$$\frac{dt}{1} = \frac{dx}{1 - 2\rho} = \frac{d\rho}{0}. \tag{3.28}$$

A characteristic curve yielded by (3.28) is

$$x - (1 - 2\rho)t = a,$$

where a and ρ are constants on each of the characteristic curves. Indeed, a is the intercept of the characteristic curve with the x-axis and ρ is the value of the density function on this curve. Note that the density function is a constant on a characteristic curve. Thus the analytic form of the general solution of (3.27) is

$$\rho = F(x - (1 - 2\rho)t), \tag{3.29}$$

where F is an arbitrary function. Here is a difference from the other cases we've discussed: ρ appears in the argument of F in (3.29). This makes the solution (3.29) an implicit one, and it is a little less tractable.

Problem for the Reader

Let the initial car density be

$$\rho(x, 0) = \frac{1}{1 + |x|}. \tag{3.30}$$

Find $\rho(x, t)$.

Solution

Setting $t = 0$ for (3.29), we have

$$\frac{1}{1+|x|} = F(x). \tag{3.31}$$

Thus we have

$$\rho = \frac{1}{1+|x-(1-2\rho)t|}.$$

To obtain ρ as an explicit function of x and t, we would need to solve the equation above. Instead of doing that, it is more instructive to study the solution graphically.

The characteristic curves satisfying the initial condition (3.30) are given by

$$x = \frac{|a|-1}{|a|+1}t + a, \tag{3.32}$$

where a is the intercept of the characteristic curve with the x-axis. We plot the characteristic curves that pass through the points $x = 0, \pm 1, \pm 2$ on the x axis in Figure 3.2.

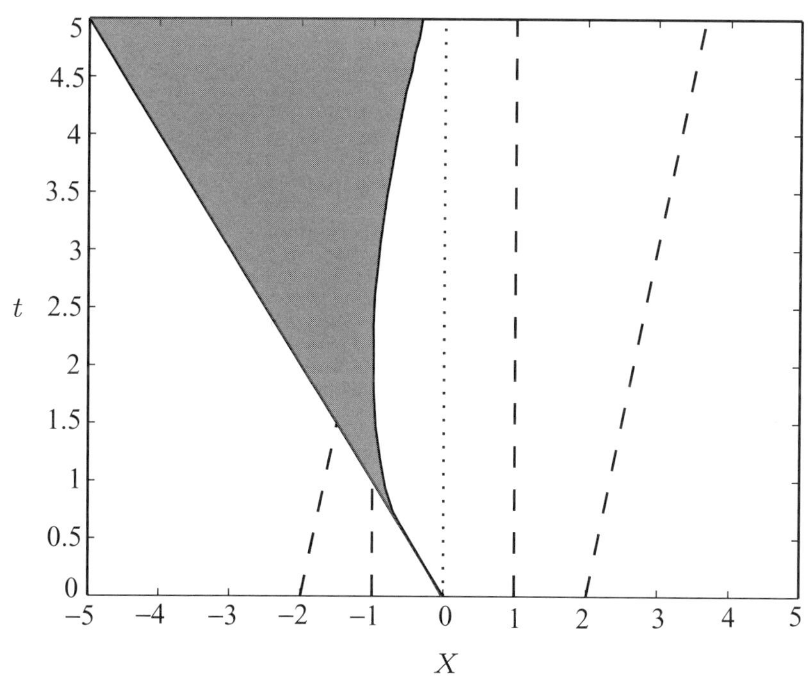

Figure 3.2

We note in particular that the characteristic curve passing through the origin, given by $x = t$, always intersects other characteristic curves that pass through the negative x axis with $a < 1$. At any of the points of intersection, there are two values of ρ. This is because by (3.26) and (3.30), a car on the former characteristic curve has zero velocity, while the cars on the latter characteristic curves have positive velocity. Thus cars collide when the curves intersect. This picture indeed holds generally for cars on characteristic curves passing through the negative x axis. Before a collision occurs, the speed of a car on one such characteristic curve is always lower than the speed of a car initially behind it. As a consequence, cars behind catch up with the cars in front at some point of time, and collisions occur. This contingency has not been anticipated by the modeling, and equation (3.27) does not describe the traffic flow after collisions occur. Thus the solution in the shaded region indicated in the figure is no good. (See homework problem 5.)

As for the characteristic curves emanating from the positive x axis, they do not intercept with one another. This is because a car on each of such curves always has a higher speed than another one behind it. Hence there are no collisions among such cars.

D. General Case

Finally, we'll discuss the most general first-order PDE expressed in the form

$$F(x, y, u, q, p) = 0, \tag{3.33}$$

where

$$p \equiv u_x, q \equiv u_y.$$

We first note that

$$p_y = q_x.$$

We also have

$$du = pdx + qdy, \quad dp = p_x dx + p_y dy, \quad dq = q_x dx + q_y d_y. \tag{3.34}$$

If the PDE is not quasi-linear, p and q are not related linearly. Thus it is not possible to eliminate p and q from du. To find a self-contained system of equations, we must enlarge the number of the dependent variables again and include p and q among them. Our goal is to choose a direction, or a relationship between dx and dy, at which p_x, p_y, q_x, and q_y disappear from (3.34). If this is done, we will have a self-contained system of ODEs, the variables being x, y, u, p, and q. The solution of these equations is, geometrically, a curve in a five-dimensional space.

Taking the partial derivative of (3.33) with respect to x, we get

$$F_x + F_u p + F_p p_x + F_q q_x = 0. \tag{3.35}$$

Since $p_y = q_x$, we may write the equation above as

$$F_p p_x + F_q p_y = -(F_x + F_u p). \tag{3.36}$$

Similarly, by taking the partial derivative of (3.33) with respect to y, we get

$$F_y + F_u q + F_p p_y + F_q q_y = 0, \tag{3.37}$$

or

$$F_p q_x + F_q q_y = -(F_y + F_u q). \tag{3.38}$$

Let us choose

$$dx = F_p \varepsilon, \; dy = F_q \varepsilon. \tag{3.39}$$

We note that if the PDE is quasi-linear, (3.39) is the same as (3.16). With the choice (3.39), (3.34) becomes

$$\begin{aligned} du &= (F_p p + F_q q)\varepsilon, \\ dp &= (F_p p_x + F_q p_y)\varepsilon = -(F_x + F_u p)\epsilon, \\ dq &= (F_p q_x + F_q q_y)\epsilon = -(F_y + F_u q)\varepsilon. \end{aligned} \tag{3.40}$$

All the equations above are independent of p_x, p_y, q_x, and q_y. Thus we have

$$\frac{dx}{F_p} = \frac{dy}{F_q} = \frac{du}{pF_p + qF_q} = -\frac{dp}{F_x + F_u p} = -\frac{dq}{F_y + F_u q}, \tag{3.41}$$

which is the system of ODEs satisfied by the characteristic curves of (3.33).

The characteristic curves given by (3.41) are curves in a five-dimensional space instead of those of a three-dimensional space. Thus one may wonder if there are additional initial conditions required. Since p and q are among the dependent variables in (3.41), the initial conditions include not only those for u, but also those for p and q. Thus, to determine an individual characteristic of (3.33), doesn't one need to be given the initial values of p and q in addition to that of u?

The answer is no, with a twist. As an example, let the initial data be on the y axis:

$$u(0, y) = f(y).$$

Then the initial value of q is given by

$$q = f^{'}(y), \tag{3.42}$$

and the initial value of p is then given by the PDE (3.33). Thus we obtain the initial values of u, p and q on any characteristic by specifying the initial condition of u on the y axis, say.

But there is a small twist. To obtain the initial value of q, we need, by (3.42), $f'(y)$, not just $f(y)$. Thus the initial value of u at the initial point of a characteristic curve is inadequate to determine the values of u on the characteristic curve. We need, in addition, the value of the tangential derivative of u at the initial point.

We have shown that the PDE (3.33) leads to the equations of the characteristic curves given by (3.40). Conversely, we may show that (3.40) together with the initial data imply (3.33). The variation of F is given by

$$\Delta F = F_x \Delta x + F_y \Delta y + F_u \Delta u + F_p \Delta p + F_q \Delta q.$$

If the variations of the variables x, y, u, p, and q are given by (3.40), then it is straightforward to show that

$$\Delta F = 0.$$

Thus (3.33) is satisfied at every point on the characteristic curve if it is satisfied at the initial point.

Problem for the Reader

Consider the PDE

$$u - px - \frac{1}{2}q^2 + x^2 = 0.$$

a. Find the equations of the characteristics for the PDE and solve these equations.
b. If

$$u(x, 0) = x^2 - \frac{x^4}{6},$$

find $u(x, y)$ satisfying this initial condition.

Solution

a. The equations for the characteristic curves are:

$$\frac{dx}{x}=\frac{dy}{q}=\frac{du}{px+q^2}=\frac{dp}{2x}=\frac{dq}{q}.$$

Thus the characteristic curves are described by

$$p=2x+a,\ q=bx, y=bx+c, u=\left(1+\frac{b^2}{2}\right)x^2+ax+d.$$

b. At the initial point $x=x_0$, $y=0$, the initial data gives

$$u=x_0^2-\frac{x_0^4}{6},\quad p=2x_0-\frac{2x_0^3}{3},$$

and by the PDE, there are two initial values of q:

$$q=\pm x_0^2.$$

We shall discuss only the case with the positive sign for the initial value of q.

The constants a, b, c, and d for the characteristic curve passing through the initial point $x=x_0$, $y=0$ are

$$a=-\frac{2x_0^3}{3},\ b=x_0,\ c=-x_0^2,\ d=0.$$

Thus this characteristic curve is

$$p=2\left(x-\frac{x_0^3}{3}\right),\ q=x_0x,\ y=x_0(x-x_0),$$

$$u=\left(1+\frac{x_0^2}{2}\right)x^2-\frac{2x_0^3}{3}x. \tag{3.43}$$

In particular, the third equation above says that the value x_0 for the characteristic curve passing through (x,y) satisfies

$$y=x_0(x-x_0),$$

or

$$x_0=\frac{x\pm\sqrt{x^2-4y}}{2}. \tag{3.44}$$

The value $u(x, y)$ is obtained from (3.43) with x_0 replaced by the expression in (3.44).

Problem for the Reader

Discuss the characteristic curves for

$$\frac{1}{2m}S_x^2 + S_t + V(x,t) = 0, \tag{3.45}$$

where m is a constant, V is a given function of x and t, and $S = S(x,t)$ is the dependent variable. The initial condition is

$$S(x,0) = f(x). \tag{3.46}$$

This equation is known as the Hamilton-Jacobi equation, and S is known as the classical action.

Solution

Let us put

$$S_x \equiv p, S_t \equiv q;$$

then the PDE is

$$\frac{p^2}{2m} + q + V(x,t) = 0. \tag{3.47}$$

The characteristic curves are given by

$$\frac{dx}{p/m} = \frac{dt}{1} = \frac{dS}{p^2/m + q} = -\frac{dp}{V_x(x,t)} = -\frac{dq}{V_t(x,t)}. \tag{3.48}$$

Using t as the independent variable, we find that the equations for x and t are

$$\frac{dx}{dt} = \frac{p}{m}, \tag{3.49}$$

$$\frac{dp}{dt} = -V_x(x,t). \tag{3.50}$$

This is a self-contained system of equations, which we may solve to obtain p and x as a function of t.

Let the initial value of x on a characteristic curve be x_0, i.e.,

$$x(0) = x_0;$$

then by (3.46), we have

$$p(0) = f'(x_0).$$

We recognize (3.49) and (3.50) as Hamilton's equations, with x and p the coordinate and the momentum, respectively, of a particle moving in a one-dimensional space under the influence of a potential $V(x,t)$, with t the time variable. Thus $x(t)$ and $p(x)$ satisfying (3.49) and (3.50) above are the position and the momentum, respectively, of a particle moving in accordance with Newton's laws under the influence of the potential $V(x,t)$.

We need to solve no differential equation to obtain q, as the PDE (3.47) gives, directly,

$$q = -\frac{p^2}{2m} - V(x,t). \tag{3.51}$$

Thus q is the negative of the energy of the particle, and is found once x and p are found. Finally, equating the second and the third terms of (3.48) gives

$$\frac{dS}{dt} = \frac{p^2}{m} + q = \frac{p^2}{2m} - V. \tag{3.52}$$

Thus on the characteristic curve under consideration, S is given by

$$S(t) - f(x_0) = \int_0^t \left[\frac{p^2(t')}{2m} - V(x(t'),t')\right] dt'. \tag{3.53}$$

After we solve Hamilton's equation together with the initial conditions, we may substitute into (3.53) the solution $p(t)$ and $x(t)$ obtained and get $S(t)$.

If the potential V is independent of t, then $q \equiv -E$ is a constant and by (3.51),

$$\frac{p^2}{2m} + V = E.$$

From the equation above we get

$$p = \sqrt{2m(E-V)}.$$

Then (3.51) becomes

$$S(t) - f(x_0) = \int_{x_0}^{x} \sqrt{2m\,[E - V(x')]}dx' - Et, \tag{3.54}$$

where x is the position, at time t, of the particle that is located at x_0 with momentum $f'(x_0)$ at $t = 0$.

The PDE (3.33) is a little simpler to tackle if we are able to solve it explicitly for p, getting

$$p = G(x, y, u, q). \tag{3.55}$$

Then the procedure above can be repeated with

$$F \equiv p - G(x, y, u, q).$$

Then we have

$$\begin{aligned} dy &= -G_q dx, \\ du &= (G - G_q q)dx, \\ dq &= (G_y + G_u q)dx, \end{aligned} \tag{3.56}$$

where we have made use of (3.55) and replaced p in the second equation above by G. The right sides of the equations above are independent of p and hence (3.56) is a self-contained system of ODEs. We may solve it to obtain the characteristic curves in the four-dimensional $xyuq$-space.

Homework Problems for This Chapter

Solutions to the Homework Problems can be found at www.lubanpress.com.

1. Obtain the equations of characteristics for the PDE

$$au_x + bu_y + cu_z = d,$$

where a, b, c, and d are functions of x, y, z, and u.

2. Find the general solution of the PDE

$$yu_x + xu_y = 0.$$

If $u(x, 0) = e^{-x}$, determine $u(x, y)$.

3. Solve

$$u_t + xu_x = 0$$

together with the initial condition $u(x, 0) = e^{-x^2}$. Show that the solution $u(x, t)$ considered as a function of x is a Gaussian at all times, with the width increasing as e^t.

4. Solve

$$(x^2 - y)u_x + (y^2 + x)u_y = 0$$

and show that the solution $u(x, y)$ is a function of

$$(y - x + 1)\exp\left(\frac{xy + 1}{y - x + 1}\right).$$

Hint: Make the change of the independent variables

$$\begin{aligned} X &\equiv \frac{x^2 + y^2}{2}, \\ Y &\equiv y - x. \end{aligned}$$

5. Find the region in the xt-plane in which the model (3.15) together with the initial condition (3.18) break down.

Hint: Find the envelope of the characteristic curves given by (3.20) originated from the negative x axis. You will find that the envelope is given by $u^2 = 4v$, where $u \equiv t - x - 1$ and $v \equiv t + x$ with $t > 1/2$. The first collision occurs at $t = 1/2$. The region where the model breaks down is the one bounded by the envelope and the characteristic originated from the origin.

6. Consider the traffic equation (3.27). Let there be a traffic light at $x = 0$ that turns green at $t = 0$, so that we have

$$\begin{aligned} \rho(x, 0) &= 0, x > 0, \\ &= 1, x < 0. \end{aligned}$$

a. Find $\rho(x, t)$.

b. Find also the speed of a particular car on the highway as a function of time.

Hint: Imagine that $\rho(x, 0)$ is not discontinuous at $x = 0$, but decreases from 1 to 0 in a very small neighborhood of $x = 0$.

7. Solve the nonlinear first-order PDE

$$xu_x + yu_y = u^2,$$

with $u(1, y) = y$. What are the characteristic curves in the xy-plane and in what region is your solution valid?

8. We have shown in the text that the characteristic curves of the quasi-linear PDE (3.15) are given by $f(x, y, u) = a$ and $g(x, y, u) = c$, and that the general solution of (3.15) can be written as $H(f, g) = 0$, where H is an arbitrary function. Consider the problem in converse: If the solutions of a PDE satisfy $H(f, g) = 0$, where H is an arbitrary function, find the PDE.

Hint: We have

$$H_f(f_x + f_u p) + H_g(g_x + g_u p) = 0,$$

and

$$H_f(f_y + f_u q) + H_g(g_y + g_u q) = 0.$$

If H_f and H_g do not both vanish, we have

$$(f_x + f_u p)(g_y + g_u q) = (g_x + g_u p)(f_y + f_u q),$$

which is quasi-linear.

9. Consider some photosensitive molecules distributed uniformly in a liquid irradiated by a light beam. For simplicity, let the liquid occupy the region $x > 0$ and the light beam propagate in the positive x direction. As the beam penetrates the emulsion, it is absorbed by the molecules, which decompose and become more transparent to light. The amount of light absorbed is proportional to the product of the light intensity and the density of the photosensitive molecules:

$$\frac{\partial I(x,t)}{\partial x} = -aN(x,t)I(x,t),$$

where I and N are the light intensity and the molecule density, respectively, and a is the constant of absorption. On the other hand, the rate of decomposition of the molecules is proportional to the amount of light absorbed:

$$\frac{\partial N(x,t)}{\partial t} = b\frac{\partial I(x,t)}{\partial x},$$

where b is a constant.

a. Solve this set of equations.

b. If the density of the molecules at $t = 0$ is uniform, i.e.,

$$N(x, 0) = N_0,$$

a constant, and the light beam at $x = 0$ is unattenuated, i.e.,

$$I(0, t) = I_0,$$

also a constant, find $I(x, t)$ and $N(x, t)$.

Answer:

$$I(x, t) = \frac{I_0}{1 + e^{-abI_0 t}(e^{aN_0 x} - 1)},$$
$$N(x, t) = \frac{N_0}{1 + e^{-aN_0 x}(e^{abI_0 t} - 1)}.$$

10. Find the characteristics for the Hamilton-Jacobi equation in two spatial dimensions

$$\frac{1}{2m}(S_x^2 + S_y^2) + S_t + V(x, y, t) = 0.$$

Find the action S if V is a function of $r \equiv \sqrt{x^2 + y^2}$ only.

11. Discuss how to solve the following quasi-linear PDE of three independent variables:

$$A(x, y, z, u)u_x + B(x, y, z, u)u_y + C(x, y, z, u)u_z = D(x, y, z, u).$$

In particular, how many parameters are there for a characteristic curve? What is the general form for the solution u? Give an example of a possible initial datum. How do you determine the arbitrary function in the general form of u from the initial data?

12. Discuss how to obtain the solution of (3.55) in an analytic form from the solutions of the characteristic curves together with the initial data.

Hint: The equations for the characteristic curves are a system of three equations. Thus there are three arbitrary constants. Therefore characteristic curves for (3.55) can be given as

$$c_n = f_n(x, y, u, q), n = 1, 2, 3.$$

Let the initial data be

$$u(0, y) = g(y);$$

then

$$q(0, y) = g'(y).$$

Let the characteristic curve with the coordinates (x, y) be associated with the initial point $(0, y_0)$. Then we have

$$f_n(0, y_0, g(y_0), g'(y_0)) = f_n(x, y, u, q), n = 1, 2, 3.$$

Eliminating y_0 and q from the equation above, we get u as a function of x and y.

13. Find the characteristic curves for the equation $xu_x + yu_y = 0$. Show that all of the characteristic curves pass through the origin. Discuss the solution of the partial differential equation with the condition $u(-1, y) = 1/(1 + y^2)$. Is the solution for $x > 0$ determined by this initial condition at $x = -1$?

Chapter 4

Second-Order Partial Differential Equations

In this chapter and in Chapter 5, we'll briefly review second-order PDEs. We'll begin with one of the simplest of such PDEs: the Laplace equation.

A. The Laplace Equation

The Laplace equation appears in many branches of physical sciences, two of them being electrostatics and fluid mechanics.

The fundamental laws of electromagnetism are Maxwell's equations. Two of these equations read

$$\vec{\nabla} \cdot \vec{E} = \rho \qquad \text{(Gauss's law)}$$

and

$$\vec{\nabla} \times \vec{E} = -\frac{\partial}{\partial t}\vec{B}, \qquad \text{(Faraday's law)}$$

where $\vec{E}$ and $\vec{B}$ are the electric field and the magnetic field, respectively, and ρ is the electric charge density. Note that the left-hand side of Gauss's law is the divergence of the electric field, and the left side of Faraday's law is the curl of the electric field. Let us express $\vec{E}$ as a Fourier integral:

$$\vec{E}(\vec{x}, t) = \iiint \frac{d^3k}{(2\pi)^3} e^{i\vec{k}\cdot\vec{x}} \vec{F}(\vec{k}, t),$$

where $\overrightarrow{F}$ is the Fourier transform of $\vec{E}$. Since $D \to m$ when D operates on e^{mx}, $\vec{\nabla} \to i\vec{k}$ when it operates on $e^{i\vec{k}\cdot\vec{x}}$. Thus Gauss's law can be considered as an equation satisfied by the longitudinal component of $\overrightarrow{F}$ (the component of $\overrightarrow{F}$ in the direction of $\vec{k}$), and Faraday's law can be considered as an equation satisfied by the transverse components of $\overrightarrow{F}$ (the components of $\overrightarrow{F}$ perpendicular to $\vec{k}$).

If the electromagnetic field is static, i.e., independent of time, then Faraday's law gives

$$\vec{\nabla} \times \vec{E} = 0.$$

Thus a static electric field is conservative and we may put

$$\vec{E} = -\vec{\nabla} u.$$

We see that the vector $\vec{E}$, which has three components in a three-dimensional space, is expressed as the gradient of one single function u. This is because two of its Fourier components, i.e., the two transverse components of the electric field, vanish. Substituting the equation above into Gauss's law, we get Poisson's equation

$$\nabla^2 u = -\rho,$$

where ∇^2 is the Laplacian.

In a region in space where there is no electric charge, the Poisson's equation is reduced to the Laplace equation

$$\nabla^2 u = 0.$$

The Laplace equation also applies to some problems in fluid mechanics. Let ρ be the density of a fluid and $\vec{v}$ be the velocity of the fluid flow. Then the rate of fluid flow out of a volume D is

$$\iint_S \rho\vec{v} \cdot d\overrightarrow{\sigma},$$

where S is the surface of the volume D. By the divergence theorem, the quantity above is equal to

$$\iiint_D (\vec{\nabla} \cdot \rho\vec{v}) dV.$$

Now the amount of fluid in the volume is given by

$$\iiint_D \rho dV.$$

As the fluid flows out, the amount of fluid left in the volume changes. The rate of change of the fluid inside D is equal to

$$\frac{d}{dt}\iiint_D \rho dV = \iiint_D \frac{\partial}{\partial t}\rho dV.$$

Equating the rate of fluid flowing out of the volume with the rate of change of the fluid left in the volume, we get

$$\frac{\partial}{\partial t}\rho = -\vec{\nabla}\cdot\rho\vec{v}.$$

The negative sign above accounts for the fact that the amount of fluid inside the volume decreases while there is a net flow going out of the surface. The equation above is called the equation of continuity. It describes the conservation of fluid, and holds if there are no sinks or sources in space. Indeed, the equation of continuity holds in electromagnetism as well, with ρ the electric charge density and $\rho\overrightarrow{v}$ the electric current, as the electric charge is conserved.

Let us return to fluid mechanics. For an incompressible fluid the density of which is a constant independent of space or time, the equation of continuity gives

$$\vec{\nabla}\cdot\vec{v} = 0.$$

If the fluid flow is, in addition, irrotational, i.e.,

$$\vec{\nabla}\times\vec{v} = 0,$$

we may put $\overrightarrow{v} = \vec{\nabla}u$. Substituting this expression of $\overrightarrow{v}$ into the equation $\vec{\nabla}\cdot\vec{v} = 0$, we find that the function u satisfies the Laplace equation $\nabla^2 u = 0$ when the fluid is both incompressible and irrotational.

The Laplace equation is one of the simplest second-order PDEs we can find, being linear, homogeneous, and of constant coefficients. That so many fundamental physical problems are described by this simple equation testifies that some of the most fundamental laws of nature are simple.

Let the Laplace equation hold in the upper half-plane in a two-dimensional space:

$$\left(\frac{\partial^2}{\partial x^2} + \frac{\partial^2}{\partial y^2}\right) u = 0, y > 0. \tag{4.1}$$

Needless to say, there are (infinitely) many solutions to this equation, corresponding to (infinitely) many different physical situations in the realms of static electricity or incompressible and irrotational fluid flow. In order for the solution of this equation to be unique, we need to impose some conditions on u.

The boundary of the upper half-plane is often thought of as an upper half-circle with center at the origin and with infinite radius. As we express the Laplace equation with Cartesian coordinates, it is conceptually clearer if we think of the upper half-plane as a square, the x axis between $-R$ and R being its lower horizontal boundary. The limit of R going to infinity is taken at the end. We shall require that the value of u be equal to zero at all sides of the square except the lower horizontal side. As R goes to infinity, we have

$$\lim_{r \to \infty} u(x, y) = 0, \tag{4.2}$$

where

$$r = \sqrt{x^2 + y^2}.$$

In addition, we require that on the lower horizontal side of the square (along the x axis),

$$u(x, 0) = f(x). \tag{4.3}$$

Equations (4.2) and (4.3) constitute a set of boundary conditions for u. To repeat, the values of u on three sides of the infinite square are required to be zero, while those on the fourth side are given by $f(x)$. Equation (4.1) together with these boundary conditions for u is a boundary-value problem.

It may help our perspective if we compare this boundary-value problem with the much simpler problem of the Laplace equation in the one-dimensional space, which is

$$\frac{d^2}{dx^2} u(x) = 0.$$

Being of the second order, the equation gives the solution

$$u(x) = ax + b,$$

which has two arbitrary constants. To determine the constants a and b, two conditions on u are required.

As an example, a meaningful problem is the one in which the value of u and the value of u' at the same point $x = 0$ are given, and we are asked to find the solution of $u'' = 0$ for $x > 0$ satisfying these initial conditions. Such a problem is called an initial-value problem, and the solution is uniquely given by $u = u(0) + u'(0)x$.

To give another example, let the range of x for the equation be $[0, 1]$. Then another meaningful problem is the one in which the values of u at the endpoints are specified, say $u(0) = 1$ and $u(1) = 2$, which are conditions imposed at two different points. This is a boundary-value problem. The solution of this problem is uniquely given by $u = x + 1$.

Let us see what happens if, in a boundary-value problem, we specify the values of u' instead of those of u. We see immediately that there is a difficulty. For, if u is a solution of $u'' = 0$, u' is a constant. Therefore, there is no solution satisfying these boundary conditions unless the two values of u' specified at the endpoints are equal. But there is another complication if the specified values of u' at the two endpoints are equal. This is because if u is a solution of this boundary-value problem, so is $u + c$, where c is an arbitrary constant. Thus the solution of such a boundary-value problem is not unique.

These difficulties or complications exist generally for the Laplace equation of any dimension. Consider, for example, the Laplace equation in three spatial dimensions:

$$\nabla^2 u = 0,$$

which we shall assume to hold in a volume D. If we think of u as the electric potential, then the right side of the Laplace equation implies that the charge density in D is zero. By Gauss's law, the total electric flux going out of the volume D is equal to zero and we have

$$\iint_S d\overrightarrow{\sigma} \cdot \overrightarrow{E} = 0,$$

where S is the surface of D. Since the electric field is equal to $-\overrightarrow{\nabla} u$, we have

$$\iint_S d\overrightarrow{\sigma} \cdot \overrightarrow{\nabla} u = 0.$$

Therefore, if the normal derivative of u at S is given as the boundary condition of the Laplace equation, there exists no solution to the boundary-value problem unless the given boundary values satisfy the equation above. And if the boundary values satisify this equation, the solution is unique up to a constant.

There are other kinds of boundary conditions for the equation $u'' = 0$. For example, we may specify the value of u at one of the boundary points and the value of u' at the other boundary point. As an example, let $u'(0) = 1$ and $u(1) = 0$. Then the solution of the boundary-value problem exists and is unique, and is given by $u(x) = (x - 1)$.

Let us return to the two-dimensional Laplace equation (4.1) with the boundary conditions (4.2) and (4.3), which specify the values of u at the boundary of the upper half-plane. To obtain the solution to this boundary-value problem, we focus on the x-dependence of $u(x,y)$ on x and denote the Fourier integral of u in the x variable by

$$\tilde{u}(k,y) \equiv \int_{-\infty}^{\infty} e^{-ikx} u(x,y)dx. \tag{4.4}$$

This Fourier transform exists because $u(\pm\infty, y) = 0$. Instead of solving for $u(x,y)$, we shall solve for $\tilde{u}(k,y)$. Once $\tilde{u}(k,y)$ is found, we immediately obtain $u(x,y)$ with the use of the Fourier inversion formula

$$u(x,y) = \int_{-\infty}^{\infty} e^{ikx} \tilde{u}(k,y) \frac{dk}{2\pi}. \tag{4.5}$$

Problem for the Reader

Substitute (4.5) into (4.1), obtain an ODE for $\tilde{u}(k,y)$, and solve this equation. Use the boundary conditions to determine $\tilde{u}(k,y)$.

Solution

Since D operating on e^{ikx} is equal to ik, where $D \equiv \frac{\partial}{\partial x}$, we get

$$\left(-k^2 + \frac{\partial^2}{\partial y^2}\right) \tilde{u}(k,y) = 0. \tag{4.6}$$

We see that with $\frac{\partial}{\partial x}$ replaced by a constant, the PDE becomes an ODE, which is easily solved. This is the advantage of expressing u as a Fourier integral. Solving the ODE, we have

$$\tilde{u}(k,y) = A(k)e^{ky} + B(k)e^{-ky}. \tag{4.7}$$

The boundary value condition (4.2) forbids the solution to blow up as $y \to \infty$. This enables us to eliminate one of the constants in the solution (4.7). More precisely, since e^{ky} for $k > 0$ blows up exponentially as $y \to \infty$, we must have

$$A(k) = 0, k > 0,$$

and hence

$$\tilde{u}(k,y) = B(k)e^{-ky}, k > 0.$$

Similarly, we may argue that

$$\tilde{u}(k, y) = A(k)e^{ky}, k < 0.$$

These two expressions for $\tilde{u}(k, y)$ can be combined into the single expression

$$\tilde{u}(k, y) = C(k)e^{-|k|y}, \tag{4.8}$$

where

$$\begin{aligned} C(k) = B(k), k > 0, \\ A(k), k < 0. \end{aligned}$$

We therefore have

$$u(x, y) = \int_{-\infty}^{\infty} e^{ikx} C(k) e^{-|k|y} \frac{dk}{2\pi}, \tag{4.9}$$

which satisfies both the Laplace equation and the boundary condition (4.2).

The condition (4.3) enables us to determine $C(k)$. By setting $y = 0$ in (4.9), we get

$$f(x) = \int_{-\infty}^{\infty} e^{ikx} C(k) \frac{dk}{2\pi}, \tag{4.10}$$

Thus $C(k)$ is equal to the Fourier transform of $f(x)$ given by

$$\tilde{f}(k) = \int_{-\infty}^{\infty} e^{-ikx} f(x) dx. \tag{4.11}$$

We therefore have

$$u(x, y) = \int_{-\infty}^{\infty} \frac{dk}{2\pi} e^{ikx - |k|y} \tilde{f}(k), \tag{4.12}$$

which satisfies both the Laplace equation (4.1) and the boundary conditions (4.2) and (4.3).

The factor $e^{-|k|y}$ provides a convergent factor for the Fourier integral in (4.12) when y is nonzero. Thus the integral of (4.12) converges nicely as one leaves the x axis, where the values of u are given, into the xy-plane, where y is nonzero. The given boundary-value problem of the Laplace equation is therefore called well posed. It is also the kind of problem that people naturally ask themselves, an example being "what is the static electric potential inside Benjamin Franklin's silver pint can?"

Substituting (4.11) into (4.12), we get, after some algebra,

$$u(x, y) = \frac{1}{\pi} \int_{-\infty}^{\infty} \frac{y}{(x - x')^2 + y^2} f(x') dx'. \tag{4.13}$$

As we recall, for the first-order PDE (3.15), all values of u on a characteristic curve are determined once an initial value on this curve is given. For the Laplace equation, we may see from (4.13) that one needs the values of u at all points of the boundary to determine the value $u(x, y)$ at any point in the interior region. Trivially, no value of the solution at any interior point is known if the value of u is given only at one point.

We mention that a useful function for many physical and engineering problems is the Dirac delta function denoted by $\delta(x)$, which vanishes for $x \neq 0$, with

$$\int_{-\infty}^{\infty} \delta(x) dx = 1 \tag{4.14a}$$

and

$$\int_{-\infty}^{\infty} f(x')\delta(x' - x) dx' = f(x). \tag{4.14b}$$

We may think of the Dirac delta function $\delta(x)$ as a function that is equal to zero except in the interval $[-\epsilon/2, \epsilon/2]$, which is an interval of width ϵ around the origin, where it is equal to $1/\epsilon$, and where ϵ is infinitesimally small. (It is to be noted that this is not the only way to define the Dirac delta function.)

Problem for the Reader

If $f(x) = \pi\delta(x - a)$, find $u(x, y)$.

Solution

By (4.13), we have

$$u(x, y) = \frac{y}{(x - a)^2 + y^2}.$$

We note that this function is finite for $y > 0$. We also note that

$$\lim_{y \to 0} \frac{y}{(x - a)^2 + y^2} = \pi\delta(x - a), \tag{4.14c}$$

as the left side of (4.14c) satisfies (4.14a) and (4.14b), and vanishes for $x \neq a$. (See homework problem 7 for other representations of the Dirac delta function.)

Problem for the Reader

If $f(x) = \dfrac{1}{1+x^2}$, find $u(x, y)$.

Solution

The Fourier transform of $(1 + x^2)^{-1}$ is equal to

$$\int_{-\infty}^{\infty} e^{-ikx} \frac{1}{1+x^2} dx = \pi e^{-|k|},$$

where the integral is evaluated by contour integration. Thus

$$u(x, y) = \int_{-\infty}^{\infty} \frac{dk}{2\pi} e^{ikx-|k|y} \pi e^{-|k|}.$$

The integral above can be easily integrated out and we get

$$u(x, y) = \frac{1+y}{x^2 + (1+y)^2}.$$

As we have discussed in Chapter 2, a function u satisfying the Laplace equation in the upper half-plane is the real part (or imaginary part) of a function of the complex variable $x + iy$ that is analytic in the upper half-plane. From the initial condition

$$u(x, 0) = \frac{1}{1+x^2},$$

a student may guess that the answer to the problem above is that $u(x, y)$ is the real part of the analytic function $1/(1+z^2)$. But this is the wrong answer, as $1/(1+z^2)$ is not analytic in the upper half-plane. Indeed, it is easy to verify that $u(x, y)$ is the real part of $i/(z + i)$.

The solution u of the Laplace equation in a region V, infinite or otherwise, with the value of u at the boundary specified, is unique. The proof of this will be left as a homework problem. (See homework problem 1.)

We next consider the initial-value problem of the Laplace equation. More precisely, let u satisfy the initial conditions

$$u(x, 0) = f(x)$$

and

$$u_y(x,0) = g(x).$$

We also require that

$$u(\pm\infty, y) = 0.$$

The last condition ensures that $u(x,y)$, considered as a function of x, has a Fourier transform. (A more precise requirement for a function $f(x)$ to have a Fourier transform is that $f(x)$ is absolutely integrable over $[-\infty, \infty]$.)

As before, we think of the upper half-plane as taking the form of an infinite square. In the problem we shall treat, two conditions are given on the lower horizontal boundary of the square, and one condition is given to each of the two vertical boundary lines, but no condition is imposed on the upper horizontal boundary line. Therefore, this is an initial-value problem with respect to the y variable and a boundary-value problem with respect to the x variable.

Problem for the Reader

Solve the Laplace equation (4.1) with the initial conditions given above.

Solution

For this problem, it is more convenient to express the solution of (4.6) as

$$A(k)\cosh(ky) + B(k)\sinh(ky).$$

In this form, the initial conditions give

$$A(k) = \tilde{f}(k), \tag{4.15}$$

and

$$kB(k) = \tilde{g}(k). \tag{4.16}$$

Thus we have

$$u(x,y) = \int_{-\infty}^{\infty} \frac{dk}{2\pi} e^{ikx} \left[\tilde{f}(k)\cosh ky + \tilde{g}(k)\frac{\sinh ky}{k} \right]. \tag{4.17}$$

Since both $\cosh ky$ and $\sinh ky$ blow up exponentially as a function of k as k goes to infinity, it is not possible for the integral in (4.17) to converge for all values of y unless both $\tilde{f}(k)$ and $\tilde{g}(k)$ vanish more rapidly than an exponential function as $|k|$ goes to infinity.

As an example, let

$$f(x) = e^{-\frac{1}{2}x^2}, \; g(x) = 0;$$

then

$$\tilde{f}(k) = \sqrt{2\pi}e^{-\frac{1}{2}k^2}, \; g(k) = 0.$$

In the above, we have made use of the formula

$$\int_{-\infty}^{\infty} e^{ixb-ax^2} dx = \sqrt{\frac{\pi}{a}} e^{-\frac{b^2}{4a}}.$$

(See homework problem 6.) Thus we get from (4.17) that

$$u(x,y) = \int_{-\infty}^{\infty} \frac{dk}{\sqrt{2\pi}} e^{ikx-\frac{1}{2}k^2} \cosh ky = e^{\frac{1}{2}(y^2-x^2)} \cos xy.$$

The function $u(x,y)$ given above has a factor e^{y^2}. Thus it blows up as $y \to \infty$ if x is kept at a finite value. On the other hand, it does vanish as $|x| \to \infty$ with $y = 0$, in consistency with the boundary conditions.

A function u satisfying the Laplace equation in the upper half-plane is called harmonic in the upper half-plane. As we have mentioned, it is the real part of a function analytic in the upper half-plane. Let the imaginary part of this analytic function be denoted by v; then u and v satisfy the Cauchy-Riemann equations

$$u_x = v_y$$

and

$$u_y = -v_x.$$

Therefore, if u_y is given on the x axis, so is v_x. This means that v is determined on the x axis up to a constant. Therefore, specifying u and u_y on the x axis is the same as specifying the analytic function $u + iv$ on the x axis up to a constant.

From this light, the result of the problem given above may appear puzzling to people who are familiar with the topic of dispersion relation. The dispersion relation, satisfied by the index of refraction and other physical quantities, is a result of the principle of causality. It is possible to prove that, by the principle of causality, the index of refraction $n(\omega)$ is an analytic function of the angular frequency ω in the upper half-plane of ω. Thus we have, by the Cauchy integral formula (2.18),

$$n(\omega) = \frac{1}{2\pi i} \int_c \frac{n(\omega')d\omega'}{(\omega' - \omega)},$$

where c is the infinite square in the upper half-plane of ω and where ω is a point in the upper half-plane. The index of refraction $n(\omega)$ is known to vanish as $|\omega|$ goes to infinity. Thus we replace the contour of integration c by the real ω axis, which is the bottom horizontal side of the infinite square, as the contribution from the other three sides vanishes. As we take ω to approach the real axis, we obtain with the Plemelj formula the dispersion relation for the index of refraction. Thus the dispersion relation says that the imaginary part of the index of refraction is determined by the real part of the index of refraction, and vice versa. In other words, there is no freedom to choose the imaginary part and the real part of the index of refraction independently.

For the solution of the initial-value problem of the Laplace equation, we do not require the solution to vanish at the upper horizontal side of the square. If a solution of such an initial-value problem exists, it need not vanish as y goes to infinity with x finite. Thus the existence of such a solution presents no contradiction.

As another example of the initial-value problem of the Laplace equation, let

$$f(x) = \frac{1}{1 + x^2} \tag{4.18}$$

and

$$g(x) = 0. \tag{4.19}$$

While these initial conditions appear innocuous enough, we will find that the solution $u(x, y)$ blows up at $y = 1$. We have

$$\tilde{f}(k) = \pi e^{-|k|}$$

and

$$\tilde{g}(k) = 0.$$

Substituting these values $\tilde{f}(k)$ and $\tilde{g}(k)$ into (4.17), we get, after carrying out the integration,

$$u(x, y) = \frac{1}{2}\left[\frac{1+y}{(1+y)^2 + x^2} + \frac{1-y}{(1-y)^2 + x^2}\right] = \mathrm{Re}\frac{1}{1+z^2}, \tag{4.20}$$

where $z = x + iy$. In particular, we have

$$u(0, y) = \frac{1}{1 - y^2},$$

which is infinite at $y = 1$. This shows that the Fourier integral (4.17) is infinite at $y = 1$. Indeed, the Fourier integral for $u(x, y)$ diverges for $y > 1$.

This means that the present initial-value problem does not have a solution. This is because since u satisfies the Laplace equation, it is the real part of a function that is analytic in the upper half-plane. At least for $y < 1$, one such function is $(z^2 + 1)$. By the uniqueness of analytic continuation, this is the only function possible. But $1/(z^2 + 1)$ is not analytic at the point $z = i$ in the upper half-plane. Thus there is no solution for the initial-value problem in question.

We call the initial-value problem of the Laplace equation not well-posed. As we have shown in an example, sometimes the initial-value problem does have a solution. But even if the solution for an initial-value problem does exist, it may be difficult to compute its values accurately away from the line on which the initial data is prescribed. For example, if we make a small error with the initial data and $f(x)$ is off by an amount ϵe^{iax}, say, where ϵ is a small number, by (4.17), the solution errs by an amount proportional to $\epsilon \cosh ay$, which becomes large as y becomes large; that is, an error of the initial data may grow exponentially as one leaves the line on which the initial data is prescribed.

Let us next consider the boundary-value problem in which $u_y(x, 0)$ rather than $u(x, 0)$ is prescribed on the x axis, i.e.,

$$u_y(x, 0) = g(x),$$

with the boundary condition at infinity remaining that in (4.2). Because of (4.2), the solution is in the form (4.9). And because of the initial condition on $u_y(x, 0)$, we have

$$kC(k) = -\tilde{g}(k).$$

Therefore,

$$u(x, y) = -\int_{-\infty}^{\infty} e^{ikx} e^{-|k|y} \frac{\tilde{g}(k)}{k} \frac{dk}{2\pi}. \tag{4.21}$$

We see from (4.21) that the integral diverges and there is no solution unless

$$\tilde{g}(0) = 0,$$

or

$$\int_{-\infty}^{\infty} g(x)dx = 0. \tag{4.22}$$

This conclusion is consistent with the general conclusion given earlier that the normal derivative of u integrated over the boundary must vanish. We may understand it with another physical consideration. The heat equation in two spatial dimensions is given by

$$\frac{\partial T}{\partial t} = \left(\frac{\partial^2}{\partial x^2} + \frac{\partial^2}{\partial y^2}\right) T.$$

(See the derivation of this equation in Section C, "The Heat Equation," below.) In the steady state, the temperature T is independent of time and the left side of the heat equation above vanishes. Thus the steady-state temperature satisfies the Laplace equation. Therefore, the function $u(x, y)$ satisfying the Laplace equation can be viewed as the steady-state temperature in a homogeneous solid. Now $u_y(0, x)$ is proportional to the heat flux entering the upper half-plane at the point $(x, 0)$ on the x axis. Since the temperature at infinity vanishes, there is net heat flux entering the xy-plane unless (4.22) is not satisfied, and the temperature cannot be time independent.

This can be formulated in a more general way. Let D be a region in a three-dimensional space and let S be the boundary of D. If u satisfies the Laplace equation, we have

$$\iiint_D \nabla^2 u dV = 0. \tag{4.23}$$

By the divergence theorem the above equation can be written as

$$\iint_S \vec{\nabla} u \cdot d\vec{\sigma} = 0.$$

B. The Wave Equation

Consider a rubber band that we idealize to be of one spatial dimension. Let the displacement of a point on this rubber band be $\phi(x, t)$. If $\phi(x, t)$ is the same for all points in this band, then it describes a uniform displacement of the entire band and there is no strain. A strain comes about when the displacement at the point x is different from that at a neighboring point $x + \Delta x$. Therefore, the strain at the point x is measured by ϕ_x. The force inside the rubber band that produces the strain is called the stress. As with a harmonic oscillator, the stress at the point x is, to a good approximation, linearly proportional to the strain at this point, and is equal to $\lambda\phi_x(x,t)$, where λ is a constant. The total force acting on an infinitesimal element of the band is equal to the difference of the stresses at the two endpoints, or

$$\lambda\phi_x(x+\Delta x,t) - \lambda\phi_x(x,t) \simeq \lambda\phi_{xx}(x,t)\Delta x,$$

where x and $x+\Delta x$ are the endpoints of this band. The mass of this infinitesimal element of the rubber band is equal to $\rho\Delta x$, where ρ is the density of the band, which we shall assume to be a constant. By Newton's law, we have

$$\rho\Delta x\ddot{\phi} = \lambda\phi_{xx}(x,t)\Delta x.$$

By choosing the dimensional units in such a way that $\rho/\lambda = 1$, we obtain the wave equation

$$\phi_{xx} - \phi_{tt} = 0. \tag{4.24}$$

The wave equation describes many physical phenomena, a notable example being the propagation of the electromagnetic wave in a vacuum. In a three-dimensional space, the wave equation reads

$$\nabla^2\phi - \phi_{tt} = 0,$$

where ∇^2 is the Laplacian.

Problem for the Reader

Let

$$u = x + t, v = x - t. \tag{4.25}$$

Express the wave equation (4.24) with the independent variables u and v.

Solution

We have

$$x = \frac{u+v}{2}, \quad y = \frac{u-v}{2}.$$

Thus we may apply the chain rule of differentiation and obtain the wave equation as

$$\frac{\partial^2 \phi}{\partial u \partial v} = 0.$$

From the equation above, we see that the solution of the wave equation is a sum of an arbitrary function of u and an arbitrary function of v:

$$\phi(x,t) = f(u) + g(v) = f(x+t) + g(x-t). \tag{4.26}$$

We note from (4.26) that the solution of the second-order wave equation has two arbitrary functions. This is to be compared with that of a first-order PDE of two independent variables, which generally has one arbitrary function. We also note that $f(x+t)$ is a wave travelling to the left with unit velocity, and $g(x-t)$ is a wave travelling to the right with unit velocity.

Problem for the Reader

Let ϕ satisfy the following initial conditions on the x axis:

$$\phi(x,0) = a(x), \phi_t(x,0) = b(x). \tag{4.27}$$

The right sides of (4.27) are known as the Cauchy data. Find $\phi(x,t)$.

Solution

It is straightforward to show that from (4.26) and (4.27)

$$\phi(x,t) = \frac{a(x+t) + B(x+t)}{2} + \frac{a(x-t) - B(x-t)}{2}, \tag{4.28}$$

where

$$B(x) \equiv \int_0^x b(t')dt'.$$

We may rewrite (4.28) as

$$\phi(x,t) = \frac{a(x+t) + a(x-t)}{2} + \frac{1}{2}\int_{x-t}^{x+t} b(x')dx'. \tag{4.29}$$

Obviously, we may also impose the Cauchy data on other lines or curves, not just the x axis. However, there is a caveat.

Problem for the Reader

Let the Cauchy data be imposed on the u axis ($v = 0$) with

$$\phi|_{v=0} = c(u), \;\; \phi_v|_{v=0} = d(u).$$

Find the solution of the wave equation.

Solution

From (4.26), we have

$$\phi_v(x, y) = g'(v).$$

Thus the second initial equation requires that

$$g'(0) = d(u).$$

If $d(u)$ is not equal to a constant, this equation cannot be satisfied and there is no solution. And if $d(u)$ is equal to a constant d, the PDE tells us that $g'(0) = d$, but provides no other information on $g(v)$ as a function of v. This is similar to the situation of assigning the initial data on a characteristic curve of a first-order PDE. Indeed, let us put

$$\phi_v \equiv \Psi.$$

Then the wave equation gives

$$\Psi_u = 0.$$

This means that v equaling a constant is a characteristic of the first-order PDE for Ψ, and that specifying the values of Ψ on such a line creates difficulties. Thus the line $v = 0$, or more generally, any line such that v is a constant is called a characteristic line for the wave equation. By exactly the same reasoning, any line such that u is a constant is also a characteristic line for the wave equation.

C. The Heat Equation

Let the density of a solid of volume ΔV be designated by ρ; then the weight of this volume is $\rho \Delta V$. Let the specific heat of this solid be s; then the thermal energy of this solid is equal to $s(\rho \Delta V)T$, where T is the temperature of the solid. Thus the thermal energy contained in a solid of finite volume D is

$$\iiint_D \rho s T dV,$$

and the change of thermal energy in the solid is

$$\frac{d}{dt}\iiint_D \rho s T dV.$$

Heat flows from points of higher temperature to points of lower temperature. Therefore, the heat flux is equal to $-c\,\vec{\nabla} T$, where c is the thermal conductivity. Consequently, the heat flux going out of the surface of a volume is equal to

$$-\iint_S c\vec{\nabla} T \cdot d\vec{\sigma} = -\iiint_D c\nabla^2 T dV,$$

where S is the surface of D and the divergence theorem has been used.

By equating the change of thermal energy with the negative of the thermal energy lost, we get the heat equation

$$\rho s T_t = c\nabla^2 T.$$

In a one-dimensional rod, the heat equation takes the form

$$\frac{\partial T}{\partial t} = \frac{\partial^2 T}{\partial x^2}, \tag{4.30}$$

where $\rho s/c$ has been set to unity.

Let the rod be of length l, and let the endpoints be kept at zero temperature at all times:

$$T(0,t) = T(l,t) = 0. \tag{4.31}$$

Also, let the initial temperature of the rod be

$$T(x,0) = f(x). \tag{4.32}$$

What is the temperature of the rod at all subsequent times?

Since the region of x is finite, we shall express $T(x,t)$ as a Fourier series rather than a Fourier integral:

$$T(x,t) = \sum_{n=1}^{\infty} c_n(t) \sin\left(\frac{n\pi}{l}x\right). \tag{4.33}$$

The Fourier sine series (instead of the Fourier cosine series, say) is chosen to represent the solution because the solution in the form (4.33) automatically satisfies the boundary conditions (4.31).

Substituting (4.33) into (4.30), we get

$$\frac{\partial c_n(t)}{\partial t} = -\frac{n^2\pi^2}{l^2} c_n(t). \tag{4.34}$$

The equation above gives

$$c_n(t) = c_n e^{-\frac{n^2\pi^2 t}{l^2}}. \tag{4.35}$$

Thus we have

$$T(x,t) = \sum_{n=1}^{\infty} c_n e^{-\frac{n^2\pi^2 t}{l^2}} \sin\left(\frac{n\pi}{l}x\right).$$

Now both the PDE (4.30) and the boundary conditions (4.31) are satisfied.

It remains to satisfy the initial condition (4.32). This is easily done by choosing

$$c_n = f_n,$$

where

$$f_n \equiv \frac{2}{l}\int_0^l f(x)\sin\left(\frac{n\pi x}{l}\right) dx$$

is the Fourier sine coefficient of $f(x)$. Thus

$$T(x,t) = \sum_{n=1}^{\infty} f_n e^{-\frac{n^2\pi^2 t}{l^2}} \sin\left(\frac{n\pi}{l}x\right). \tag{4.36}$$

We note that the series in (4.36) converges very well when t is nonzero, as its Fourier sine coefficient has a factor

$$e^{-\frac{n^2\pi^2 t}{l^2}}.$$

The problem is well posed.

We observe that the heat equation is of the first order in t. Thus only one initial condition is required. It is of the second order in x. Thus a set of boundary conditions is imposed. A well-posed problem for the heat equation consists of an initial condition in the time variable and two boundary conditions in the space variable.

We also observe that the right side of (4.36) vanishes at the endpoint $x = 0$ for all values of t. If $f(0)$ does not vanish, then the two sides of (4.36) are not equal at $x = t = 0$. This is a well-known difficulty with expanding a function into a Fourier sine series. If we do a numerical computation, adding up a large (but of course finite) number of terms of the series for $f(x)$, the truncated Fourier sine series vanishes at the endpoint $x = 0$, while by assumption $f(x)$ does not. Thus the result we get represents $f(x)$ accurately only when $x \neq 0$. But with x close to zero, one finds that the sum exhibits a rapidly oscillatory behavior, eventually vanishing at $x = 0$. This is the Gibbs phenomenon discussed in Chapter 2. This phenomenon reflects the fact that the Fourier sine coefficient for a function $f(x)$ with $f(0) \neq 0$ does not go to zero very rapidly as $n \to \infty$, as we shall discuss in Chapter 8. In a numerical computation, one must add up a larger and larger number of terms in the series for the series to converge as x becomes smaller and smaller. Happily, because of the factor

$$e^{-\left(\frac{n\pi}{l}\right)^2 t}$$

in (4.36), this difficulty does not exist for $t > 0$.

Similar remarks apply to the endpoint $x = l$.

Next let us consider the heat equation in a rod of infinite length rather than of finite length. The initial condition at $t = 0$ is

$$T(x, 0) = f(x), \quad -\infty < x < \infty, \tag{4.37}$$

and the boundary conditions are that the temperature of the rod is zero at infinity:

$$T(\pm\infty, t) = 0. \tag{4.38}$$

We wish to find the temperature in the rod at all points of x and at time $t > 0$. Thus the region where the heat equation holds is the upper half of the xt-plane. As before, we think of this upper half-plane as an infinite square. Thus the conditions imposed are one initial condition in the t variable and two boundary conditions in the x variable.

Problem for the Reader

Find $T(x, t)$.

Solution

Since the region of x is from $-\infty$ to ∞, we shall represent $T(x,t)$ as a Fourier integral of the variable x:

$$T(x,t) = \int_{-\infty}^{\infty} \frac{dk}{2\pi} e^{ikx} \tilde{T}(k,t).$$

Substituting this integral into the heat equation (4.30), we get

$$\frac{\partial}{\partial t}\tilde{T}(k,t) = -k^2 \tilde{T}(k,t),$$

which gives

$$\tilde{T}(k,t) = A(k) e^{-k^2 t}.$$

Thus we have

$$T(x,t) = \int_{-\infty}^{\infty} \frac{dk}{2\pi} e^{ikx - k^2 t} A(k). \tag{4.39}$$

The expression (4.39) satisfies the PDE as well as the boundary condition (4.38). To satisfy the initial condition (4.37), we put

$$A(k) = \tilde{f}(k),$$

where $\tilde{f}(k)$ is the Fourier transform of $f(x)$ given explicitly by (4.11). Thus we have

$$T(x,t) = \int_{-\infty}^{\infty} \frac{dk}{2\pi} e^{ikx - k^2 t} \tilde{f}(k). \tag{4.40}$$

Let us substitute (4.11) into (4.40) and carry out the integration over k. Since

$$\int_{-\infty}^{\infty} e^{-ak^2} dk = \sqrt{\frac{\pi}{a}},$$

we have

$$\begin{aligned}
\int_{-\infty}^{\infty} \frac{dk}{2\pi} e^{ik(x - x') - k^2 t} &= e^{-\dfrac{(x-x')^2}{4t}} \int_{-\infty}^{\infty} \frac{dk}{2\pi} \exp\left[-t\left(k - i\frac{x-x'}{2t}\right)^2\right] \\
&= \sqrt{\frac{1}{4\pi t}} e^{\dfrac{-(x-x')^2}{4t}}.
\end{aligned}$$

Thus (4.40) becomes

$$T(x,t) = \int_{-\infty}^{\infty} G(x-x',t)f(x')dx', \tag{4.41}$$

where

$$G(x,t) \equiv \sqrt{\frac{1}{4\pi t}}e^{-\frac{x^2}{4t}}. \tag{4.42}$$

Equation (4.41) can be used to solve any initial condition $f(x)$ of the heat equation. In particular, let

$$f(x) = \delta(x-a),$$

i.e., the initial temperature $T(x,0)$ vanishes everywhere except at the point a, where it is infinite, with the area

$$\int_{-\infty}^{\infty} T(x,0)dx = 1.$$

It follows from (4.41) that the subsequent temperature $T(x,t)$ is simply the function $G(x-a,t)$ given by (4.42), which is called the Green function for the heat equation.

Equation (4.41) also says that the solution of the heat equation satisfying the initial condition in the general form (4.37) is a linear superposition of the Green function $G(x-x',t)$ over all values of x'. This is because the heat equation is linear. The initial condition $T(x,0) = f(x)$ can be expressed as

$$T(x,0) = f(x) = \int_{-\infty}^{\infty} dx' f(x')\delta(x-x').$$

The rightmost term is a linear superposition of the delta function $\delta(x-x')$ with the weight factor $f(x')$. Thus the temperature $T(x,t)$ is a linear superposition of $G(x-x',t)$ with the weight factor $f(x')$.

The Green function $G(x-x',t)$ tells us qualitatively how heat is being dispersed. If the temperature is initially nonzero only at the point x', we see by (4.42) that the temperature spreads into a region of the size

$$|x-x'| = 2\sqrt{t},$$

at which $G(x-x',t)$ is down by a factor e of its peak value at time t. The square root in the above expression is indeed implied by the form of the PDE. Let

$$t \to ct;$$

then the heat equation (4.30) is unchanged if

$$x \to \sqrt{c}x.$$

This means that it takes four times as much time for heat to diffuse to two times the distance.

In this chapter we have studied three partial differential equations. They are among the simplest PDEs of the second order, being linear, homogeneous, and of constant coefficients. More generally, a linear and homogeneous second-order PDE is of the form

$$A(x,y)u_{xx} + 2B(x,y)u_{xy} + C(x,y)u_{yy} = 0. \tag{4.43}$$

We call the PDE above elliptic, hyperbolic, or parabolic according to

$$\Delta < 0, \quad \Delta > 0, \quad or \quad \Delta = 0,$$

respectively, where

$$\Delta \equiv B^2 - AC.$$

The Laplace equation is elliptic, with $A = C = 1$ and $B = 0$. If we substitute the solution $e^{imx+int}$ into the Laplace equation, we get $m^2 + n^2 = 0$, which means that if n is chosen to be real, then m is imaginary. Thus if the solution is chosen to be oscillatory in the x variable, then it must be exponential in the t variable.

The wave equation in one spatial dimension is hyperbolic, with $A = -C = 1$ and $B = 0$. Consequently, if we substitute the solution $e^{imx+int}$ into the wave equation, we get $m^2 - n^2 = 0$.This means that the solution can be chosen to be an oscillatory function of x times an oscillatory function of t.

The heat equation in one spatial dimension is parabolic. If we substitute the solution $e^{imx+int}$ into the heat equation, we get a solution that is oscillatory in the x variable and is exponentially decaying in the t variable.

Consider the case in which the coefficients A, B, and C are constants. If we substitute $e^{imx+iny}$ into (4.43), we get

$$Am^2 + 2Bnm + Cn^2 = 0.$$

From this quadratic equation, it is clear why the sign of Δ is relevant.

In particular, for elliptic PDEs, Δ is negative. If we choose n to be real, then both roots of m have imaginary parts. Thus if the solution is oscillatory in one of the variables, then it has two solutions in the other variable, one being exponentially growing and the other being exponentially decaying.

For hyperbolic PDEs, Δ is positive. Thus the ratio m/n is real, and the solution can be oscillatory in both variables.

For parabolic PDEs, Δ is zero. If $A \neq 0$, we have

$$C = \frac{B^2}{A},$$

and the PDE is

$$\left(A\frac{\partial}{\partial x} + B\frac{\partial}{\partial y}\right)^2 u = 0.$$

If A, B, C, and D are constants, we make the change of independent variables

$$X \equiv Ax + By$$

and

$$Y \equiv -Bx + Ay,$$

and the PDE becomes

$$\frac{\partial^2}{\partial X^2} u = 0.$$

The solution is

$$u(x, y) = F(Bx - Ay) + (Ax + By)G(Bx - Ay).$$

The heat equation is of the parabolic type but there is an additional first derivative term. For such a case, we have $m = in^2$, and there are solutions that are oscillatory in the x variable and exponentially decaying in the t variable, as given by (4.40).

The heat equation in three spatial dimensions is

$$\frac{\partial T}{\partial t} = \left(\frac{\partial^2}{\partial x^2} + \frac{\partial^2}{\partial y^2} + \frac{\partial^2}{\partial z^2}\right) T, \tag{4.44}$$

which is reduced to the Laplace equation for steady-state (time-independent) solutions. Equation (4.44) is parabolic between t and one of the spatial variables, and elliptic between two spatial variables. A well-posed problem for this PDE is to impose an initial condition for the variable t and boundary conditions on the spatial variables. This is also physically meaningful. For if the temperature at the boundary of the solid as well as the initial temperature of the solid are given, we expect the temperature of the solid at all space-time points to be well defined.

In problems of the heat equation treated so far, the temperature at the spatial boundary is always zero. We shall treat a case in which the temperature does not vanish at the spatial boundary.

Problem for the Reader

Solve the heat equation (4.30) that holds for $0 < x < l$ with the initial condition (4.32) and the boundary conditions

$$T(0,t) = g(t) \text{ and } T(l,t) = h(t). \tag{4.45}$$

Solution

We note that in this problem, the temperature at the boundary is not zero but a function of t. To make the boundary values vanish, we put

$$T(x,t) \equiv u(x,t) + g(t)(1 - x/l) + h(t)x/l; \tag{4.46}$$

then $u(x,t)$ vanishes at the endpoints $x = 0$ and $x = l$. The new dependent variable $u(x,t)$ satisfies the inhomogeneous PDE

$$u_t - u_{xx} = -g'(t)(1 - x/l) - h'(t)x/l. \tag{4.47}$$

Also, the initial condition for u is

$$u(x,o) = f(x) - g(0)(1 - x/l) - h(0)x/l. \tag{4.48}$$

Now that $u(x,t)$ vanishes at the endpoints, we may express $u(x,t)$ by a Fourier sine series:

$$u(x,t) = \sum_{n=1}^{\infty} c_n(t) \sin\left(\frac{n\pi}{l}x\right). \tag{4.49}$$

First, we express the functions

$$\frac{x}{l}$$

and

$$\frac{1-x}{l}$$

as the Fourier sine series

$$\frac{x}{l} = -\sum_{n=1}^{\infty} \frac{2}{n\pi}(-1)^n \sin\left(\frac{n\pi}{l}x\right),$$

and

$$\frac{1-x}{l} = \sum_{n=1} \frac{2}{n\pi} \sin\left(\frac{n\pi}{l}x\right).$$

Then we get

$$\left(\frac{d}{dt} + \frac{n^2\pi^2}{l^2}\right) c_n(t) = -\frac{2}{n\pi}g'(t) + \frac{2(-1)^n}{n\pi}h'(t).$$

By (4.48), the initial condition for $c_n(t)$ is

$$c_n(0) = f_n - \frac{2}{n\pi}g(0) + \frac{2(-1)^n}{n\pi}h(0). \tag{4.50}$$

This is a first-order linear ODE and the solution is

$$\begin{aligned} c_n(t) &= e^{-\frac{n^2\pi^2}{l^2}t} c_n(0) \\ &\quad - \frac{2}{n\pi}\int_0^t e^{-\frac{n^2\pi^2}{l^2}(t-t')} \left[g'(t') - (-1)^n h'(t')\right] dt'. \end{aligned} \tag{4.51}$$

By expressing

$$e^{-\frac{n^2\pi^2}{l^2}(t-t')} dt' = \frac{l^2}{n^2\pi^2} d\left(e^{-\frac{n^2\pi^2}{l^2}(t-t')}\right)$$

and performing an integration by parts, it is easy to show that for $t > 0$, $c_n(t)$ vanishes like

$$\frac{1}{n^3}$$

as n becomes very large. Thus the series in (4.49) converges reasonably well.

We may reduce eq. (4.51) to

$$\begin{aligned} c_n(t) &= f_n e^{-\frac{n^2\pi^2}{l^2}t} - \frac{2}{n\pi}\left[g(t) - (-1)^n h(t)\right] \\ &\quad + \frac{2n\pi}{l^2}\int_0^t e^{-\frac{n^2\pi^2}{l^2}(t-t')} \left[g(t') - (-1)^n h(t')\right] dt'. \end{aligned} \tag{4.52}$$

The second term on the right side of (4.52) is precisely the Fourier sine coefficient of $-g(t)(1-x/l)-h(t)x/l$. By (4.46), we have

$$T(x,t)=\sum_{n=1}^{\infty} C_n(t)\sin\left(\frac{n\pi}{l}x\right), \tag{4.53}$$

where

$$C_n(t)=f_n e^{-\frac{n^2\pi^2}{l^2}t} + \frac{2n\pi}{l^2}\int_0^t e^{-\frac{n^2\pi^2}{l^2}(t-t')}\left[g(t')-(-1)^n h(t')\right]dt'. \tag{4.54}$$

The series (4.53) vanishes at $x=0,l$, and does not satisfy the boundary conditions (4.45). The truncated series will give approximately the boundary value if x is chosen to be close to, but not exactly equal to, one of the endpoints, provided that the truncation is made at a sufficiently large number. As we have mentioned, such a truncated series exhibits the Gibbs phenomenon as the endpoints are approached. This is reflected by the fact that $C_n(t)$ vanishes as slowly as $1/n$ as n becomes large, which also makes the series in (4.53) converge more slowly than the series in (4.49). For numerical computations, it is better to use (4.49) together with (4.46) than to use (4.53).

Homework Problems for This Chapter

Solutions to the Homework Problems can be found at www.lubanpress.com.

1. Let $\nabla^2 u=0$ in a region V, and let the value of u at the boundary of V be specified. Show that u is unique.

   Hint: Let u_1 and u_2 be two of the solutions of this boundary-value problem. Consider the integral

   $$\int_V (\vec{\nabla}u)^2 dV,$$

   where

   $$u=u_1-u_2.$$

2. Consider the Laplace equation (4.1) with the condition (4.2) together with

   $$u_y(x,0)=\frac{x}{(1+x^2)^2}.$$

   Does a solution exist? If so, find this solution.

3. Solve the initial-value problem for the Laplace equation in the upper half-plane with

$$f(x) = 0$$

and

$$g(x) = e^{-\frac{1}{2}x^2}.$$

4. Consider the wave equation (4.24).

a. Let the wave equation hold in the spatial region $0 < x < \infty$, and let the solution $\phi(x, t)$ satisfy the initial conditions (4.27) as well as the boundary condition $\phi(0, t) = 0$. Find $\phi(x, t)$.

Hint: Use the method of images.

b. Let the wave equation hold in the spatial region $-L < x < L$, and let the solution $\phi(x, t)$ satisfy the initial conditions (4.27) as well as the boundary condition $\phi(\pm L, t) = 0$. Find $\phi(x, t)$.

5. Verify that at $t = 0$, the Green function for the heat equation is equal to the Dirac delta function. Also, verify directly that for $t > 0$, it satisfies the heat equation.

6. While it is not possible to express the indefinite integral

$$\int^x e^{-x'^2} dx'$$

in an analytic form, the definite integral

$$I \equiv \int_{-\infty}^{\infty} e^{-x^2} dx$$

is equal to $\sqrt{\pi}$. Prove it.

Hint: It is not easy to prove that something is equal to the square root of π. Thus, try to prove $I^2 = \pi$ by expressing I^2 as a double integral over a complete two-dimensional plane.

7. Instead of thinking of the Dirac delta function as a rectangular function of width ϵ and height $1/\epsilon$, we may, alternatively, think of the Dirac delta function as

a. The left side of (4.14c);

b. $\lim_{\epsilon\to 0} \sqrt{\frac{1}{\pi\epsilon}} e^{-\frac{x^2}{\epsilon}}$; or

c. $\lim_{K\to\infty} \dfrac{\sin(Kx)}{\pi x}$.

Show that the functions in (a) and (c) satisfy (4.14a) and (4.14b).

8. Consider the second-order linear and homogeneous PDE

$$Au_{xx} + 2Bu_{xy} + Cu_{yy} + 2Du_x + 2Eu_y + Fu = 0,$$

where A, B, C, D, E, and F are constants. Show that unless $B^2 - AC = 0$, we may make a change of the dependent variable

$$u(x, y) = e^{ax+by}U(x, y)$$

and eliminate the first-order partial derivatives of U from the partial differential equation satisfied by U. Show also that we may further make linear transformations of the independent variables from x and y to ξ and η and cast the PDE into the forms

$$\begin{aligned} U_{\xi\xi} + U_{\eta\eta} + fU &= 0, \Delta < 0, \\ U_{\xi\xi} - U_{\eta\eta} + fU &= 0, \Delta > 0, \\ U_{\xi\xi} + 2dU_\eta + fU &= 0, \Delta = 0. \end{aligned}$$

where $\Delta \equiv B^2 - AC$.

9. Consider the coupled first-order partial differential equations

$$\begin{aligned} A_1u_x + B_1u_y + C_1v_x + D_1v_y &= 0, \\ A_2u_x + B_2u_y + C_2v_x + D_2v_y &= 0, \end{aligned}$$

where u and v are the dependent variables. This set of coupled equations is equivalent to a single second-order partial differential equation. Give the condition that determines whether this set of equations is elliptic, hyperbolic, or parabolic.

Chapter 5

Separation of Variables

Compared to first-order partial differential equations, solvable second-order partial differential equations are much more limited in number. Many such second-order partial differential equations are solved with the method of separation of variables. In particular, the method of separation of variables can be used to solve all the partial differential equations discussed in the preceding chapter, which are linear, homogeneous, and of constant coefficient. The method of separation of variables may also be useful for a number of linear partial differential equations with non-constant coefficients; I'll give an example of this at the end of the chapter.

As a matter of fact, we have already used the method of separation of variables implicitly to solve the Laplace equation and the heat equation in Chapter 4. We have shown, by making a Fourier transform, that the Laplace equation (4.1) is satisfied by the solution

$$e^{ikx}e^{\pm ky},$$

which is in the form of a function of x times a function of y. The constant k in this solution may take any real value. Since the Laplace equation is linear and homogeneous, a superposition of such solutions also satisfies the Laplace equation. We use such a superposition to fulfill the boundary condition. Similarly, the solution

$$e^{ikx}e^{-k^2t},$$

which is in the form of a function of x times a function of t, solves the heat equation (4.30) for all values of k. We use a superposition of such solutions to satisfy the initial condition (4.37).

In this chapter, we shall discuss the method of separation of variables and demonstrate this method with several PDE examples.

A. The Laplace Equation

We shall start again with the Laplace equation in two spatial dimensions:

$$u_{xx} + u_{yy} = 0, \tag{5.1}$$

which holds inside the unit disk

$$\sqrt{x^2 + y^2} < 1. \tag{5.2}$$

Let us adopt the polar coordinates r and θ. We shall require that on the boundary of this disk, $u(r, \theta)$ satisfies the condition

$$u(1, \theta) = g(\theta), \tag{5.3}$$

where

$$\theta = \tan^{-1} \frac{y}{x} \tag{5.4}$$

and

$$r \equiv \sqrt{x^2 + y^2}. \tag{5.5}$$

This is a boundary-value problem for the Laplace equation. (Note that, because the geometry of this problem is cylindrically symmetrical, it is more convenient to use as the independent variables the polar coordinates r and θ instead of the Cartesian variables x and y.)

Problem for the Reader

Calculate $\dfrac{\partial r}{\partial x}$ and $\dfrac{\partial \theta}{\partial x}$.

Solution

We have, from (5.4) and (5.5),

$$\frac{\partial r}{\partial x} = \cos\theta, \frac{\partial \theta}{\partial x} = -\frac{\sin\theta}{r}.$$

Problem for the Reader

Calculate the Laplacian $\dfrac{\partial^2}{\partial x^2} + \dfrac{\partial^2}{\partial y^2}$ in the polar coordinates.

Solution

$$\frac{\partial}{\partial x} = \frac{\partial r}{\partial x}\frac{\partial}{\partial r} + \frac{\partial \theta}{\partial x}\frac{\partial}{\partial \theta} = \cos\theta\frac{\partial}{\partial r} - \frac{\sin\theta}{r}\frac{\partial}{\partial \theta}. \tag{5.6}$$

Hence

$$\begin{aligned}\frac{\partial^2}{\partial x^2} &= \left(\cos\theta\frac{\partial}{\partial r} - \frac{\sin\theta}{r}\frac{\partial}{\partial \theta}\right)\left(\cos\theta\frac{\partial}{\partial r} - \frac{\sin\theta}{r}\frac{\partial}{\partial \theta}\right)\\ &= \cos^2\theta\frac{\partial^2}{\partial r^2} - \frac{\sin\theta}{r}\frac{\partial}{\partial \theta}\cos\theta\frac{\partial}{\partial r}\\ &\quad - \cos\theta\sin\theta\frac{\partial}{\partial r}\frac{1}{r}\frac{\partial}{\partial \theta} + \frac{\sin\theta}{r^2}\frac{\partial}{\partial \theta}\sin\theta\frac{\partial}{\partial \theta}.\end{aligned} \tag{5.7}$$

In order to streamline the calculation, we will make use of the formula

$$D(au) = a\left(D + \frac{a'}{a}\right)u. \tag{5.8}$$

This means that moving a function a from the right of D to the left of D is compensated for by replacing D with

$$D + \frac{a'}{a}.$$

Equation (5.8) is another form of (1.3).

Thus (5.7) can be calculated to be

$$\begin{aligned}\frac{\partial^2}{\partial x^2} &= \cos^2\theta\frac{\partial^2}{\partial r^2} - \frac{\sin\theta\cos\theta}{r}\left(\frac{\partial}{\partial \theta} - \frac{\sin\theta}{\cos\theta}\right)\frac{\partial}{\partial r}\\ &\quad - \frac{\cos\theta\sin\theta}{r}\left(\frac{\partial}{\partial r} - \frac{1}{r}\right)\frac{\partial}{\partial \theta} + \frac{\sin^2\theta}{r^2}\left(\frac{\partial}{\partial \theta} + \frac{\cos\theta}{\sin\theta}\right)\frac{\partial}{\partial \theta}.\end{aligned} \tag{5.9}$$

In (5.9), all functions of θ and r are located to the left of the differential operators.

Next we calculate

$$\frac{\partial^2}{\partial y^2}.$$

Instead of repeating this dreary process again, we notice that if we make the replacement

$$\theta \to \frac{\pi}{2} - \theta, \tag{5.10}$$

then

$$x \leftrightarrow y.$$

With this replacement, we have from (5.9) that

$$\begin{aligned}\frac{\partial^2}{\partial y^2} &= \sin^2\theta \frac{\partial^2}{\partial r^2} + \frac{\sin\theta\cos\theta}{r}\left(\frac{\partial}{\partial\theta} + \frac{\cos\theta}{\sin\theta}\right)\frac{\partial}{\partial r} \\ &\quad + \sin\theta\cos\theta\left(\frac{\partial}{\partial r} - \frac{1}{r}\right)\frac{\partial}{\partial\theta} + \frac{\cos^2\theta}{r^2}\left(\frac{\partial}{\partial\theta} - \frac{\sin\theta}{\cos\theta}\right)\frac{\partial}{\partial\theta},\end{aligned} \tag{5.11}$$

where, by (5.10), we have made the replacement

$$\frac{\partial}{\partial\theta} \to -\frac{\partial}{\partial\theta}. \tag{5.12}$$

From (5.9) and (5.11), we get

$$\nabla^2 \equiv \frac{\partial^2}{\partial x^2} + \frac{\partial^2}{\partial y^2} = \frac{\partial^2}{\partial r^2} + \frac{1}{r}\frac{\partial}{\partial r} + \frac{1}{r^2}\frac{\partial^2}{\partial\theta^2}. \tag{5.13}$$

With (5.13), we find that the Laplace equation in the polar coordinates is

$$\left(\frac{\partial^2}{\partial r^2} + \frac{1}{r}\frac{\partial}{\partial r} + \frac{1}{r^2}\frac{\partial^2}{\partial\theta^2}\right)u = 0. \tag{5.14}$$

We shall find the solution of the Laplace equation in the form

$$u = R(r)\Theta(\theta), \tag{5.15}$$

which is a product of a function of r and a function of θ.

Problem for the Reader

Substitute (5.15) into (5.14) and divide the resulting equation by $R(r)\Theta(\theta)/r^2$.

Solution

Carrying out this simple procedure, we find

$$\frac{r^2 R''(r) + rR'(r)}{R(r)} = -\frac{\Theta''(\theta)}{\Theta(\theta)}. \tag{5.16}$$

Note that we replace $\frac{\partial}{\partial r}R(r)$ by $R'(r)$ as $R(r)$ depends on r only. We also replace

$$\frac{\partial}{\partial \theta}\Theta(\theta)$$

by $\Theta'(\theta)$.

Let us rewrite (5.16) as

$$\frac{r^2 R''(r) + rR'(r)}{R(r)} = -\frac{\Theta''(\theta)}{\Theta(\theta)} \equiv c. \tag{5.17}$$

Since c is equal to

$$-\frac{\Theta''(\theta)}{\Theta(\theta)},$$

it is not a function of r. And since c is also equal to

$$\frac{r^2 R''(r) + rR'(r)}{R(r)},$$

it is not a function of θ. Thus c is a constant.

Problem for the Reader

Equate the second and the third quantities in (5.17) and find the solution of the equation obtained.

Solution

Equating the second and the third quantities in (5.17), we get

$$\frac{d^2}{d\theta^2}\Theta(\theta) \equiv -c\Theta(\theta). \tag{5.18}$$

Thus

$$\Theta(\theta) = e^{\pm i\sqrt{c}\theta}. \tag{5.19}$$

Since θ and $\theta + 2\pi$ are physically the same angle, the value of $\Theta(\theta)$ and that of $\Theta(\theta + 2\pi)$ are required to be the same, i.e., the function $\Theta\,(\theta)$ is required to be periodic, with period 2π. Thus we impose on the solution of (5.18) the boundary conditions

$$\Theta(0) = \Theta(2\pi). \tag{5.20}$$

We shall also require

$$\Theta'(0) = \Theta'(2\pi). \tag{5.21}$$

These conditions are satisfied only if

$$e^{\pm i\sqrt{c}2\pi} = 1,$$

which requires that

$$\sqrt{c} = n,\ n = 0, 1, 2\cdots,$$

or

$$c = n^2,\ n = 0, 1, 2\cdots. \tag{5.22}$$

The solutions

$$e^{\pm in\theta}$$

are of period 2π.

This is an example of a Sturm-Liouville problem. In such a problem, there is a linear and homogeneous differential equation with a parameter. The solution of this equation is required to satisfy a set of homogeneous boundary conditions, in which all terms are linear in u. In the present problem, the equation is (5.18), the parameter in the equation is c, and the homogeneous boundary conditions are (5.20) and (5.21). To give other examples of homogeneous boundary conditions, we mention that the boundary conditions (4.2), (5.20), and (5.21) are homogeneous. We also mention that (5.3) is not homogeneous.

A homogeneous boundary condition is always satisfied by the solution $u = 0$. The solution $u = 0$ also satisfies a homogeneous linear differential equation. Thus

$u = 0$ satisfies both the differential equation and the boundary conditions, provided that both are homogeneous. Such a solution is obviously trivial, and is indeed called a trivial solution. A trivial solution is not useful.

Nontrivial solutions exist for such a problem only if the parameter in the equation takes some special values called eigenvalues. The nontrivial solutions corresponding to such eigenvalues are called eigenfunctions. For example, in the eigenvalue problem we just discussed, the eigenvalues are

$$n^2, \; n = 0, 1, 2 \cdots$$

and the eigenfunctions are

$$\Theta_n(\theta) = e^{\pm in\theta}.$$

These eigenfunctions form a complete set for any function $f(\theta)$ with θ in the region $0 \leq \theta \leq 2\pi$. This means that any $f(\theta)$ can be expressed as a superposition of such eigenfunctions.

Problem for the Reader

Equate the first and the third quantities in (5.17) and solve the equation obtained.

Solution

With c chosen to be the eigenvalue n^2, we get

$$\left(r^2 \frac{d^2}{dr^2} + r \frac{d}{dr}\right) R_n(r) = n^2 R_n(r), \tag{5.23}$$

which is an equidimensional ODE. The solutions of (5.23) are

$$R_n(r) = r^{\pm n}, \; n = 1, 2, \cdots \tag{5.24}$$

and

$$R_0(r) = 1 \text{ or } \ln r. \tag{5.25}$$

Any solution of the form

$$R_n(r) e^{\pm in\theta}$$

satisfies the Laplace equation. Since the Laplace equation is linear, a linear superposition of them also satisfies the Laplace equation. We shall use a linear superposition of these solutions to fulfill the boundary condition (5.3).

We mention that we do not always employ all solutions found above to satisfy the boundary condition. For the Laplace equation (5.1), which holds inside the unit disk $r < 1$, the solutions r^{-n} and $\ln r$ must be excluded. This is because they blow up at $r = 0$. Therefore, we put

$$u(r, \theta) = A_0 + \sum_1^\infty (A_n \cos n\theta + B_n \sin n\theta) r^n. \tag{5.26}$$

This solution satisfies the Laplace equation inside the unit disk. And since 1, $\cos n\theta$, $\sin n\theta$ form a complete set for a function of θ, with $-\pi < \theta < \pi$, the solution in the form (5.26) can be made to satisfy the boundary condition (5.3). Specifically, the boundary condition (5.3) means that

$$g(\theta) = A_0 + \sum_1^\infty (A_n \cos n\theta + B_n \sin n\theta).$$

Thus the coefficients A_0, A_n, and B_n are the Fourier coefficients of $g(\theta)$.

The method of separation of variables can be applied to a number of other problems. Generally, this method, when applied to a linear and homogeneous PDE, yields infinitely many solutions of the PDE. In a physical problem, we seek a solution of the PDE that satisfies a set of boundary conditions or initial conditions. While an individual solution obtained with the method of separation of variables may not satisfy these conditions, a linear superposition of these solutions often does. Actually, the method of separation of variables often yields more solutions than we need. For example, we argued just a moment ago that we must exclude

$$r^{-n} e^{\pm i\theta}$$

and $\ln r$ from the expansion series of the solution of the boundary-value problem we just treated. We shall now address the general question of how to select the set of functions useful for a particular problem.

First, an almost trivial point. A PDE has more than one independent variable. And we are allowed to make a change of these variables. Therefore, before we perform the separation of variables, we must first decide what are the best variables to use. The point is to ensure that the independent variables we use are the most suitable for the problem. This is mostly a matter of geometry. For example, if the region under consideration is the three-dimensional sphere with its center at the origin, one should consider using the spherical variables. But if the region in which the PDE is required to hold is the box $0 < x < a$, $0 < y < b$, and $0 < z < c$, then one should try the Cartesian coordinates. Change the variables to the ones most suitable for the problem.

With the variables chosen, one expresses the solution as a product of functions of one variable. Substitute this solution into the equation and obtain an equation one side of which depends on only one of the variables, an example being (5.36) below. We then argue that both sides of the equation must be equal to a constant, which is an eigenvalue.

We emphasize that the eigenvalues such as c in (5.17) are determined from boundary conditions. In a problem in which there are two or more sets of boundary conditions, it is often best to start with the boundary conditions that are homogeneous.

As an example, let us revisit the boundary-value problem in the Laplace equation (4.1) in the upper half-plane. We put

$$u = X(x)Y(y),$$

and the Laplace equation gives

$$\frac{X''}{X} = -\frac{Y''}{Y} = c. \tag{5.27}$$

The boundary condition (4.2)

$$\lim_{r\to\infty} u(x, y) = 0$$

is homogeneous, and we make use of it first. The boundary condition (4.3) is not homogeneous, and will be used later.

We argue that if c in (5.27) is positive, all solutions of the equation grow exponentially either as $x \to \infty$ or as $x \to -\infty$, which is not allowed by (4.2). So the eigenvalues are restricted to

$$c = -k^2.$$

From (5.27) we then get

$$X(x) = e^{\pm ikx},$$

and

$$Y(y) = e^{\pm ky}.$$

If we require $u(x, y)$ to approach zero as y approaches ∞, then the solution $Y(y) = e^{|k|y}$ must be excluded. Thus the basis of functions to be used in the superposition is chosen to be the functions

$$e^{\pm ikx} e^{-|k|y},$$

with k any real number. The solution of the Laplace equation satisfying the boundary conditions (4.2) and (4.3) has been given in Chapter 4.

On the other hand, if u is not required to vanish at $y = \infty$, then both

$$e^{ky}$$

and

$$e^{-ky}$$

must be admitted. As an example, consider the boundary-value problem for the Laplace equation held in the infinite strip between $y = 0$ and $y = a$. Let the boundary conditions for $u(x, y)$ be

$$u(\pm\infty, y) = 0$$

and

$$u(x, 0) = f(x), \; u(x, a) = g(x).$$

We make use of the homogeneous conditions $u(\pm\infty, y) = 0$ and conclude again that c must be negative. As we have mentioned, now both independent solutions of $Y(y)$ are allowed. We shall choose the independent solutions to be $\sinh ky$ and $\sinh k(a - y)$ instead of

$$e^{\pm ky}.$$

Note that the solution $\sinh ky$ vanishes at the boundary $y = 0$, and the solution $\sinh k(a - y)$ vanishes at the boundary $y = a$. We have

$$u(x, y) = \int_{-\infty}^{\infty} e^{ikx} \left[A(k) \sinh ky + B(k) \sinh k(a - y)\right].$$

Problem for the Reader

Determine $A(k)$ from the boundary condition at $y = a$, and $B(k)$ from the boundary condition at $y = 0$.

Solution

From the boundary condition at $y = a$, we get

$$A(k) = \tilde{g}(k)/\sinh ka.$$

From the boundary condition at $y = 0$, we get

$$B(k) = \tilde{f}(k)/\sinh ka.$$

In the above, $\tilde{f}(k)$ and $\tilde{g}(k)$ are the Fourier transforms of $f(x)$ and $g(x)$, respectively.

Problem for the Reader

Let the Laplace equation be valid in the rectangle with the length of its sides equal to b and a, respectively, and with the lower-left corner at the origin, and let the boundary conditions be

$$u(0, y) = u(b, y) = 0, \tag{5.28}$$

$$u(x, 0) = f(x), \text{ and } u(x, a) = g(x). \tag{5.29}$$

Solution

The eigenvalues are determined from the homogeneous boundary conditions (5.28). We get

$$c = -(n\pi/b)^2, \tag{5.30}$$

where n is a positive integer. From (5.27), we get

$$X_n(x) = \sin(n\pi x/b),$$

which satisfies the boundary conditions (5.28). From the equation for Y in (5.27), we get

$$Y_n(y) = A_n \sinh \frac{n\pi y}{b} + B_n \sinh \frac{n\pi(a-y)}{b}.$$

Thus we have

$$u(x, y) = \sum_{n=1}^{\infty} \left[A_n \sinh \frac{n\pi y}{b} + B_n \sinh \frac{n\pi(a-y)}{b} \right] \sin(n\pi x/b). \tag{5.31}$$

Setting $y = 0$ in (5.31), we find that

$$B_n \sinh\left(\frac{n\pi a}{b}\right)$$

is equal to the Fourier coefficient of $f(x)$. Similarly,

$$A_n \sinh\left(\frac{n\pi a}{b}\right)$$

is equal to the Fourier coefficient of $g(x)$.

Problem for the Reader

If the boundary conditions for $u(x, y)$ are

$$u(x,0) = f(x),\ u(x,a) = g(x),$$
$$u(0,y) = F(y),\ \text{and}\ u(b,y) = G(y),$$

how would you seek the solution of this boundary-value problem?

Solution

There are no homogeneous boundary conditions. Thus we split the solution into the linear superposition

$$u = u_1 + u_2.$$

We require u_1 and u_2 to satisfy the Laplace equation individually. We also require u_1 to vanish at $x = 0$ and $x = b$, while u_2 vanishes at $y = 0$ and $y = a$.

B. The Wave Equation with Two Spatial Variables

Next we turn to the wave equation in two spatial dimensions. This equation is satisfied by the displacement of the surface of a circular drum, which we'll study in this section. The surface of the drum has circular symmetry; therefore, for the vibration of the drum surface, the spatial variables we shall use are the polar coordinates r and θ. The equation is

$$\frac{\partial^2 u}{\partial t^2} - \frac{\partial^2 u}{\partial r^2} - \frac{1}{r}\frac{\partial u}{\partial r} - \frac{1}{r^2}\frac{\partial^2 u}{\partial^2 \theta} = 0,$$

where

$$u = u(r, \theta, t),$$

a function of three variables.

The PDE is of second order in each of the independent variables. We note that in this equation, the sign of the coefficient of u_{tt} is the opposite of that of u_{rr}. Therefore, the PDE is hyperbolic between the variables t and r. Similarly, it is hyperbolic between t and θ. Thus we impose two initial conditions at $t = 0$:

$$u(r, \theta, 0) = f(r, \theta) \tag{5.32}$$

and

$$u_t(r, \theta, 0) = g(r, \theta). \tag{5.33}$$

The PDE is elliptic in the spatial variables; thus we impose the boundary condition

$$u(1, \theta, t) = 0, \tag{5.34}$$

which is well posed.

The conditions above are physically meaningful as well. Equation (5.34) says that the circumference of the drum is kept fixed at all times, and (5.32) and (5.33) give the initial displacement and the initial velocity of the surface of the drum. We expect from experience that a unique solution of the wave equation will result under these conditions. It is satisfying that such a set of conditions on the solution makes the problem well posed.

Problem for the Reader

Apply the method of separation of variables to the wave equation by putting

$$u = T(t)F(r, \theta), \tag{5.35}$$

and obtain the equations for $T(t)$ and $F(r, \theta)$.

Solution

We have

$$\frac{T''(t)}{T(t)} = \frac{1}{F(r,\theta)}\left(\frac{\partial^2}{\partial r^2} + \frac{1}{r}\frac{\partial}{\partial r} + \frac{1}{r^2}\frac{\partial^2}{\partial \theta^2}\right)F(r, \theta) = c. \tag{5.36}$$

By (5.36), c is neither a function of t nor a function of r and θ, and hence must be a constant. Thus we have

$$\left(\frac{\partial^2}{\partial r^2}+\frac{1}{r}\frac{\partial}{\partial r}+\frac{1}{r^2}\frac{\partial^2}{\partial \theta^2}\right)F(r,\theta)=cF(r,\theta) \tag{5.37}$$

and

$$T''(t)=cT(t). \tag{5.38}$$

The eigenvalues c are determined from the homogeneous boundary conditions (5.34). We shall first show that, as a consequence of (5.34), c must be a negative real constant.

We multiply (5.37) by F and integrate over the unit disk, getting

$$\iint F\nabla^2 F dA = c\iint F^2 dA. \tag{5.39}$$

Performing an integration by parts for the left side of (5.39), and making use of the fact that F vanishes at the boundary, we get

$$-\iint (\vec{\nabla}F)^2 dA = c\iint F^2 dA. \tag{5.40}$$

If F is not a constant, c is equal to the negative of the ratio of two positive integrals and hence must be negative. (If F is a constant, then it must be identically zero if it is equal to zero at $r=a$, and is a trivial solution.) Thus we put

$$c=-\omega^2, \tag{5.41}$$

and (5.38) gives

$$T(t)=e^{\pm i\omega t}, \tag{5.42}$$

which is oscillatory if $\omega \neq 0$. We see that the boundary condition (5.34) guarantees that the displacement of the drum will not grow exponentially and the drum will not eventually explode. Nor will the vibration of the drum surface be damped with time if the wave equation is strictly obeyed.

We shall next show that the values of ω, the frequencies of oscillation of the drum, are discrete, and are indeed related to the roots of a special function called the Bessel function.

Problem for the Reader

Solve the PDE (5.37) by doing another separation of variables:

$$F(r,\theta) = R(r)\Theta(\theta). \tag{5.43}$$

Solution

Substituting (5.43) into (5.37) with c replaced by $-\omega^2$, we get

$$\frac{r^2}{R(r)}\left(\frac{d^2}{dr^2} + \frac{1}{r}\frac{d}{dr} + \omega^2\right)R(r) = -\frac{1}{\Theta(\theta)}\frac{d^2}{d\theta^2}\Theta(\theta) = c'. \tag{5.44}$$

Equating the second quantity and the third quantity in (5.44), we get

$$c' = n^2,$$

which is a second eigenvalue for this problem. The corresponding eigenfunction $\Theta(\theta)$ is given by

$$\Theta_n(\theta) = e^{\pm in\theta}.$$

Equating the first quantity and the third quantity in (5.44), we get

$$\left(r^2\frac{d^2}{dr^2} + r\frac{d}{dr} + \omega^2 r^2\right)R_n(r) - n^2 R_n(r) = 0. \tag{5.45}$$

It remains to satisfy the boundary condition (5.34), which requires that

$$R_n(1) = 0. \tag{5.46}$$

Equation (5.45) is known as the Bessel equation, and the solutions of (5.45) are known as the Bessel functions. Since the Bessel equation is of the second order, it has two independent solutions. As we shall discuss in the next chapter, it turns out that one of them is infinite at $r = 0$. Since the solution is required not to blow up at any point inside the unit circle, this solution is excluded. The one that is finite at $r = 0$ is denoted by $J_n(\omega r)$. By requiring that (5.46) be satisfied, we get

$$J_n(\omega) = 0.$$

This constitutes a Sturm-Liouville problem. Let us denote the roots of the Bessel function of order n as ω_{mn}, i.e.,

$$J_n(\omega_{mn}) = 0.$$

Indeed, the natural angular frequencies of a drum of radius a are equal to ω_{mn} divided by a. The corresponding eigenfunctions, $J_n(\omega_{mn}r)$, $m = 1, 2\cdots$, constitute a complete set. We may use a linear superposition of such functions to satisfy the initial conditions. Thus we represent the general solution of the two-dimensional wave equation with circular geometry by a linear superposition of the solutions obtained by

$$u(r,\theta,t) = \sum_{n,m} \Big[e^{i\omega_{mn}t}(a_{mn}e^{in\theta} + b_{mn}e^{-in\theta}) \\ + e^{-i\omega_{mn}t}(c_{mn}e^{in\theta} + d_{mn}e^{-in\theta}) \Big] J_n(\omega_{mn}r). \tag{5.47}$$

This solution satisfies the wave equation as well as the boundary condition (5.34). And the coefficients a_{mn}, b_{mn}, c_{mn}, and d_{mn} are so chosen that the initial conditions (5.32) and (5.33) are satisfied.

We point out that the wave equation for the drum surface has three independent variables: r, θ, and t. The application of the method of separation of variables is done in two steps: separate the t variable from r and θ, and separate θ from r. A constant is introduced at each of these steps. Hence there are two eigenvalues in the problem: n and ω_{mn}. These eigenvalues are determined from the homogeneous boundary conditions (5.20), (5.21), and (5.34). Each set of eigenvalues n and ω_{mn} is associated with the eigenfunctions

$$e^{\pm i\omega_{mn}t} J_n(\omega_{mn}r)e^{\pm in\theta}.$$

A superposition of these functions is a double sum over n and m.

Finally, let me add a few more words about the Sturm-Liouville problem. A Sturm-Liouville problem is an eigenvalue problem, a solution of which satisfies the differential equation

$$\frac{d}{dx}\left[p(x)\frac{dy}{dx}\right] + [q(x) + \lambda w(x)]\, y = 0,$$

$$a < x < b,$$

as well as a homogeneous boundary condition at each of the endpoints a and b. In the above, $p(x)$, $q(x)$, and $w(x)$ are given functions of x, and λ is the eigenvalue to be determined.

Let $y_1(x)$ and $y_2(x)$ be eigenfunctions of the Sturm-Liouville problem given above with the eigenvalues λ_1 and λ_2, respectively. This means that

$$\frac{d}{dx}\left[p(x)\frac{dy_1}{dx}\right] + [q(x) + \lambda_1 w(x)]\, y_1 = 0$$

and

$$\frac{d}{dx}\left[p(x)\frac{dy_2}{dx}\right] + [q(x) + \lambda_2 w(x)]\, y_2 = 0.$$

We multiply the first equation above by y_2 and the second equation above by y_1, and take the difference of these two resultant equations. We get

$$y_2\frac{d}{dx}\left(p\frac{dy_1}{dx}\right) - y_1\frac{d}{dx}\left(p\frac{dy_2}{dx}\right) + (\lambda_1 - \lambda_2)wy_1y_2 = 0.$$

The sum of the first two terms of the equation above is equal to the derivative of

$$\left(y_2p\frac{dy_1}{dx} - y_1p\frac{dy_2}{dx}\right).$$

Therefore, by integrating the equation above from a to b, we get

$$\left[y_2p\frac{dy_1}{dx} - y_1p\frac{dy_2}{dx}\right]_a^b + (\lambda_1 - \lambda_2)\int_a^b wy_1y_2dx = 0.$$

If the eigenfunctions satisfy homogeneous boundary conditions, such as

$$y + c\frac{dy}{dx} = 0,$$

we easily show that the first term in the equation above vanishes, and we get

$$\int_a^b wy_1y_2dx = 0 \quad \text{if } \lambda_1 \neq \lambda_2.$$

Thus the eigenfunctions y_1 and y_2 are orthogonal to each other with the weighting function $r(x)$.

We shall state without proof that the eigenfunctions of a Sturm-Liouville problem form a complete set. Therefore, one may use these eigenfunctions to make an expansion for a function defined in the interval $[a, b]$.

Take the example of eq. (5.18), where $p = w = 1$, $q = 0$, and $\lambda = c$. Let us require the eigenfunctions be periodic with the period 2π, thus ensuring that

$$\left[y_2\frac{dy_1}{dx} - y_1\frac{dy_2}{dx}\right]_0^{2\pi} = 0,$$

where y_1 and y_2 are solutions of this Sturm-Liouville problem. As we have shown, the eigenfunctions of this Sturm-Liouville problem are

$$e^{\pm in\theta},$$

which are orthogonal to one another with the weighting function 1. If we use these eigenfunctions to make an expansion for a function, the series obtained is the Fourier series of this function.

Another example is offered by (5.45), which we shall write in the Sturm-Liouville form

$$\frac{d}{dr}\left(r\frac{dR_n}{dr}\right)+\left(\omega^2 r-\frac{n^2}{r}\right)R_n=0.$$

From this form we identify $w = r$. Let us require the eigenfunctions to vanish at $r = a$ and not to blow up at $r = 0$. These boundary conditions ensure that

$$\left[y_2 r\frac{dy_1}{dr}-y_1 r\frac{dy_2}{dr}\right]_0^a=0,$$

where y_1 and y_2 are eigenfunctions of this Sturm-Liouville problem. These eigenfunctions have been found to be $J_n(\omega_{mn}r)$, and we here conclude that

$$\int_0^a J_n(\omega_{mn}r)J_n(\omega_{ln}r)rdr=0 \quad \text{if } \omega_{mn}\neq\omega_{ln}.$$

We may use these eigenfunctions to make expansions for functions defined in the interval $[0, a]$. Let $f(r)$ be such a function and let the expansion for $f(r)$ be

$$f(r)=\sum_{m=1}^{\infty}c_m J_n(\omega_{mn}r),\ \ 0<r<a.$$

The series above is called the Fourier-Bessel series for $f(r)$. By multiplying the equation above by $J_n(\omega_{mn}r)r$ and integrating from 0 to a, we get the Fourier-Bessel coefficients c_m as

$$c_m=\frac{\int_0^a f(r)J_n(\omega_{mn}r)rdr}{\int_0^a\left[J_n(\omega_{mn}r)\right]^2 rdr}.$$

C. The Schrödinger Equation

We'll now discuss the Schrdinger equation of quantum mechanics. We shall first explain how this equation was first derived.

A plane wave of frequency

$$\frac{\omega}{2\pi}$$

and wave number k travelling in the x direction is described by

$$e^{-i\omega t + ikx}.$$

According to the principle of particle-wave duality of de Broglie, this wave is also a particle with energy

$$E = h\frac{\omega}{2\pi}$$

and momentum

$$p = h\frac{k}{2\pi},$$

where h is the Planck constant 6.6238×10^{-27} erg-sec.

When it operates on the plane wave

$$e^{-i\omega t + ikx},$$

the differential operator

$$\frac{\partial}{\partial t} \to -i\omega;$$

thus

$$i\frac{h}{2\pi}\frac{\partial}{\partial t} \to E. \qquad (5.48)$$

Similarly, when it operates on the plane wave

$$e^{-i\omega t + ikx}$$

$$\frac{\partial}{\partial x} \to ik;$$

thus

$$-i\frac{h}{2\pi}\frac{\partial}{\partial x} \to p. \qquad (5.49)$$

Schrödinger took this one step further. He assumed that a particle in a three-dimemsional space is described by a complex-valued function $\Psi(\overrightarrow{x}, t)$, the square of the absolute value of which is the probability density of finding the particle at position $\overrightarrow{x}$ and at time t. Schrödinger also assumed that the correspondence (5.48) and (5.49) holds. As we know, the Hamiltonian of a free particle moving in a three-dimensional space is given by

$$H = E = \frac{\overrightarrow{p}^2}{2m}. \tag{5.50}$$

Thus, substituting (5.49) into (5.50) and requiring that (5.48) hold, Schrödinger suggested that Ψ satisfies the PDE

$$i\frac{h}{2\pi}\frac{\partial}{\partial t}\Psi(\overrightarrow{x}, t) = -\frac{1}{2m}\left(\frac{h}{2\pi}\right)^2 \nabla^2\Psi(\overrightarrow{x}, t). \tag{5.51}$$

Equation (5.51) is the Schrdinger equation for a free particle.

The Hamiltonian for a particle in an electromagnetic field $(A_0, \overrightarrow{A})$ is

$$H = \frac{(\overrightarrow{p} - e\overrightarrow{A})^2}{2m} + eA_0,$$

where A_0 and $\overrightarrow{A}$ are functions of space and time. Schrödinger assumes that his equation holds not only for a free particle but also for a particle in a field. Thus the Schrdinger equation for a particle in an electromagnetic field is

$$i\frac{h}{2\pi}\frac{\partial}{\partial t}\Psi = -\frac{1}{2m}\left(\frac{h}{2\pi}\right)^2\left(\overrightarrow{\nabla} - \frac{2\pi ie}{h}\overrightarrow{A}\right)^2\Psi + eA_0\Psi. \tag{5.52}$$

This is the basic equation in quantum mechanics, which supersedes Newtonian mechanics; Newtonian mechanics fails when one gets to the small atomic dimensions. In situations in which the Planck constant h can be regarded as a small parameter, the solutions of the Schrdinger equation are approximated by the WKB solutions, which we will discuss in Chapter 7. In the WKB approximation, the exponent of the wave function obeys the Hamilton-Jacobi equation of classical mechanics. Thus quantum mechanics is a higher law, the laws of Newtonian mechanics being approximations of it.

We shall choose the unit so that $2m = 1$ and $h = 2\pi$. We shall also restrict ourselves here to the case $\overrightarrow{A} = 0$. Then we have

$$i\frac{\partial}{\partial t}\Psi = -\nabla^2\Psi + V\Psi, \tag{5.53}$$

where

$$V(\overrightarrow{x}, t) = eA_0(\overrightarrow{x}, t).$$

Consider the Schrdinger equation in one spatial dimension with the potential independent of time, i.e.,

$$V = V(x).$$

Problem for the Reader

Put

$$\Psi(x, t) = X(x)T(t)$$

and perform separation of variables for the Schrödinger equation (5.53).

Solution

We get

$$i\frac{dT(t)}{dt} = ET(t), \tag{5.54}$$

and

$$-\frac{d^2X(x)}{d^2x} + V(x)X(x) = EX(x). \tag{5.55}$$

We shall require that (5.55) hold for all x, with

$$X(\pm\infty) = 0. \tag{5.56}$$

The energy eigenvalues are determined from the homogeneous condition (5.56). Indeed, while $X(x)$ is a complex value function, we may prove from (5.56) that the eigenvalues E are real. (See homework problem 1). Thus $T(t)$ is oscillatory.

The eigenfunctions $X_n(x)$ of (5.55) together with (5.56) form a complete set. Specifically, let

$$\Psi(x,t) = \sum_n a_n X_n(x) e^{-iE_n t}.$$

Then the initial-value problem

$$\Psi(x,0) = f(x)$$

is solved by choosing

$$a_n = \int_{-\infty}^{\infty} X_n^*(x) f(x) dx,$$

where we have normalized the eigenfunctions so that

$$\int_{-\infty}^{\infty} X_n^*(x) X_m(x) dx = \delta_{nm}.$$

We'll give a few specific examples.

In the case of a linear potential,

$$V(x) = x.$$

Then (5.55) becomes

$$\frac{d^2 X(x)}{d^2 x} + (E - x) X(x) = 0, \tag{5.57}$$

which, with a shift of the dependent variable x, is known as the Airy equation.

If $V(x)$ is the potential of a harmonic oscillator,

$$V(x) = \frac{1}{2}\kappa x^2, \kappa \text{ a constant},$$

the Schrdinger equation (5.55) is the parabolic cylinder equation

$$-\frac{d^2 X(x)}{d^2 x} + \frac{1}{2}\kappa x^2 X(x) = E X(x). \tag{5.58}$$

We shall discuss these equations in later chapters.

Before we close this chapter, let me say a few words about the gauge principle of quantum mechanics, which is one of the most important physical principles discovered in the twentieth century. As we have mentioned, the square of the absolute value of the wavefunction Ψ is the probability density, which leaves the phase of the complex number Ψ without any physical meaning. Let us change the wavefunction Ψ to Ψ', where

$$\Psi = e^{i\chi}\Psi', \tag{5.59}$$

with χ a real constant. We call this change of the wavefunction a gauge transformation of the first kind. By (5.59), we have

$$|\Psi|^2 = |\Psi'|^2.$$

Thus Ψ' is just as good as Ψ in the role of representing the quantum state of the particle. Indeed, if we substitute (5.59) into (5.51), then after cancelling the common factor $e^{i\chi}$ from the terms of the resulting equation, we find that Ψ' satisfies exactly the same equation as that satisfied by Ψ.

It is gratifying to see that as all wavefunctions differing by a phase factor have the same physical contents, they also satisfy the same equation. There is a complete democracy in the choice of these wavefunctions. This is the gauge principle of quantum mechanics.

But this is true only if χ is a constant. Let χ be a function of space and time. This is to say that we add one phase factor to the wavefunction at the space-time point A, and a different phase factor to the wavefunction at the space-time point B. This is called making a gauge transformation of the second kind.

Substituting (5.59) into (5.51), we find that, since χ depends on $\overrightarrow{x}$ and t, we cannot move the factor $e^{i\chi}$ to the left of the differential operators in the equation. Thus we cannot cancel out the factor $e^{i\chi}$ from the equation, and the equation satisfied by Ψ' is different from that satisfied by Ψ.

But what happens if the electron is under the influence of an electromagnetic field? Let us substitute (5.59) into (5.52). We get

$$i\left(\frac{\partial}{\partial t} + ieA_0\right)e^{i\chi}\Psi' = -(\overrightarrow{\nabla} - ie\overrightarrow{A})^2 e^{i\chi}\Psi', \tag{5.60}$$

where we have set h to be 2π. We move the phase factor $e^{i\chi}$ to the left of the differential factor, and compensate by replacing the differential operator with itself plus the derivative of $i\chi$.

We get

$$i\left(\frac{\partial}{\partial t} + i\chi_t + ieA_0\right)\Psi' = -(\overrightarrow{\nabla} + i\overrightarrow{\nabla}\chi - ie\overrightarrow{A})^2\Psi'. \tag{5.61}$$

Let us denote

$$A_0' \equiv A_0 + \frac{\dot{\chi}}{e} \tag{5.62}$$

and

$$\overrightarrow{A'} \equiv \overrightarrow{A} - \overrightarrow{\nabla}\frac{\chi}{e}, \tag{5.63}$$

where $\dot{\chi}$ *is* $\frac{\partial \chi}{\partial t}$. Then (5.61) is exactly the same as (5.52) with Ψ, A_0, and $\vec{A}$ replaced by Ψ', A_0', and $\vec{A'}$. Democracy is now restored.

What is more, since the magnetic field $\vec{B}$ is equal to $\vec{\nabla} \times \vec{A}$, we have by (5.63) that

$$\vec{\nabla} \times \vec{A'} = \vec{\nabla} \times \left(\vec{A} - \vec{\nabla}\frac{\chi}{e}\right) = \vec{\nabla} \times \vec{A} = \vec{B}.$$

Thus, $\vec{A'}$ and $\vec{A}$ give exactly the same $\vec{B}$ field. Also, the electric field $\vec{E}$ is equal to

$$-\vec{\nabla} A_0 - \frac{\partial \vec{A}}{\partial t}.$$

We have

$$-\vec{\nabla} A_0' - \frac{\partial \vec{A'}}{\partial t} = -\vec{\nabla}\left(A_0 + \frac{\dot{\chi}}{e}\right) - \frac{\partial\left(\vec{A} - \vec{\nabla}\frac{\chi}{e}\right)}{\partial t} = -\vec{\nabla} A_0 - \frac{\partial \vec{A}}{\partial t} = \vec{E}.$$

Thus A_0' and $\vec{A'}$ give the same $\vec{E}$ field as that given by A_0 and $\vec{A}$. Indeed, (5.62) and (5.63) had been known, long before quantum mechanics was discovered, to be the gauge transformation that leaves the electric field $\vec{E}$ and the magnetic field $\vec{B}$ unchanged.

It is striking how nature takes advantage of a beautiful property of the differential operator and turns it into one of the fundamental laws of nature. Now we understand the motivation behind the command "Let there be light." Light is the electromagnetic field, the existence of which is necessary for the physical laws to possess invariance under gauge transformation of the second kind.

Homework Problems for This Chapter

Solutions to the Homework Problems can be found at www.lubanpress.com.

1. Consider the eigenvalue problem (5.55) together with the boundary condition $X(\pm\infty) = 0$. Show that the eigenvalue E is real.
   Hint: Multiply (5.55) by $X^*(x)$ and integrate from $-\infty \to \infty$.

2. Express the Laplacian in three spatial dimensions

$$\nabla^2 \equiv \frac{\partial^2}{\partial x^2} + \frac{\partial^2}{\partial y^2} + \frac{\partial^2}{\partial z^2}$$

   in terms of spherical coordinates.
   Hint: It reduces the algebra if you make use of (5.13).

3. Find the Green function for the Laplace equation

$$\nabla^2 G(x, x') = \delta(\vec{x} - \vec{x}')$$

in the three-dimensional space as well as in the two-dimensional space. What is the behavior of the Green function at infinity?

4. Discuss how to apply separation of variables to the Schrdinger equation in two spatial dimensions

$$\left[\frac{\partial^2}{\partial x^2} + \frac{\partial^2}{\partial y^2} - U(x, y)\right] \Psi(x, y, t) = i\frac{\partial}{\partial t}\Psi(x, y, t)$$

when

$$U(x, y) = U_1(x) + U_2(y).$$

Is the equation separable if

$$U(x, y) = U_1(x)U_2(y)?$$

What if $U(x, y) = xy$? Can you change the independent variables and make the equation separable?

5. Solve the Hamilton-Jacobi equation (last example in Chapter 3) with the method of separation of variables in the following cases:

a. One spatial dimension with $V = V(x)$.

b. Two spatial dimensions with $V = V(r)$, where $r = \sqrt{x^2 + y^2}$.

c. Three spatial dimensions with $V = a(r) + \dfrac{b(\phi)}{\rho^2} + \dfrac{c(\theta)}{\rho^2 \sin^2 \phi}$, where $\theta \equiv \tan^{-1} \dfrac{y}{x}$, $\rho \equiv \sqrt{x^2 + y^2 + z^2}$, and $\phi \equiv \tan^{-1} \dfrac{r}{z}$.

6. Using any variables you deem appropriate, solve by separation of variables the equation

$$\left(\frac{\partial^2}{\partial x^2} + \frac{\partial^2}{\partial y^2}\right) u(x, y) = \frac{2xy}{(x^2 + y^2)^2} u(x, y).$$

Do you recognize the resulting ordinary differential equations?

Chapter 6

Singular Points of Ordinary Differential Equations

In the last chapter, we applied the method of separation of variables to various PDEs and found a number of transcendental equations. Two of them are

$$\frac{d^2y}{dx^2} - xy = 0 \qquad \text{(Airy eq.)}$$

and

$$x^2\frac{d^2y}{dx^2} + x\frac{dy}{dx} + (x^2 - p^2)y = 0. \qquad \text{(Bessel eq.)}$$

These equations cannot be solved in closed forms.

As one handles problems in the real world, one commonly encounters equations that cannot be solved in closed form. In such cases it's helpful if one is able to find approximate solutions for them. In this chapter, we'll focus on making approximations for equations that cannot be solved in closed form.

A. Taylor Series Solutions

We first demonstrate with the example of the Airy equation

$$y'' - xy = 0. \tag{6.1}$$

Consider the initial-value problem in which the values $y(0)$ and $y'(0)$ of the solution of this equation are given.

While it is clear that when x is small,

$$y(x) \simeq y(0) + y'(0)x$$

is a reasonable approximation of $y(x)$, we shall show how to get better approximations than that.

We see from eq. (6.1) that y'' is equal to xy. Thus it enables us to obtain $y''(0)$ once $y(0)$ is given. Since xy vanishes at $x = 0$, the value $y''(0)$ is equal to zero.

By differentiating (6.1) we find

$$y''' = xy' + y. \tag{6.2}$$

Thus $y'''(0) = y(0)$. Repeating this process successively, we may evalaute all $y^{(n)}(0)$. With all the derivatives of y at $x = 0$ calculated, we obtain the solution of the Airy equation in the form of the Maclaurin series

$$y(x) = y(0) + y'(0)x + \frac{y''(0)}{2!}x^2 + \cdots + \frac{y^{(n)}(0)}{n!}x^n + \cdots . \tag{6.3}$$

While it is perfectly correct to obtain the Maclaurin series for the solution of the Airy equation in this manner, there is a more efficient way to do it.

Let's write the Maclaurin series as

$$y(x) = \sum a_n x^n, \tag{6.4}$$

where the summation ostensibly covers all positive and negative integers n. We will understand that

$$a_{-n} = 0, \; n = 1, 2, \cdots .$$

Thus (6.4) is indeed a Maclaurin series.

We have

$$y''(x) = \sum_n a_n n(n-1)x^{n-2} \tag{6.5}$$

and

$$xy(x) = \sum_n a_n x^{n+1} = \sum_n a_{n-3} x^{n-2}. \tag{6.6}$$

In the last step above, we redesignate n to be $(n-3)$. The purpose for doing this is to cause the power function inside the summation to be x^{n-2}, the same as that in (6.5). Note that adopting the notation (6.4) saves us the labor of respecifying the range of the summation index n as we redesignate n.

Equation (6.1) is satisfied if the coefficient of x^{n-2} (6.5) is equal to that of (6.6), or

$$a_n n(n-1) = a_{n-3}. \tag{6.7}$$

Equation (6.7) is a recurrence formula that relates two Taylor coefficients with indices differing by 3.

Problem for the Reader

Set $n = 0, 1, 2, 3$, and 4 in (6.7) and see what you can conclude.

Solution

Setting $n = 0$, we find that the left side is zero, but so is the right side, as a_{-3} is zero. Thus the recurrence formula is self-fulfilling.
Setting $n = 1$ in (6.7), we again find that the two sides are zero.
Setting $n = 2$ in (6.7), we get $a_2 = 0$.
Setting $n = 3$ in (6.7), we get $a_3 = a_0/6$.
Setting $n = 4$ in (6.7), we get $a_4 = a_1/12$.

Generally, once a_0 is given, the recurrence formula gives us a_3, which begets a_6, which begets a_9, and successively all a_{3m}.

Similarly, we may express a_{3m+1} in terms of a_1.

And since a_2 is zero, we find from (6.7) that a_{3m+2} is zero.

The recurrence formula (6.7) has only two terms, which makes it easy to solve. But not all recurrence formulae are so simple. Indeed, the number of coefficients in the recurrence formula for an ODE has to do with the dimensions of operators in the equation. We define the dimension of x to be 1, as x changes linearly when the unit of measurement changes, say from centimeters to inches. Now if we make the transformation $x \to cx$, then $D \to c^{-1}D$. Thus we will assign the dimension -1 to the operator D. More generally, the operator $x^m D^n$ has the dimension $(m-n)$. As we apply the operator $x^m D^n$ on the power function x^p, where p is a constant, we get $(x^m D^n)x^p = p(p-1)\cdots(p-n+1)x^{p+m-n}$. The relevant point here is that the power of x increases by $m-n$, which is precisely the dimension of the operator $x^m D^n$. For example, applying D and $1/x$ to the power function x^p leads to the same power function x^{p-1} multiplied by different constants.

It is now easy to understand why the recurrence formula (6.7) for the Airy equation has only two terms. It is because the operator $(D^2 - x)$ in the Airy equation has only two terms. Since the dimension of D^2 and that of $-x$ differ by 3, the exponent of the power function for the series of $D^2 y$ and that for $-xy$ differ by three. As we redesignate the exponent of $-xy$, making the exponents of the power functions of the two series the same, the recurrence formula (6.7) obtained has two terms relating a_n with a_{n-3}.

If the terms in the differential equation have m different dimensions, the recurrence formula for this equation has m terms.

Counting the number of dimensions in an ODE sometimes gives us hints to find the transformation that may help in solving the ODE (see, for example, homework problems 1c and 1e) or in streamlining the calculations of series solutions, as we shall soon demonstrate.

We set $n = 3m$ in (6.7); we get

$$a_{3m} = \frac{a_{3(m-1)}}{3^2 m(m - \frac{1}{3})}. \tag{6.8}$$

Note that we have taken care to write

$$(3m - 1) = 3(m - 1/3),$$

making the coefficient of m inside the parenthesis unity. As we shall see, this makes it convenient to express a_{3m} in terms of Gamma functions.

Problem for the Reader

Keep applying (6.8), each time reducing m by unity, until you succeed in expressing a_{3m} in terms of a_0.

Solution

Since there is a factor 3^2 in the denominator of (6.8), we get an additional factor 3^2 in the denominator each time we apply (6.8). Thus, applying (6.8) m times, we get a factor 3^{2m} in the denominator. Similarly, the factor m in the denominator of

(6.8) generates, as we apply (6.8) successively, the factors $m(m-1)(m-2)\cdots$, and similarly for the factor $(m-\frac{1}{3})$ in the denominator of (6.8). Thus we get

$$a_{3m} = \frac{a_0}{3^{2m}\left[m(m-1)\cdots 1\right]\left[(m-\frac{1}{3})(m-\frac{4}{3})\cdots(\frac{2}{3})\right]}. \tag{6.9}$$

The expression (6.9) can be written in a more compact form. First of all, we note that the quantity inside the first square bracket above is

$$m(m-1)\cdots 1 = m!.$$

The quantity in the second square bracket of (6.9) can be expressed more neatly too, with the use of the Gamma function $\Gamma(z)$, defined by the integral

$$\Gamma(z) \equiv \int_0^\infty e^{-t} t^{z-1} dt. \tag{6.10}$$

Note in particular that as we set $z = 1$ in the formula above, the integral is easily evaluated and we get

$$\Gamma(1) = 1.$$

We shall show in the Appendix the identity

$$z(z-1)(z-2)\cdots(z-n+1) = \Gamma(z+1)/\Gamma(z-n+1). \tag{6.11}$$

By setting $z = n$ in (6.11), we get

$$n! = \Gamma(n+1), \tag{6.12}$$

as $\Gamma(1) = 1$. Thus the Gamma function is simply the factorial function when its argument is an integer.

Note that $(z-n+1)$, the argument of the rightmost Gamma function in (6.11), is equal to the last factor on the left side of (6.11), while $(z+1)$, the argument of the other Gamma function in (6.11), is 1 greater than the first factor on the left side of (6.11). Consequently, the quantity in the second square bracket of (6.9) is equal to $\Gamma(m+2/3)/\Gamma(2/3)$, and (6.9) is

$$a_{3m} = \frac{\Gamma(2/3)}{3^{2m} m! \Gamma(m+2/3)} a_0.$$

Problem for the Reader

Express a_{3m+1} in terms of a_{3m-2}.

Solution

Setting $n = 3m + 1$ in (6.7), we get

$$a_{3m+1} = \frac{a_{3(m-1)+1}}{3^2\left(m+\frac{1}{3}\right)(m)}. \tag{6.13}$$

Problem for the Reader

Express a_{3m+1} in terms of a_1.

Solution

We may derive from (6.13) and (6.10) that

$$a_{3m+1} = \frac{a_1\Gamma(4/3)}{3^{2m}m!\Gamma(m+4/3)}. \tag{6.14}$$

Thus the general solution of the Airy equation is

$$y(x) = C_1 y_1(x) + C_2 y_2(x),$$

where

$$y_1(x) = \sum_m \frac{x^{3m}}{3^{2m}m!\Gamma\left(m+\frac{2}{3}\right)} \tag{6.15}$$

and

$$y_2(x) = \sum_m \frac{x^{3m+1}}{3^{2m}m!\Gamma\left(m+\frac{4}{3}\right)}, \tag{6.16}$$

with

$$C_1 \equiv a_0\Gamma(2/3) \text{ and } C_2 \equiv a_1\Gamma(4/3).$$

If x is small, taking a few terms of (6.15) and (6.16) gives a good numerical approximation of the solution of the Airy equation.

But (6.15) and (6.16) are more useful than that. Indeed, the series can be used for any finite value of x, not just for small x. This is because the series is convergent even for large x.

Problem for the Reader

For what values of x is the series (6.15) convergent?

Solution

The series is convergent if the magnitude of the ratio of the mth term with its preceding term is less than unity. This needs to be true only when m is very large, as a finite number of terms being disagreeably large affects the value but not the convergence of the series. Thus we calculate with the use of (6.8) the ratio

$$\frac{a_{3m}x^{3m}}{a_{3(m-1)}x^{3(m-1)}} = \frac{x^3}{9m\left(m-\frac{1}{3}\right)}. \tag{6.17}$$

We see that the ratio in (6.17) vanishes as $m \to \infty$, no matter how large x is. Thus the series (6.15) converges for all finite values of x, and is useful for all finite values of x.

However, if x is not small, we may have to take a large number of terms in the series to approximate the solution well. How many terms of the series should we take in a practical calculation? Very roughly, stopping at the value m when the ratio (6.17) becomes much less than unity means that the mth term we neglect is much less than the $(m-1)^{\mathrm{th}}$ term, which we keep. In the present example, this value m is much larger than $\frac{1}{3}x^{\frac{3}{2}}$. When x is large, m is a large number. Thus the series solution is inefficient to use if x is large. What is more, the series hardly tells us anything about the qualitative behavior of the Airy functions for large values of x.

If we want to find the behavior of the solution when x is very large, it will be far easier if we define

$$t \equiv \frac{1}{x}.$$

Note that t is very small when x is very large, and vice versa. Thus, to study the behavior of the solution of the Airy equation when x is large, we try to express this solution in the form of an expansion series of power functions of t, rather than those of x. As it turns out, this series is not a Taylor series of t. We will return to this problem in Section C of this chapter.

B. Frobenius method

Consider next the ODE

$$x^2y'' - y = 0. \tag{6.18}$$

Problem for the Reader

Find the solution of (6.18) in the form of a Maclaurin series.

Solution

$$y(x) = \sum a_n x^n,$$

$$x^2y''(x) = \sum a_n n(n-1)x^n.$$

Thus we have

$$(n^2 - n - 1)a^n = 0. \tag{6.19}$$

There is only one term in this recurrence formula, as both x^2D^2 and -1 are of the same dimension. Since $(n^2 - n - 1) \neq 0$ for any integer n, we conclude $a_n = 0$ for all n.

Thus the only solution is the trivial solution $y = 0$. We obtain nothing useful because we seek the solution in the form of a Maclaurin series, and the fact that such a solution is identically zero means that there is no solution of (6.18) that is expressible in the form of a Maclaurin series.

Indeed, an astute reader may recognize that (6.18) is equidimensional and can be solved exactly. Thus we substitute $y = x^s$ into (6.18) and get

$$s(s-1) = 1, \tag{6.20}$$

the two roots of which are

$$s_1 = \frac{1 + i\sqrt{5}}{2}, \; s_2 = \frac{1 - i\sqrt{5}}{2}, \qquad (6.21)$$

and the two independent solutions are

$$y_1(x) = x^{s_1} \text{ and } y_2(x) = x^{s_2}.$$

Since neither s_1 nor s_2 is an integer, we see that none of the solutions of (6.18) are analytic at the origin. This is why we get the trivial solution $y = 0$ when we try to find the solution of (6.18) in the form of a Maclaurin series. The general solution of (6.18) is

$$y(x) = c_1 y_1(x) + c_2 y_2(x),$$

where c_1 and c_2 are arbitrary constants.

Problem for the Reader

Let $y(0) = 1$ and $y'(0) = 2$; determine c_1 and c_2.

Solution

Since

$$y_1(0) = y_2(0) = 0,$$

no values of c_1 and c_2 can be chosen to make $y(0) = 1$. Also, since

$$y_1'(0) = y_2'(0) = \infty,$$

no values of c_1 and c_2 can be chosen to make $y'(0) = 2$. (It is often impossible to impose initial conditions at a singular point of a differential equation.)

Equation (6.18) is perhaps a little too trivial. Let us consider the Bessel equation

$$x^2 \frac{d^2 y}{dx^2} + x \frac{dy}{dx} - p^2 y = -x^2 y. \qquad (6.22)$$

Let us seek the solution of the Bessel equation in the form of some kind of expansion around the point $x = 0$.

When x^2 is much smaller than p^2, we make the approximation of neglecting x^2y, as it is much smaller than p^2y. In this approximation we set the right side of (6.22) to zero. We note that the resulting ODE is equidimensional. Thus the solution of the resulting equation is of the form x^s. Substituting this solution into the differential equation, we get

$$s(s-1)+s-p^2=0,$$

or

$$s = p \text{ or } -p. \tag{6.23}$$

Thus the approximate solutions are

$$y(x) \approx x^p \text{ or } x^{-p}, \tag{6.24}$$

which are good approximations when x is small.

We see once again from (6.24) that it is not meaningful to assign initial values $y(0)$ and $y'(0)$ to the solution of (6.22). This is because at the point $x = 0$, one of the solutions of (6.24) vanishes and the other blows up unless $\mathrm{Re}\, p = 0$. Even if $p = 0$, it is still not possible to assign two initial values of y at the origin. This is because while both x^p and x^{-p} with p set to zero are finite at $x = 0$, these two solutions are actually the same solution and are not independent of each other. The second independent solution is obtained by differentiating x^p with respect to p before setting p to zero, and is equal to $\ln x$. This solution blows up at $x = 0$.

We have seen that it is not possible to assign initial values $y(0)$ and $y'(0)$ to the solution of the Bessel equation. Indeed, we have encountered this initial value problem before, when we treated in Chapter 5 the motion of the surface of a drum. As you may recall, the solution $R(r)$ obtained with the method of separation of variables is a sum of two solutions of the Bessel equation of order n, where n is an integer. One of the imposed conditions is based on physical considerations. This condition requires the solution to be finite at $x = 0$, but does not require the solution to take any specific value. Such a condition eliminates one of the solutions of the Bessel equation, i.e., the solution that blows up at the origin.

Let us seek an improvement of the approximate solutions of (6.24). Take one of the approximate solutions, say x^p, and better it by improving the right side of (6.22). Specifically, instead of setting it to zero, we replace y in the right side of (6.22) by x^p. We get

$$\left(x^2\frac{d^2}{dx^2}+x\frac{d}{dx}-p^2\right)y\approx -x^{p+2}.$$

Problem for the Reader

Find a particular solution of the equation above.

Solution

Since all the terms in the operator

$$\left(x^2\frac{d^2}{dx^2}+x\frac{d}{dx}-p^2\right)$$

are of zero dimension, and since the right side of the equation above is proportional to x^{p+2}, a particular solution is of the form

$$y_P = Ax^{p+2}.$$

We easily find that

$$A=-\frac{1}{4(p+1)}.$$

Thus an approximation better than x^p is

$$y\approx x^p-\frac{x^{p+2}}{4(p+1)}. \tag{6.25}$$

If we go one step further and replace y in the right side of (6.22) by the two terms of (6.25) instead of by just one, we get an additional term, which is a constant times x^{p+4}. We may continue this process and seek successive terms in this approximation, obtaining a solution in the form of the infinite series $x^p\sum a_nx^n$. If instead of starting with the approximate solution x^p, we start with the approximate solution x^{-p} and repeat the same procedure, we obtain a solution in the form of the infinite series $x^{-p}\sum a_nx^n$.

This motivates us to express y as

$$y(x) = \sum a_n x^{n+s}. \tag{6.26}$$

As before, we shall understand that

$$a_{-1} = a_{-2} = \cdots = 0.$$

We shall require that

$$a_0 \neq 0.$$

This means that $a_0 x^s$ is by assumption the first nonvanishing term in the series solution of the Bessel equation. The series in (6.26) is called the Frobenius series. We shall show that as we substitute (6.26) into the Bessel equation, we determine the index s as well as the coefficients a_n, $n = 1, 2 \cdots$. The values of the coefficients a_n, $n = 1, 2 \cdots$, depend on the value of s as well as a_0. There are two possible values of s, and hence we obtain, except in some special cases, two Frobenius series solutions of the Bessel equation. The general solution of the Bessel equation is a linear superposition of these two solutions. This method is known as that of Frobenius.

Problem for the Reader

Find the recurrence formula for the Bessel equation.

Solution

Determining the dimensions of the terms in the equation will help to reduce the chore of calculation. This is because as we apply two operators of the same dimension to a Frobenius series, the two resulting series can be added together without reassigning the index of summation, both operators changing the power of x^{n+s} by the same amount. Thus it will save both space and labor if we handle the operators with the same dimension at the same time.

The only term in the Bessel operator $(x^2D^2 + xD + x^2 - p^2)$ of dimension 2 is x^2. We have

$$x^2 y = \sum_n a_n x^{n+s+2} = \sum_n a_{n-2} x^{n+s}.$$

The rest of the terms in the Bessel operator are of dimension zero, and we will take care of all of them at the same step. We have

$$(x^2D^2 + xD - p^2)y = \sum \left[(n+s)(n+s-1) + (n+s) - p^2\right] a_n x^{n+s-2}.$$

Thus the recurrence formula for the Bessel equation is

$$(n+s+p)(n+s-p)a_n = -a_{n-2}. \tag{6.27}$$

We note that there are only two coefficients in (6.27): a_n and a_{n-2} . This is because there are only two different dimensions for the operators in the Bessel equation.

To obtain the solution of the Bessel equation, we first set $n = 0$ in (6.27). Since a_{-2} is zero and since $a_0 \neq 0$, we get

$$(s+p)(s-p) = 0. \tag{6.28}$$

Equation (6.28) is called the indicial equation. It determines the index s to be either s_1 or s_2. We get

$$s_1 = p, \ s_2 = -p,$$

in agreement with (6.23). Without loss of generality, we may assume that $\mathrm{Re}\, p \geq 0$, and hence $\mathrm{Re}\, s_1 \geq \mathrm{Re}\, s_2$.

Next we write (6.27) as

$$a_n = -\frac{a_{n-2}}{(n+s+p)(n+s-p)}, n > 0. \tag{6.29}$$

Problem for the Reader

Solve the recurrence formula (6.29).

Solution

The recurrence formula enables one to obtain a_n from a_{n-2}. Thus once a_0 is given, we can calculate a_2, from which we can calculate a_4, etc.

Let us therefore set $n = 2m$ in (6.29). We get

$$a_{2m} = -\frac{a_{2(m-1)}}{4\left(m + \dfrac{s+p}{2}\right)\left(m + \dfrac{s-p}{2}\right)}.$$

Note that we factor out the factor 4 so that the coefficient of m in each of the denominator factors is unity.

By applying the above recurrence formula m times, we get with the help of (6.10) that

$$a_{2m} = (-1)^m \frac{a_0\Gamma(1 + s/2 + p/2)\Gamma(1 + s/2 - p/2)}{4^m\Gamma(m + s/2 + p/2 + 1)\Gamma(m + s/2 - p/2 + 1)}. \tag{6.30}$$

The right side of (6.30) depends on s, with the two values of s determined from the indicial equation (6.28) given by (6.23).

Problem for the Reader

Find a solution of the Bessel equation.

Solution

Setting $s = p$ in (6.30), we get

$$a_{2m} = (-1)^m \frac{\Gamma(1+p)a_0}{4^m\Gamma(m+p+1)\Gamma(m+1)}. \tag{6.31}$$

Thus one of the solutions of the Bessel equation is

$$J_p(x) = \sum_{m\geq 0} (-1)^m \frac{\left(\dfrac{x}{2}\right)^{2m+p}}{\Gamma(m+p+1)\Gamma(m+1)}, \tag{6.32}$$

where the immaterial constant multiple $a_0\Gamma(1+p)$ has been ignored. The function $J_p(x)$ is known to be the Bessel function of order p.

We may get a second solution of the Bessel equation by choosing s in (6.30) to be the other index $-p$. But there is a slightly easier way. The Bessel equation is unchanged as we replace p by $-p$. Thus another solution of the Bessel equation is obtained by replacing p in (6.32) by $-p$. The solution obtained is $J_{-p}(x)$.

By (6.29), the ratio a_n/a_{n-2} vanishes like $1/n^2$ as $n \to \infty$. Thus the series (6.32) converges for all values of x. Therefore, the series is meaningful as an approximate series not only when x is small, but even if x is not small. However, when x is large, it is inefficient to calculate the values of the Bessel functions $J_p(x)$ with the series (6.32). Also, the series hardly tells us anything about the qualitative behavior of the Bessel functions for large values of x. To find the behavior of the Bessel functions for large values of x, it is a good idea to obtain its expansion in the variable $t = x^{-1}$, rather than the variable x. We will discuss this in more detail in the next section.

There are special cases in which $J_p(x)$ and $J_{-p}(x)$ are not independent. In particular, when p is equal to zero, $J_p(x)$ and $J_{-p}(x)$ are both equal to $J_0(x)$, and the second independent solution for the Bessel equation of order zero is still missing. In this case, we take the function

$$y(x) = \frac{J_p(x) - J_{-p}(x)}{2p}, \tag{6.33}$$

which is a solution of the Bessel equation of order p. Let us take the limit as $p \to 0$. The expression in (6.33), if it approaches a limit, satisfies the Bessel equation of order zero. Thus it is possible to find the second solution of the Bessel equation of order zero by applying l'Hopital's rule to (6.33). This solution is

$$\left.\frac{\partial J_p(x)}{\partial p}\right|_{p=0}. \tag{6.34}$$

Problem for the Reader

Evaluate the expression (6.34) with the use of (6.32).

Solution

Differentiating (6.32) with respect to p, and then setting p to zero, we get

$$\left.\frac{\partial J_p(x)}{\partial p}\right|_{p=0} = \log\left(\frac{x}{2}\right) J_0(x) - \sum_{m\geq 0}(-1)^m \frac{\Gamma'(m+1)\left(\frac{x}{2}\right)^{2m}}{\Gamma^3(m+1)}, \tag{6.35}$$

which is a solution of the Bessel equation of order zero.

Problem for the Reader

Show that

$$J_{-n}(z) = (-1)^n J_n(z), n = 0, 1, 2, \cdots . \tag{6.36}$$

Solution

Make use of (6.32) and the fact that $\Gamma(z)$ is infinite as $z = 0, 1, 2 \cdots$.

The second independent solution of the Bessel equation with the order p equal to n can be obtained by taking the limit $p \to n$ of the ratio

$$\frac{J_p(x) \cos \pi p - J_{-p}(x)}{\sin \pi p}. \tag{6.37}$$

I will leave this as an exercise for the reader. (See homework problem 3.)

Consider now the general second-order linear homogeneous differential equation

$$y'' + c(x)y' + d(x)y = 0. \tag{6.38}$$

If $c(x)$ and $d(x)$ are both analytic at x_0, x_0 is called an ordinary point of (6.38). And if either $c(x)$ or $d(x)$ or both have a singularity at x_0, x_0 is called a singular point of (6.38).

If x_0 is an ordinary point of (6.38), we may obtain the two independent solutions of (6.38) by seeking them in the form of a Taylor series expanded around x_0:

$$y(x) = \sum a_n (x - x_0)^n, a_{-1} = a_{-2} = \cdots = 0.$$

The coefficients a_n, $n = 2, 3, \cdots$, are determined from the recurrence formula obtained by plugging the series into the differential equation. These coefficients are dependent on the coefficients a_0 and a_1, which are arbitrary constants. The latter coefficients are indeed the values of $y(x_0)$ and $y'(x_0)$, respectively. Thus the method of series expansion gives us two independent solutions, the first being obtained with the choice of $a_0 = 1$ and $a_1 = 0$, and the second being obtained with the choice of $a_0 = 0$ and $a_1 = 1$. Each of these series is convergent at least up to the nearest singular point x_1 of (6.38) or, more precisely, for x satisfying $|x - x_0| < |x_1 - x_0|$. If the variable x is regarded to be complex, this is the circle with center at x_0 and a radius equal to the distance between x_0 and x_1. For example, the Airy equation has no finite singular point; thus its series solutions converge for all finite values of x.

Let x_0 be a singular point of (6.38). There are two kinds of singular points. If x_0 is a singular point of (6.38) with $(x - x_0)c(x)$ and $(x - x_0)^2 d(x)$ both analytic at x_0, then x_0 is called a regular singular point of (6.38).

The Bessel equation (6.22) offers an example of a differential equation with a regular singular point. We first note that this equation is not in the form of (6.38). This is because the coefficient of y'' in (6.22) is x^2, while that in (6.38) is unity. It is, of course, easy to cast (6.22) into the form (6.38). We divide the equation by x^2 and the Bessel equation takes the form

$$y'' + \frac{1}{x}y' + \left(1 - \frac{p^2}{x^2}\right) y = 0.$$

Comparing with (6.38), we find that the coefficients $c(x)$ and $d(x)$ for the Bessel equation are x^{-1} and $(1 - p^2/x^2)$, respectively. The point $x = 0$ is a regular singular point of the Bessel equation, as $c(x)$ has a simple pole and $d(x)$ has a double pole at this point. Consider the Taylor series of a solution of the Bessel equation expanded around $x = 1$, say. To find the region in which this Taylor series is convergent, we draw a circle with the center at $x = 1$ and with unit radius. Since $x = 0$ is a singular point of the Bessel equation, we are certain of the convergence of the series only if x is inside the circle. Indeed, if p is not an integer, we see from (6.32) that both solutions of the Bessel equation have a branch point at $x = 0$. Thus, if p is not an integer, all solutions of the Bessel equation have a singularity at $x = 0$, and hence their expansions around $x = 1$ must diverge at $x = 0$. If p is equal to an integer n, the solution $J_n(x)$ is analytic at $x = 0$. Since the Bessel equation has no other finite singular points, $J_n(x)$ is an entire function. Thus the Taylor series of $J_n(x)$ expanded around any point x_0 converges for all values of x. However, the other independent solution of the Bessel equation of integral order has a singularity at $x = 0$; thus its Taylor series expanded around x_0 converges at x only if the distance between x_0 and x is less than the distance between x_0 and the origin.

Let us return to (6.38). We may obtain the solutions of (6.38) expanded around a regular singular point x_0 in the same way we obtained the solutions of the Bessel equation expanded around $x = 0$. More precisely, we seek these solutions in the form of the Frobenius series

$$y(x) = \sum_n a_n (x - x_0)^{n+s}, \tag{6.39}$$

with

$$a_{-1} = a_{-2} = \cdots = 0.$$

The series is convergent at least up to the next singular point of (6.38). We plug (6.39) into (6.38) and determine s as well as a_n with $n = 1, 2 \cdots$. Note that a_0 is the only arbitrary constant in the Frobenius series, as all other a_n are, except in some exceptional cases, determined from the recurrence formula. This is different from the Taylor series solution expanded around an ordinary point of (6.38), for which there are two arbitrary constants: a_0 and a_1. Note also that for the Frobenius series, there is one more constant to determine: the index s. The index s is determined from the indicial equation, which has two roots. Since an independent solution of (6.38) is obtained from each of the roots, this procedure often yields two independent solutions of the equation. Some of the exceptional cases in which this is not true will be discussed below.

To understand why the number of arbitrary constants in a Frobenius series solution differs from that in a Taylor series solution, let us consider the following example. Let

$$y'' + \frac{A}{x}y' - y = 0.$$

If $A \neq 0$, the point $x = 0$ is a regular singular point of the differential equation, and if $A = 0$, it is an ordinary point of the differential equation.

Problem for the Reader

Solve the differential equation above with the Frobenius method with $x_0 = 0$. Find the recurrence formula satisfied by the coefficients of the Frobenius series. Obtain the indicial equation and determine the value of a_1.

Solution

Let

$$y = \sum a_n x^{n+s}.$$

The recurrence formula is

$$(n+s)(n+s-1+A)a_n = a_{n-2}.$$

The indicial equation is obtained by setting $n = 0$ in the recurrence formula. We get

$$s(s-1+A) = 0,$$

which has the roots

$$s = 0, 1 - A.$$

Consider first the solution with the index $s = 0$. The recurrence formula for the coefficients a_n with $s = 0$ is

$$n(n - 1 + A)a_n = a_{n-2}.$$

Setting $n = 1$ in the recurrence formula above, we get

$$Aa_1 = 0.$$

Therefore,

$$a_1 = 0 \quad \text{for} \quad A \neq 0$$

$$a_1 \quad \text{arbitrary for } A = 0.$$

Thus the coefficient a_1 is determined to be zero if $A \neq 0$. Remember that when $A \neq 0$, the point $x_0 = 0$ is a regular singular point of the differential equation. In this case, the Frobenius series solution corresponding to the index $s = 0$ has only one arbitrary constant: a_0.

But as we have also shown above, a_1 is arbitrary if A is equal to zero. When $A = 0$, the point $x_0 = 0$ is an ordinary point of the differential equation. In this case, the Frobenius series has two arbitrary constants: a_0 and a_1. Note that since $s = 0$, the Frobenius series is a Maclaurin series.

In brief, if the point of expansion x_0 is an ordinary point of the differential equation, the recurrence formula for $n = 1$ is satisfied by any value of a_1. Therefore, the series solution has two arbitrary constants and is the general solution of the differential equation.

Sometimes the value of a_1 is arbitrary even if the point of expansion x_0 is a regular singular point of the differential equation (6.38). An example is offered by the second Frobenius series solution of the present problem. This solution is the one of the index $s = (1 - A)$, the other root of the indicial equation. The recurrence formula satisfied by a_n with $s = (1 - A)$ is

$$(n + 1 - A)na_n = -a_{n-2}.$$

Setting $n = 1$, we get

$$(2 - A)a_1 = 0.$$

If A happens to be 2, a_1 is arbitrary. Therefore, if A takes the special value 2, the second Frobenius series solution has two arbitrary constants: a_0 and a_1. Note that the index $(1 - A)$ with $A = 2$ is equal to -1. Having two arbitrary constants, the Frobenius series with $s = -1$ is the general solution of the differential equation. The first Frobenius series solution with $s = 0$ is obtained from this general solution with the choice of $a_0 = 0$ and $a_1 = 1$.

There are other special cases in which the second arbitrary constant of the Frobenius series is a_2, while the value of a_1 is fixed by the recurrence formula of $n = 1$ and is not arbitrary. Indeed, it is possible for the second arbitrary constant to be a_k with k any positive integer.

We mention that x_0 being a regular singular point of the ODE does not imply that both of the solutions of the ODE are necessarily singular at x_0. An example is the Bessel equation of order zero. This equation has a regular singular point at the origin, yet $J_0(x)$ is analytic at the origin. But as long as one of the independent solutions blows up at x_0, there is no solution satisfying the general initial conditions $y(x_0) = a$ and $y'(x_0) = b$.

Finally, if either $(x - x_0)c(x)$ or $(x - x_0)^2 d(x)$ or both are not analytic at x_0, then x_0 is called an irregular singular point of (6.38).

C. Solutions Near an Irregular Singular Point

Solutions expanded around an irregular singular point are distinctive in one aspect: They are usually in the form of an exponential function times a Frobenius series. Due to the factor of the exponential function, a solution near an irregular singular point behaves differently from that near a regular singular point. It may blow up exponentially, vanish exponentially or oscillate wildly.

Let us start with the discussion of irregular singular points with the simpler case of the first-order linear and homogeneous equation

$$y' + p(x)y = 0. \tag{6.40}$$

We begin with the example

$$p(x) = \frac{b_1}{x},$$

where b_1 is constant. Then (6.40) is

$$y' + \frac{b_1}{x}y = 0. \tag{6.41}$$

We observe that the coefficient of y in the differential equation (6.41) has a simple pole at $x = 0$. We also observe that eq. (6.41) is equidimensional. Thus the solution of (6.40) is of the form x^s. Indeed, s is equal to $-b_1$.

This result can be slightly extended. As we know, the solution of (6.40) is

$$y(x) = ce^{-P(x)}, \tag{6.42}$$

where

$$P(x) = \int p(x)dx$$

and c is a constant. Now if $p(x)$ has a simple pole at $x = 0$, then it has the Laurent series expansion

$$p(x) = \frac{b_1}{x} + a_0 + a_1 x + \cdots .$$

Thus

$$P(x) = b_1 \ln x + T(x),$$

where $T(x)$ is a Maclaurin series. Since $e^{-T(x)}$ is a Maclaurin series if $T(x)$ is a Maclaurin series, we find by (6.42) that the solution of (6.40) is of the form

$$x^{-b_1} M(x)$$

where

$$M(x) = e^{-T(x)}.$$

Note that this solution is in the form of a Frobenius series (6.39) with $x_0 = 0$ and $s = -b_1$. For this reason, eq. (6.40) is said to have a regular singular point at $x = 0$ if $p(x)$ has a simple pole at $x = 0$.

Let us next consider the case in which $p(x)$ has a double pole rather than a simple pole at $x = 0$. We shall show that the solution of (6.40) has a new feature.

If $p(x)$ has a double pole at the origin, then it has the Laurent series expansion

$$p(x) = \frac{b_2}{x^2} + \frac{b_1}{x} + a_0 + a_1 x + \cdots .$$

Thus

$$P(x) = -\frac{b_2}{x} + b_1 \ln x + \cdots . \tag{6.43}$$

By (6.42), the solution of (6.40) is in the form of

$$\exp\left(\frac{b_2}{x}\right) x^{-b_1} M(x). \tag{6.44}$$

Again, $M(x)$ is a Maclaurin series, and hence $x^{-b_1} M(x)$ is a Frobenius series. But note the additional factor of the exponential function in (6.44). We emphasize that this additional factor appears in the solution of (6.40) when the singularity of $p(x)$ at $z = 0$ is a double pole rather than a simple pole. In such a case, eq. (6.40) is said to have an irregular singular point of rank 1 at the point $x = 0$.

Near the singular point $z = 0$, the behavior of the function

$$\exp\left(\frac{b_2}{z}\right)$$

is drastically different from that of a power function z^s. Let us, for simplicity, set $b_2 = 1$ and consider the function

$$\exp\left(\frac{1}{z}\right).$$

We first let z be real and be equal to x. If x is small and positive, $1/x$ is large and positive, and

$$\exp\left(\frac{1}{x}\right)$$

is very large, approaching infinity as x approaches zero. But when x is small and negative, $1/x$ is large in magnitude but is negative, and

$$\exp\left(\frac{1}{x}\right)$$

is very small, approaching zero as x approaches zero. Next, let z be purely imaginary and equal to iy. In this case, the magnitude of

$$\exp\left(\frac{1}{iy}\right)$$

is always equal to unity, but the phase of

$$\exp\left(\frac{1}{iy}\right)$$

goes to infinity as y approaches zero. This behavior of

$$\exp\left(\frac{1}{z}\right)$$

is in contrast to that of a function having a pole at $z = 0$. For example, the function $1/z$ always blows up as z approaches zero, no matter whether z is positive, negative, or imaginary. Note also that while the series expansion

$$\exp\left(\frac{1}{z}\right) = 1 + \frac{1}{z} + \frac{1}{2!}\left(\frac{1}{z}\right)^2 + \cdots$$

is convergent no matter how small z is, this expansion gives little hint of the behavior of the function

$$\exp\left(\frac{1}{z}\right)$$

near $z = 0$ as described above.

Equation (6.40) is said to have an irregular singular point of rank 1 at $x = 0$ if $p(x)$ has a double pole at $x = 0$. More generally, if $p(x)$ has a pole of order $(k+1)$ at $x = 0$ with the Laurent series expansion

$$p(x) = \frac{b_{k+1}}{x^{k+1}} + \cdots + \frac{b_1}{x} + a_0 + a_1 x + \cdots,$$

eq. (6.40) is said to have an irregular singular point of rank k at $x = 0$. Note that in this case, $x^{k+1}p(x)$ is analytic at $x = 0$ and can be expressed as a convergent Maclaurin series.

Problem for the Reader

Find the solution of (6.40) if this equation has an irregular singular point of rank k at $x = 0$.

Solution

By integrating $p(x)$, we have

$$P(x) = -\frac{b_{k+1}}{kx^k} + \cdots + b_1 \ln x + T(x),$$

where $T(x)$ is a Maclaurin series. Thus the solution of (6.40) is of the form

$$\exp\left(\frac{b_{k+1}}{kx^k}+\cdots\right)x^{-b_1}M(x),$$

where $M(x)$ is a Maclaurin series. Note that the exponent above is a kth-order polynomial of x^{-1}, where k is the rank of the singular point at the origin.

If $p(x)$ has a pole of order $(k+1)$ at x_0 and x_0 is not equal to zero, and if we are interested in the behavior of the solution near x_0, we can make the change of variable

$$X = x - x_0.$$

The solution near x_0 is obtained from the one given above with x replaced by X. If the singular point is at infinity, we define $t = x^{-1}$, and the solution near the infinity of x is obtained from the one above with x replaced by t.

We are now ready to discuss the second-order linear homogeneous differential equations (6.38) with irregular singular points. Few such equations have closed-form solutions, and our purpose here is to find the solutions in the form of expansion series that are useful when x is near an irregular singular point of the differential equation.

Let x_0 be an irregular singular point of (6.38). If

$$(x-x_0)^{k+1}c(x) = c_0 + c_1(x-x_0) + \cdots$$

and

$$(x-x_0)^{2k+2}d(x) = d_0 + d_1(x-x_0) + \cdots$$

are both convergent Taylor series, and if at least one of c_0 and d_0 are not zero, then x_0 is called an irregular singular point of rank k. If the order of the pole of $c(x)$ and that of $d(x)$ at x_0 are (k_1+1) and $(2k_2+2)$, respectively, with k_1 not equal to k_2, then the rank k is equal to the greater of k_1 and k_2. Note that if $k=0$, then x_0 is a regular singular point of (6.38).

We shall show that if x_0 is an irregular singular point of rank k, where k is a positive integer, then the solutions of (6.38) are of the form

$$y(x) = \exp\left[F(x)\right]Y(x), \tag{6.45}$$

where

$$F(x) = \frac{A_k}{(x-x_0)^k} + \frac{A_{k-1}}{(x-x_0)^{k-1}} + \cdots \frac{A_1}{x-x_0}$$

and

$$Y(x) = \sum a_n (x - x_0)^{n+s}, \quad a_0 \neq 0, \quad a_{-1} = a_{-2} = \cdots = 0 \tag{6.46}$$

is a Frobenius series. Thus the general solution of (6.38) near its irregular singular point x_0 is a linear combination of two expansion solutions, each of which is in the form of an exponential function e^F, where F is a polynomial of $(x - x_0)^{-1}$, times a Frobenius series.

To prove this, we substitute (6.45) into (6.38), obtaining the differential equation satisfied by Y:

$$(D + F')^2 Y + c(D + F')Y + dY = 0,$$

or

$$Y'' + (c + 2F')Y' + (d + cF' + F'^2 + F'')Y = 0. \tag{6.47}$$

We shall choose the constants A_n, $n = 1, \cdots, k$, to eliminate the most singular terms in the coefficient of Y in (6.47). Since the number of A_n available is k, we will be able to eliminate from the coefficient of Y in (6.47) all poles of order greater than or equal to $(k + 3)$. This determines the values of the coefficients A_n. In particular, we find that A_k must satisfy a quadratic equation the roots of which are

$$A_k = \frac{c_0 \pm \sqrt{c_0^2 - 4d_0}}{2k}. \tag{6.48}$$

These roots are distinct unless

$$c_0^2 - 4d_0 = 0. \tag{6.49}$$

We choose A_k to be one of the roots; then (6.47) is of the form

$$Y'' + \frac{e_0 + e_1(x - x_0) + \cdots}{(x - x_0)^{k+1}} Y' + \frac{f_0 + f_1(x - x_0) + \cdots}{(x - x_0)^{k+2}} Y = 0, \tag{6.50}$$

where e_n and f_n are constants. In particular,

$$e_0 = \mp\sqrt{c_0^2 - 4d_0}.$$

Problem for the Reader

Show that (6.50) admits a Frobenius series as one of its solutions. What is the index s for this Frobenius series?

Solution

We shall show that, while x_0 is still an irregular singular point of (6.50), the exponential function for one of the solutions of (6.50) is unity. Thus this solution can be expressed as a Frobenius series expanded around x_0. To show this, we note that there are two terms in (6.50) that are of the lowest dimension. They are

$$\frac{e_0 Y'}{(x-x_0)^{k+1}} \text{ and } \frac{f_0 Y}{(x-x_0)^{k+2}}. \tag{6.51}$$

By equating the most singular terms in them, we shall find one index and the corresponding Frobenius solution. Indeed, let us substitute

$$Y(x) = \sum_n a_n (x-x_0)^{n+s}$$

into (6.50). The indicial equation comes from the two terms in (6.51) with $Y(x)$ replaced by the first term in the series above. We get

$$a_0(e_0 s + f_0) = 0. \tag{6.52}$$

By assumption

$$a_0 \neq 0;$$

thus (6.52) gives the indicial equation

$$e_0 s + f_0 = 0.$$

This equation has the single root

$$s = -f_0/e_0. \tag{6.53}$$

The index s is finite provided that $e_0 \neq 0$. Under this condition we obtain two independent solutions of (6.38), each associated with a value of A_k given by (6.48).

If $e_0 = 0$ and $f_0 \neq 0$, eq. (6.50) has only one term with the lowest dimension. In this case, (6.52) cannot be satisfied and hence no Frobenius solution for (6.50) expanded around x_0 exists. To find a series solution for (6.50), we note that with $e_0 = 0$ and $f_0 \neq 0$, the order of the pole at x_0 for the coefficient of Y' in (6.50) is k. Also, the order of the pole at x_0 for the coefficient of Y in (6.50) is $(k+2)$. Thus the rank of the differential equation (6.50) at the point x_0 is equal to the greater of $k/2$ and $(k-1)$. If k is an even integer, both $k/2$ and $(k-1)$ are integers, and we may apply the method we just described to solve this problem. If k is an odd integer, the change of variable

$$x - x_0 = \rho^2$$

transforms the differential equation into one with an irregular singular point of integral rank. (See homework problem 5.)

Because of the exponential factor e^F, the function in (6.45) is known to have an essential singularity at x_0. If x_0 is an irregular singular point of (6.38), at least one of the solutions of (6.38) has an essential singularity at x_0. (See homework problem 6.)

We mention that if one is interested in the behavior of the solution of (6.38) at very large values of x, rather than that at values of x close to a finite point x_0, one makes the change of variable

$$x = \frac{1}{t}.$$

When x is very large, t is very small. Indeed, $x = \infty$ corresponds to $t = 0$. By making an expansion of the solution around $t = 0$ in the way we just outlined, we obtain an expansion of the solution around $x = \infty$. Keeping the first few terms of this expansion is often a good approximation of the solution at very large values of x.

The change of the independent variable from x to t is done with

$$\frac{d}{dx} = \frac{d}{d\left(\frac{1}{t}\right)} = -t^2\frac{d}{dt},$$

and

$$\frac{d^2}{dx^2} = t^2\frac{d}{dt}\left(t^2\frac{d}{dt}\right) = t^4\frac{d^2}{dt^2} + 2t^3\frac{d}{dt}.$$

Thus (6.38) becomes

$$\frac{d^2y}{dt^2} + \frac{2t - c\left(\frac{1}{t}\right)}{t^2}\frac{dy}{dt} + \frac{d\left(\frac{1}{t}\right)}{t^4}y = 0. \tag{6.54}$$

We call $x = \infty$ an ordinary point, a regular singular point, or an irregular singular point of rank k of eq. (6.38), respectively, depending on whether $t = 0$ is an ordinary point, a regular singular point, or an irregular singular point of rank k of (6.54). In particular, we see from (6.54) that infinity is an irregular singular point of rank k if either $d(x)$ blows up like x^{2k-2} or $c(x)$ blows up like x^{k-1} as x goes to infinity.

If infinity is an irregular singular point of rank k, then the solution of (6.38) for very large x is of the form

$$y(x) = \exp\left[A_k x^k + A_{k-1}x^{k-1} + \cdots + A_1 x\right] Y(x), \tag{6.55}$$

where

$$Y(x) = \sum a_n x^{-n-s}, \quad a_0 \neq 0, \quad a_{-1} = a_{-2} = \cdots = 0, \tag{6.56}$$

which is easily deduced from (6.45) and (6.46) by identifying $(x - x_0)$ with t or $1/x$. Note that the terms in the series (6.56) are in decreasing powers of x, and hence the series is a useful approximation for x large.

Problem for the Reader

Find the series solution of

$$y'' + x^{-4} y = 0, \tag{6.57}$$

which is useful for very small values of x.

Solution

The point $x = 0$ is an irregular singular point with rank 1. Thus the series solution near the origin is of the form

$$y = \exp\left(\frac{A_1}{x}\right) Y(x), \tag{6.58}$$

with

$$Y(x) = \sum a_n x^{n+s}, \quad a_0 \neq 0, \quad a_{-1} = a_{-2} = \cdots = 0. \tag{6.59}$$

We substitute (6.58) into (6.57). Since

$$\begin{aligned}\frac{d^2}{dx^2} \exp\left(\frac{A_1}{x}\right) Y(x) &= \exp\left(\frac{A_1}{x}\right)\left(\frac{d}{dx} - \frac{A_1}{x^2}\right)\left(\frac{d}{dx} - \frac{A_1}{x^2}\right) Y(x) \\ &= \exp\left(\frac{A_1}{x}\right)\left(\frac{d^2}{dx^2} - \frac{2A_1}{x^2}\frac{d}{dx} + \frac{2A_1}{x^3} + \frac{A_1^2}{x^4}\right) Y(x),\end{aligned}$$

eq. (6.57) becomes

$$\frac{d^2Y}{dx^2} - \frac{2A_1}{x^2}\frac{dY}{dx} + \left(\frac{2A_1}{x^3} + \frac{A_1^2+1}{x^4}\right)Y = 0. \tag{6.60}$$

The value of A_1 is determined by requiring the most divergent term in the coefficient of Y in (6.60) to vanish. There is, in eq. (6.60), a term in the coefficient of Y that is a fourth-order pole at the point of expansion $x_0 = 0$. Thus we require

$$A_1^2 + 1 = 0,$$

or

$$A_1 = \pm i.$$

Let us choose the root

$$A_1 = i.$$

Then (6.60) becomes

$$\left(\frac{d^2}{dx^2} - \frac{2i}{x^2}\frac{d}{dx} + \frac{2i}{x^3}\right)Y = 0.$$

One of the solutions can be obtained by inspection. We note that the operator

$$\left(\frac{d^2}{dx^2} - \frac{2i}{x^2}\frac{d}{dx} + \frac{2i}{x^3}\right)$$

has a term with dimension -2, and two terms of dimension -3. The sum of the latter two terms,

$$\left(-\frac{2i}{x^2}\frac{d}{dx} + \frac{2i}{x^3}\right),$$

vanishes as it operates on x. The former term,

$$\frac{d^2}{dx^2},$$

also vanishes as it operates on x. Thus one of the solutions of the equation for Y is simply

$$Y(x) = x.$$

This result can also be obtained if we go through the grind to obtain the recurrence formula

$$(n+s-1)(n+s-2)a_{n-1} = 2i(n+s-1)a_n.$$

From the recurrence formula with $n = 0$, we get

$$s = 1.$$

Thus the recurrence formula is

$$n(n-1)a_{n-1} = 2ina_n.$$

Setting $n = 1$, we obtain from the recurrence formula that

$$a_1 = 0,$$

and hence $a_2 = a_3 = \cdots = 0$. Therefore, the series solution for $Y(x)$ terminates after one term. Thus we have obtained the closed form expression of one of the independent solutions of (6.57) as

$$y_1(x) = x \exp\left(\frac{i}{x}\right).$$

Taking the complex conjugate $y_1(x)$, we get the second independent solution of (6.57) as

$$y_2(x) = x \exp\left(-\frac{i}{x}\right).$$

Problem for the Reader

Find the series solution of the Bessel equation

$$y'' + x^{-1}y' + (1 - p^2/x^2)y = 0, \tag{6.61}$$

which is useful for very large values of x.

Solution

Let us make a change of the independent variable from x to t. Acoording (6.54), we get

$$\frac{d^2y}{dt^2} + \frac{1}{t}\frac{dy}{dt} + \frac{1 - p^2t^2}{t^4}y = 0.$$

We find that the rank of the irregular singular point at $x_0 = \infty$ is unity.

Alternately, we may read off the rank at infinity from the Bessel equation (6.61) directly. As x goes to infinity, the coefficient of y goes to a constant, or x^0. Therefore, according to this coefficient,

$$2k - 2 = 0,$$

or

$$k = 1.$$

The coefficient of y' is x^{-1}. Thus, according to this coefficient,

$$k - 1 = -1,$$

or

$$k = 0.$$

The rank of the Bessel equation at $x_0 = \infty$ is the greater of the two, and is unity.

Thus we put

$$y(x) = \exp(A_1 x) Y(x),$$

where

$$Y(x) = \sum a_n x^{-n-s}, a_0 = 1, a_{-1} = a_{-2} = \cdots = 0.$$

The Bessel equation becomes

$$(D^2 + 2A_1 D + x^{-1}D + A_1^2 + 1 + A_1/x - p^2/x^2)Y = 0.$$

We require the sum of the most divergent terms in the coefficient of Y in the equation above to vanish. As $x \to \infty$, the coefficient of Y in the equation above is $(A_1^2 + 1)$. Thus we get

$$A_1 = \pm i.$$

This determines the two roots of A_1.

Take the root

$$A_1 = i.$$

Then the equation for Y becomes

$$Y'' + (2i + x^{-1})Y' + (i/x - p^2/x^2)Y = 0.$$

We have

$$\begin{aligned}&(D^2 + x^{-1}D - p^2/x^2)Y(x)\\&= \sum \left[(n+s)(n+s+1) - (n+s) - p^2\right] a_n x^{-n-s-2}\end{aligned}$$

and

$$(2iD + i/x)Y(x) = \sum \left[-2i(n+s) + i\right] a_n x^{-n-s-1}.$$

Thus we have

$$2i(n+s-1/2)a_n = \left[(n+s-1)^2 - p^2\right] a_{n-1}.$$

Setting $n = 0$, we get

$$s = 1/2.$$

For $n \neq 0$, the recurrence formula is

$$a_n = \frac{(n+p-1/2)(n-p-1/2)}{2in} a_{n-1}. \tag{6.62}$$

Thus we have

$$a_n = \frac{1}{(2i)^n} \frac{\Gamma(n+p+1/2)\Gamma(n-p+1/2)}{n!\Gamma(p+1/2)\Gamma(-p+1/2)} a_0.$$

Therefore, one of the series solutions for the Bessel equation for very large values of x is

$$y_1(x) = x^{-1/2}\exp(ix) \sum \frac{1}{(2ix)^n} \frac{\Gamma(n+p+1/2)\Gamma(n-p+1/2)}{\Gamma(p+1/2)\Gamma(-p+1/2)n!}. \tag{6.63}$$

The other series solution for the Bessel equation for very large values of x is

$$y_2(x) = x^{-1/2}\exp(-ix) \sum_{n=0}^{\infty} \frac{1}{(-2ix)^n} \frac{\Gamma(n+p+1/2)\Gamma(n-p+1/2)}{\Gamma(p+1/2)\Gamma(-p+1/2)n!}. \tag{6.64}$$

Equations (6.63) and (6.64) give the solutions of the Bessel equation useful when $|x|$ is large. This is to be compared with the series solution (6.32) useful when $|x|$ is small. Since $J_p(x)$ is a solution of the Bessel equation, it must be a linear superposition of y_1 and y_2. Thus we have

$$J_p(x) = cy_1(x) + c^\star y_2(x).$$

The coefficient c cannot be determined with the present analysis, but we will be able to show in Chapter 8 that

$$c = \frac{e^{-i\pi(1/4+p/2)}}{\sqrt{2\pi}}.$$

Thus

$$J_p(x) \approx \sqrt{\frac{2}{\pi x}} \cos(x - \pi/4 - p\pi/2), \ x \gg 1.$$

The series (6.63) and (6.64) differ from the series (6.32) in one important respect: The former are divergent for all values of x! This can be seen from (6.62), which shows that when n is very large, the ratio a_n/a_{n-1} goes to infinity.

That these series are divergent does not imply that they are not useful for the purpose of approximation. Indeed, they are called asymptotic series and give very good approximations to the solutions when x is large if the number of terms is chosen appropriately.

The classic example of an asymptotic series is

$$I(x) = \int_0^\infty \frac{e^{-t}}{1+tx} dt, \ x > 0. \tag{6.65}$$

We shall find an approximation of this integral when x is small. We have

$$I(0) = \int_0^\infty e^{-t} dt = 1.$$

Thus we expect

$$I(x) \approx 1$$

when x is small.

To seek an approximation better than this, we approximate the factor $(1+tx)^{-1}$ in the integrand of $I(t)$ by

$$\frac{1}{1+tx} \approx 1 - tx + t^2x^2 + \cdots + (-tx)^n.$$

Using this approximation for the integrand of $I(x)$, we get

$$\begin{aligned} I(x) &\approx \int_0^\infty e^{-t}\left[1 - tx + t^2x^2 + \cdots + (-tx)^n\right] dt \\ &= 1 - x + 2!x^2 + \cdots + n!(-x)^n. \end{aligned} \tag{6.66}$$

The first term of this series is unity, an approximation we have already obtained. However, we note that the series in (6.66) diverges for all values of x as we let $n \to \infty$.

Nevertheless, (6.66) is a useful formula for approximation, as we shall prove. We have

$$\frac{1}{1+tx} = 1 - tx + t^2x^2 + \cdots + (-tx)^n + \frac{(-tx)^{n+1}}{1+tx},$$

which is exact. Thus we have

$$I(x) = \int_0^\infty e^{-t}\left[1 - tx + t^2x^2 + \cdots + (-tx)^n\right] dt + R_n, \tag{6.67}$$

where

$$R_n = \int_0^\infty e^{-t}\frac{(-tx)^{n+1}}{1+tx}dt.$$

Equation (6.67) is exact. This is to be compared with (6.66), which is obtained by dropping R_n.

How big is the term dropped? We have

$$|R_n| \le \int_0^\infty e^{-t}(tx)^{n+1}dt = x^{n+1}(n+1)!.$$

Let $x = 0.1$. Then the sum of the first five terms of (6.66), which is of the order of unity, differs from the exact value by R_4, which is estimated to be about one tenth of one percent. We may improve this approximation somewhat by taking more terms. But as the series eventually diverges, it is obviously not a great idea to be too enthused and take a large number of terms in (6.66). Where do we stop? Very roughly, we should stop when the upper bound of R_n no longer goes down as we increase n by unity. For $x = 0.1$, this happens when $n = 10$.

If $x = 0.01$, and if we take 100 terms in the series (6.57), the error is as small as 10^{-42}.

Appendix: The Gamma Function

We first prove that the integral in (6.10) converges only if $\mathrm{Re}\, z > 0$. Since the integrand is finite for all values of t in the range $0 < t < \infty$, we need to examine the possible divergence of this integral only at the endpoints $t = 0$ and $t = \infty$.

The integral is always convergent at the endpoint $t = \infty$, no matter what z is. This is because while t^{z-1} may blow up as $t \to \infty$, which happens when the value of $\mathrm{Re}(z - 1)$ is greater than zero, the exponentially vanishing factor e^{-t} always beats t^{z-1}, and the integrand $e^{-t}t^{z-1}$ vanishes sufficiently rapidly to make the integral converge at the upper endpoint $t = \infty$.

Near the endpoint $t = 0$, the integrand is approximately equal to t^{z-1}, as e^{-t} is approximately equal to unity. Let us first examine the special case of $z = 0$. If $z = 0$, the integrand is approximately equal to t^{-1} near $t = 0$. Since $\int dt/t$ diverges logarithmically at the point $t = 0$, the integral with z set to zero diverges at the endpoint $t = 0$. Next we shall show that the integral is convergent as long as $\mathrm{Re}\, z > 0$. We have, with $\mathrm{Re}\, z > 0$,

$$\int_0^{\epsilon} t^{z-1} dt = \frac{t^z}{z}\Big|_{t=0}^{t=\epsilon} = \frac{\epsilon^z}{z},$$

as

$$\lim_{t \to 0} t^z = 0.$$

In the above, ϵ is a positive real number much smaller than unity. Since the integral above is finite for $\mathrm{Re}\, z > 0$, the integral of (6.10) is finite for $\mathrm{Re}\, z > 0$. Therefore, the function $\Gamma(z)$ defined by (6.10) is meaningful for $\mathrm{Re}\, z > 0$. Indeed, since the integrand of (6.10) is an analytic function of z, $\Gamma(z)$ is an analytic function of z provided that the integral of (6.10) is convergent. Thus $\Gamma(z)$ is an analytic function of z in the right half-plane $\mathrm{Re}\, z > 0$.

With the Gamma function $\Gamma(z)$ defined by (6.10) for $\mathrm{Re}\, z > 0$, it is possible to analytically continue it to all complex values z. To see this, we will need to first derive an identity for the Gamma function. From (6.10), we have

$$\Gamma(z+1) = \int_0^{\infty} e^{-t} t^z dt = -\int_0^{\infty} t^z d(e^{-t}) = -t^z e^{-t}|_0^{\infty} + z\int_0^{\infty} e^{-t} t^{z-1} dt.$$

The integrated term in the above vanishes if $\mathrm{Re}\, z > 0$. Thus we have

$$\Gamma(z+1) = z\Gamma(z), \mathrm{Re}\, z > 0. \tag{A.1}$$

We may write (6A.1) as

$$\Gamma(z) = \Gamma(z+1)/z, \ \mathrm{Re}\, z > 0. \tag{A.2}$$

The identity (6A.2) has been proven to hold only in the region $\mathrm{Re}\, z > 0$. Indeed, the left side of (6A.2), which is $\Gamma(z)$, has yet to be defined when the real part of z is less than zero. We note, however, that the Gamma function in the right side of (6A.2), which is $\Gamma(z+1)$, is defined as long as $\mathrm{Re}(z+1) > 0$, or $\mathrm{Re}\, z > -1$, a region larger than the region $\mathrm{Re}\, z > 0$. We shall define $\Gamma(z)$ to be the right side of (6A.2) in the region where the latter is defined. Therefore, with the values of $\Gamma(z+1)$ in the strip $0 \geq \mathrm{Re}\, z > -1$ given by the integral of (6.11), we may use (6A.2) to determine the values of $\Gamma(z)$ in the strip $0 \geq \mathrm{Re}\, z > -1$. Generally, (6A.2) enables us to determine the value of $\Gamma(z)$ from the value of $\Gamma(z+1)$. Therefore, we may further determine the values of $\Gamma(z)$ in the strip $-1 \geq \mathrm{Re}\, z > -2$ from the values of $\Gamma(z)$ in the strip $0 \geq \mathrm{Re}\, z > -1$ we just succeeded in defining. By repeating this process, we obtain the values of $\Gamma(z)$ for any z in the left half-plane.

The Gamma function $\Gamma(z)$ so obtained is analytic everywhere except at the points $z = 0, -1, -2 \cdots$, where it has a simple pole. For example, we have, near $z = 0$,

$$\Gamma(z) = \frac{\Gamma(z+1)}{z} \approx \frac{1}{z}, \ z \approx 0.$$

Therefore, $\Gamma(z)$ has a simple pole at $z = 0$ with the residue unity. Similarly, we may prove from (6A.2) that $\Gamma(z)$ has a simple pole at $z = -1, -2 \cdots$. The Gamma function in the left half-plane so obtained is the analytic continuation of the Gamma function defined in the right half-plane by the integral of (6.10).

Let us write (6A.1) as

$$z = \Gamma(z+1)/\Gamma(z). \tag{A.3}$$

Replacing z by $z-1$ in (6A.3), we get

$$z - 1 = \Gamma(z)/\Gamma(z-1). \tag{A.4}$$

From (6A.3) and (6A.4), we get

$$z(z-1) = \Gamma(z+1)/\Gamma(z-1).$$

Repeating this procedure n times and reducing the argument of the Gamma function by unity each time, we get

$$z(z-1)(z-2)\cdots(z-n+1) = \Gamma(z+1)/\Gamma(z-n+1). \tag{A.5}$$

Problem for the Reader

Show that $\Gamma(z)$ at $z = 0, 1, \cdots$ is infinite.

Solution

Let us replace z in (6A.5) by $(z + n - 1)$. We get

$$\Gamma(z) = \frac{\Gamma(z+n)}{(z+n-1)(z+n-2)\cdots(z)}. \tag{A.6}$$

Setting $z = 0$ in (6A.6) with $n = 1$, we find that $\Gamma(z)$ has a simple pole at $z = 0$ with the residue unity. Setting $z = -1$ in (6A.6) with $n = 2$, we find that $\Gamma(z)$ has a simple pole at $z = -1$ with the residue -1. Similarly, we may find that $\Gamma(z)$ has a simple pole at $z = -n$, where n is a nonpositive integer, with the residue $(-1)^n/n!$.

Homework Problems for This Chapter

Solutions to the Homework Problems can be found at www.lubanpress.com.

1. Find and classify the singular points (∞ included) for the following ODE:

a. $xy'' + (c - x)y' - ay = 0.$ (Confluent hypergeometric equation)

b. $x(1-x)y''+[c - (a + b + 1)x]\,y'-aby = 0.$ (Hypergeometric equation)

c. $y'' - (x^4 - 3x^{-2}/16)y = 0.$

d. $y'' + (x^2 + 3x^{-2}/16)y = 0.$

e. $y''+\left(\nu + \frac{1}{2} - \frac{1}{4}x^2\right) y = 0$, ν a constant. (Parabolic cylinder equation)

The recurrence formula for the last equation involves three different a_n. Find a change of the dependent variable so that the recurrence formula for the transformed equation involves only two different a_n.

2. Find the Maclaurin series solutions or Frobenius series solutions expanded around $x_0 = 0$ for the equations in problem 1. In what region is each of the series convergent?

3. Calculate the quantity in (6.37) as $p \to n$. Show that this solution of the Bessel equation blows up at the origin.

4. Let $y_1(x) = (1+x)$ and $y_2(x) = x^2$ be the two independent solutions of eq. (6.38). Find $c(x)$ and $d(x)$ and show that both of them have a simple pole at $x = 0$. What do you learn from this example?

5. Consider (6.50) with $e_0 = 0$ and with k an odd integer. Show that with the change of variable $x - x_0 = \rho^2$, eq. (6.50) becomes a differential equation of integral rank at x_0. What is this rank?

6. Show that if x_0 is an irregular singular point of (6.38), at least one of the solutions of (6.38) has an essential singularity at x_0.

7. Consider the integral $I(x)$ given in (6.65). Let $x = z$, a complex variable.

   a. Give the asymptotic form for $I(z)$ when $|z|$ is small, with $-\pi/2 \leq \arg z \leq \pi/2$.

   b. Give the asymptotic form for $I(z)$ when $|z|$ is small, with $-\pi < \arg z < \pi$.
   Hint: For (b), rotate the contour of integration.

8. Find the entire asymptotic series for the solutions of the equations in problem 1.

Chapter 7

The WKB Approximation

The WKB method is a powerful tool to obtain solutions for many physical problems. It is generally applicable to problems of wave propagation in which the wave number of the wave is very high or, equivalently, the wavelength of the wave is very small. The WKB solutions are approximate solutions, but sometimes they are surprisingly accurate. In this chapter we'll discuss this method, which is applicable to linear equations only.

A. WKB in the Zeroth and the First Order

Consider the first-order linear differential equation

$$y' = ipy.$$

If p is a constant, the solution is simply

$$y = e^{ipx},$$

which describes a wave of propagation number p.

But if p is a function of x, the solution of this equation is

$$y(x) = \exp\left(i\int p(x)dx\right).$$

If we regard $y(x)$ as a wave, then the exponent of $y(x)$ is the phase of a wave with the wave number $p(x)$. We see that at the point x, the phase is equal to the waveshift accumulated over the path of propagation, not to x times the local wave number $p(x)$—a perfectly logical result.

We mention that a change of the lower limit of the integral above merely changes the integral by an additive constant, and hence y by a multiplicative constant. For a linear homogeneous equation, the freedom of multiplying a solution by a constant is always understood. Thus we shall leave the lower limit of integration unspecified.

Next we consider the second-order differential equation

$$y'' + p^2 y = 0. \tag{7.1}$$

If p is a constant, the two independent solutions of (7.1) are $\exp(\pm ipx)$, waves with wave number p travelling in opposite directions of the x axis.

If p is a function of x, it appears reasonable that the solutions are two waves with the phase $\pm \int p(x)dx$. Thus we may surmise that the independent solutions of (7.1) are

$$e^{\pm i \int p(x)dx}, \tag{7.2}$$

which are called the zeroth-order WKB solutions.

Let us see if these solutions satisfy (7.1). It is straightforward to show that

$$\left(\frac{d^2}{dx^2} + p^2\right) e^{\pm i \int p(x)dx} = \pm ip' e^{\pm i \int p(x)dx}. \tag{7.3}$$

Therefore, the WKB solutions (7.2) do not satisfy (7.1) unless $p' = 0$, or p is independent of x.

While we get a negative answer, (7.3) suggests that $\exp(\pm i \int p(x)dx)$ are good approximate solutions of (7.1) provided that $\pm ip'$ is negligible, or, more precisely, if

$$|p'| \ll p^2,$$

the right side of this inequality being a term inside the parentheses of (7.3). The inequality above can be written as

$$\left| \frac{d}{dx} \frac{1}{p(x)} \right| \ll 1. \tag{7.4}$$

In particular, this condition is satisfied if the wave number $p(x)$ is of the form

$$p(x) \equiv \lambda P(x), \tag{7.5}$$

where λ is a large constant, i.e.,

$$\lambda \gg 1, \tag{7.6}$$

and $P(x)$ is of the order of unity. We have assumed that none of the constant parameters in $P(x)$ are large. Indeed, if $p(x)$ is given by (7.5), the inequality (7.4) is

$$\left| \frac{1}{\lambda} \frac{d}{dx} \frac{1}{P(x)} \right| \ll 1. \tag{7.7}$$

Clearly, (7.7) is satisfied if $\lambda \gg 1$, provided that x is not near a zero of $P(x)$.

As a side remark, while it is easy to accept that we may drop a term in the equation that is much smaller than a term in the equation being kept, we shall see in later chapters that this is not always a valid procedure. For example, the effects from a small term of the differential equation may add up, and as the solution evolves over a long interval of the independent variable, small perturbations may accumulate into a large correction. A justification of the WKB solutions will be given later.

We note that the phases of the solutions $\exp(\pm i \int p(x)dx)$ are functions of x, but the magnitudes of these approximate solutions are independent of x. Let us remember that, in Chapter 1, we have shown that if y_1 and y_2 are two independent solutions of (7.1), then the Wronskian $W(x) \equiv y_1 y_2' - y_1' y_2$ is independent of x. Now the Wronskian of $\exp(i \int p(x)dx)$ and $\exp(-i \int p(x)dx)$ is easily shown to be equal to $2ip(x)$, not a constant unless $p(x)$ is a constant. This suggests that these approximate solutions still leave something to be desired. Because the Wronskian of the approximate solutions miss by a factor of $p(x)$, let us try to fix it by adding an additional factor $1/\sqrt{p(x)}$ to each of the approximate solutions. The resulting approximate solutions are

$$y^{\pm}_{WKB}(x) = \frac{1}{\sqrt{p(x)}} e^{\pm i \int p(x)dx}, \tag{7.8}$$

which is said to include both the zeroth-order and the first-order terms of the WKB approximations.

The magnitude of these solutions varies with x like $1/\sqrt{p(x)}$. The Wronskian of $y^{\pm}_{WKB}(x)$ is now exactly a constant. (See homework problem 1.) It is therefore tempting to surmise that under the condition (7.4) or, equivalently, (7.7), $y^{\pm}_{WKB}(x)$ are even better approximations than

$$e^{\pm i \int p(x)dx}.$$

To see if this is true, we put

$$y \equiv e^{\pm i \int p(x) dx} v, \tag{7.9}$$

and substitute this expression of y into (7.1). We get

$$(D \pm ip)(D \pm ip)v + p^2 v = 0,$$

or

$$\left(\frac{d^2}{dx^2} \pm 2ip \frac{d}{dx} \pm ip' \right) v = 0.$$

We shall write the equation above as

$$v' + \frac{p'}{2p} v = \pm \frac{i}{2p} v''. \tag{7.10}$$

By (7.5), (7.10) can be written as

$$v' + \frac{P'}{2P} v = \pm \epsilon \frac{i}{2P} v'', \tag{7.11}$$

where

$$\epsilon \equiv 1/\lambda$$

is a small number. In the first-order approximation, we ignore the right side of (7.11) and we get

$$v' + \frac{P'}{2P} v \approx 0, \tag{7.12}$$

which gives

$$v(x) \approx \frac{1}{\sqrt{P(x)}}. \tag{7.13}$$

Thus (7.9) and (7.13) give, aside from an immaterial overall constant, the WKB solutions (7.8).

The more traditional way to derive the WKB solutions is given in homework problem 2 in this chapter.

We have mentioned that if $P(x)$ has a zero at x_0, the inequality (7.4) does not hold at x_0. To see how far away from x_0 it must be for the WKB approximation to hold, let $P(x)$ near x_0 be approximately given by

$$P(x) \approx a(x - x_0)^n, \quad x \approx x_0. \tag{7.14}$$

Equation (7.7) requires

$$|x - x_0| \gg \left(\frac{n}{\lambda a}\right)^{1/(n+1)}. \tag{7.15}$$

Equation (7.15) tells us how far away from x_0 it must be for the WKB approximate solutions to be valid.

We mention that if $P(x)$ vanishes in the way given by (7.14), we say that $P(x)$ has an nth-order zero at x_0. Not all zeroes of $P(x)$ are of finite order; an example of $P(x)$ having a zero of infinite order is given by homework problem 7.

As we have stated at the beginning of this chapter, the WKB approximation is useful for describing the propagation of waves with very small wavelengths. We shall now explain what this means more precisely. The wave number for the WKB solutions (7.2) is $p(x)$, and the corresponding wavelength $\Omega(x)$ is $2\pi/p(x)$. When λ of (7.5) is large, $p(x)$ is large and the wavelength $\Omega(x)$ is small. In the meantime, the inequality (7.4) is satisfied, justifying the WKB approximation.

While all of this is straightforward in mathematical terms, we deem it useful to clarify it further in physical terms. This is because the wavelength has the dimension of the distance, and the numerical value of a wavelength depends on the distance unit we choose. For example, 10^{-3} centimeters is exactly the same as 10^4 nanometers. While 10^{-3} is a small number, 10^4 is a large number. Is the wavelength of such a value small or large?

The fact is that it is not meaningful to classify a quantity as either small or large unless we compare it with another quantity of the same dimension. Let us first examine (7.4) in this light. Expressed in terms of $\Omega(x)$, (7.4) is

$$\left|\frac{d}{dx}\frac{\Omega(x)}{2\pi}\right| \ll 1.$$

As both $d\Omega(x)$ and dx have the dimension of length, $d\Omega(x)/dx$ is dimensionless. Since it is meaningful to say that a dimensionless quantity is much less than the dimensionless constant unity, (7.4) is meaningful.

Problem for the Reader

What is the dimension of $\int p(x)dx$?

Solution

The wave number $p(x)$ has the dimension of the inverse of the distance. Since dx has the dimension of distance, the integral $\int p(x)dx$ is dimensionless. We may add that the integral $\int p(x)dx$ appears in the exponent of the WKB solutions, and an exponent should always be a dimensionless quantity.

We shall now explain what is the physical quantity we must compare the wavelength to. Let $p(x)$ in (7.1) be equal to $a^{-1}P(x/L)$, where both a and L have the dimension of length, and where $P(x/L)$ is dimensionless and is of the order of unity. Then $\Omega(x) = 2\pi a/P(x/L)$. This says that the magnitude of $\Omega(x)$ is of the order of $2\pi a$. Note that x/L is dimensionless and by $\Omega(x)$ being a function of x/L we imply that L is the scale characterizing the variation of the wavelength as a function of x. This means that the derivative of $\Omega(x)$ is of the order of $1/L$ times $\Omega(x)$. Therefore, $d\Omega(x)/dx$ is of the order of $2\pi a/L$. As a result, the inequality (7.4) is satisfied if

$$a/L \ll 1,$$

or

$$a \ll L.$$

This says that the WKB approximation is valid if the wavelength is small compared to $2\pi L$, where L is the length characterizing the scale of the variation of the wavelength.

As a trivial example, if p is a constant, the wavelength does not change no matter how much x varies. Thus L is equal to infinity and the wavelength is much smaller than L. Indeed, if p is a constant, the WKB solutions are not only good approximate solutions of (7.1), but the exact solutions of (7.1).

Let us revisit the case of (7.5), for which the WKB approximation has been justified. In this case, we have $a = \lambda^{-1}$and $L = 1$. Hence a is much less than L.

Problem for the Reader

Consider (7.1) with $p(x) = P(\epsilon x)$, where $\epsilon \ll 1$ and $P(\epsilon x)$ is of the order of unity, i.e.,

$$\frac{d^2y}{dx^2} + P^2(\epsilon x)y = 0.$$

Note that the coefficient of y in this equation is not large, but varies slowly with x. Can we apply the WKB approximation to this equation?

Solution

The WKB approximation is valid if (7.4) is satisfied. Since

$$\frac{d}{dx}\frac{1}{P(\epsilon x)} = -\epsilon\frac{P'(\epsilon x)}{P^2(\epsilon x)},$$

(7.4) is satisfied if ϵ is very small and P does not vanish. Therefore, we conclude that we may apply the WKB approximation to (7.1) with $p(x) = P(\epsilon x)$ when x is not near a zero of $P(\epsilon x)$. We note that in this case, $a = 1$ and $L = \epsilon^{-1}$. Thus the wavelength is again much smaller than the characteristic length L.

The case of $p(x) = P(\epsilon x)$ and the case of $p(x) = \lambda P(x)$ are actually related by a change of the scale of the independent variable. To wit, let

$$X = \epsilon x;$$

then eq. (7.1) with $p(x) = P(\epsilon x)$ becomes

$$\frac{d^2y}{dX^2} + \lambda^2 P^2(X)y = 0,$$

where the large parameter λ is equal to ϵ^{-1}. This says that if we use X as the independent variable, p is in the form $\lambda P(X)$.

We mention a couple of physical problems in which the WKB approximation is useful. Consider the problem of determining the shadow cast on a wall by a point light source in front of a screen. To obtain the exact solution of this problem, one solves the wave equation and makes the solution satisfy the boundary conditions imposed by the presence of the screen. This is a difficult boundary-value problem. On the other hand, the shadow on the wall is very accurately determined simply by drawing straight lines from the light source to the edges of the screen. This is because when the wavelength of light is very small compared to the dimensions of the screen, the WKB approximation can be used to justify the results obtained with the use of geometric optics.[1] As another example, we know that Newtonian mechanics is an approximation of quantum mechanics. However, the behavior of atoms obeying the rules of wave mechanics is drastically different from that of particles obeying the rules of Newtonian mechanics. How does one reconcile these two sets of rules? The answer again lies in the WKB approximation, in which the Schrödinger equation is reduced to the Hamilton-Jacobi equation satisfied by the classical action of Newtonian mechanics.

The WKB approximation can also be used to solve problems in which the functional behavior is rapidly growing or rapidly decaying rather than rapidly oscillatory, an example being the problems of boundary layer, which we will discuss in Chapter 9. Consider the equation

$$y'' - \eta^2 y = 0. \tag{7.16}$$

The WKB solutions are given by

$$y^{\pm}_{WKB}(x) = \frac{1}{\sqrt{\eta(x)}} e^{\pm \int \eta(x)dx}. \tag{7.17}$$

These solutions are good approximations of the solutions of (7.16) if

$$\left| \frac{d}{dx} \frac{1}{\eta(x)} \right| \ll 1. \tag{7.18}$$

The counterpart of (7.5) is

$$\eta(x) = \lambda N(x), \tag{7.19}$$

where $\lambda \gg 1$ and $N(x)$ is of order unity. As before, if η is in the form (7.19), the inequality (7.18) is always satisfied unless x is near a zero of $N(x)$.

B. Solutions Near an Irregular Singular Point

In some mathematical problems, the large parameter λ is not explicitly exhibited. As an example, consider the problem of solving the equation $y'' + xy = 0$ when x

[1] S. I. Rubinow and T. T. Wu, *Journal of Applied Physics* 27:1032 (1956); T. T. Wu, *Physical Review* 104:1201 (1956).

is very large. In this problem, x inherently contains a large parameter. Indeed, let x be of the order of Λ, with $\Lambda \gg 1$. We may put

$$x \equiv \Lambda X,$$

where X is of the order of unity. Then the Airy equation is

$$\left(\frac{d^2}{dX^2} + \Lambda^3 X\right) y = 0.$$

This is in the canonical form for which the WKB method can be applied.

Thus the WKB approximation is useful for obtaining the asymptotic solutions near an irregular singular point of a second-order linear homogeneous equation. While we have already given a method in the preceding section to obtain these solutions, it applies only when the rank of the singular point is an integer. The WKB method has no such restriction. In addition, the use of the WKB method makes it easy to obtain the leading terms of the asymptotic series.

Let us consider the leading asymptotic terms for the solutions of the equation

$$y'' + xy = 0. \tag{7.20}$$

For $x < 0$, we have, comparing with (7.16),

$$\eta = (-x)^{1/2}.$$

Thus the WKB solutions are

$$|x|^{-1/4} e^{\pm 2|x|^{3/2}/3}. \tag{7.21}$$

We conclude immediately that when x is large and negative, one of these solutions is an exponentially increasing function of x and the other is an exponentially decreasing function of x.

When x is positive, we have, comparing with (7.1),

$$p = x^{1/2}.$$

Thus the WKB solutions are

$$x^{-1/4} e^{\pm i2x^{3/2}/3}, \tag{7.22}$$

both being oscillatory functions of x.

Next we will give the entire asymptotic series for when x is positive and very large in magnitude.

Problem for the Reader

Find the asymptotic series for the solutions of the equation (7.20) for $x \gg 1$.

Solution

We note from the power function of the exponent in the WKB solutions that the rank of a singular point at infinity is $3/2$. Since the rank is not an integer, the method given in the preceding chapter cannot be directly applied. By (7.4), these WKB solutions are good approximate solutions if

$$x \gg (1/2)^{2/3}.$$

To find corrections of the WKB solutions, we put

$$y = \exp\left(\pm\frac{2x^{3/2}}{3}i\right)v.$$

Then v satisfies the equation

$$\left(\frac{d}{dx}+\frac{1}{4x}\right)v = \pm\frac{i}{2x^{1/2}}\frac{d^2}{dx^2}v. \tag{7.23}$$

The dimension of the operator on the left side of the equation above is -1, while the dimension of the operator on the right side is $-5/2$. These two dimensions differ by $3/2$, which is not an integer. Thus we make the change of variable

$$\rho \equiv x^{3/2}.$$

In terms of the variable ρ, these two dimensions differ by unity, an integer. Then (7.23) becomes

$$\left(\frac{d}{d\rho}+\frac{1}{6\rho}\right)v = \pm\frac{3i}{4}\left(\frac{d}{d\rho}+\frac{1}{3\rho}\right)\frac{dv}{d\rho}. \tag{7.24}$$

Let

$$v = \sum a_n\rho^{-n-s},\ a_0 \neq 0,\ \text{ and } a_{-1} = a_{-2} = \cdots = 0. \tag{7.25}$$

We have

$$\frac{dv}{d\rho}+\frac{v}{6\rho} = \sum -\left(n+s-\frac{1}{6}\right)a_n\rho^{-n-s-1},$$

and

$$\pm\frac{3i}{4}\left(\frac{d}{d\rho}+\frac{1}{3\rho}\right)\frac{dv}{d\rho} = \pm\frac{3i}{4}\sum\left[(n+s)(n+s+1)-1/3(n+s)\right]a_n\rho^{-n-s-2}$$
$$= \pm\frac{3i}{4}\sum(n+s-1)(n+s-1/3)a_{n-1}\rho^{-n-s-1}.$$

Thus we get

$$-\left(n+s-\frac{1}{6}\right)a_n = \pm\frac{3i}{4}(n+s-1)(n+s-1/3)a_{n-1}. \tag{7.26}$$

Setting $n=0$ in the equation above, we get

$$s = 1/6.$$

We set $s=1/6$ in (7.26) and get, for $n>0$,

$$a_n = \mp\frac{3i}{4}\frac{(n-5/6)(n-1/6)}{n}a_{n-1} = \left(\mp\frac{3i}{4}\right)^n\frac{\Gamma(n+1/6)\Gamma(n+5/6)}{\Gamma(1/6)\Gamma(5/6)n!}a_0. \tag{7.27}$$

Therefore, for $x \gg 1$, the two asymptotic solutions of (7.20) are

$$x^{-1/4}\exp\left(\pm\frac{2x^{3/2}}{3}i\right)\sum_{n=0}^{\infty}\left(\mp\frac{3i}{4}\right)^n\frac{\Gamma(n+1/6)\Gamma(n+5/6)}{n!}x^{-3n/2}. \tag{7.28}$$

Problem for the Reader

Find the WKB solutions for the Bessel equation

$$\left(\rho^2\frac{d^2}{d\rho^2}+\rho\frac{d}{d\rho}+\rho^2-p^2\right)Y(\rho) = 0. \tag{7.29}$$

Solution

First we transform eq. (7.29) into the form of (7.1). This is done by putting

$$Y(\rho) = \rho^{-1/2}y(\rho),$$

getting

$$\left(\frac{d^2}{d\rho^2} + 1 - \frac{p^2 - 1/4}{\rho^2}\right) y = 0.$$

Identifying

$$p = \sqrt{1 - \frac{p^2 - 1/4}{\rho^2}},$$

we have, when ρ is large,

$$p \simeq 1.$$

Thus we have

$$y_{WKB}(\rho) = e^{\pm i\rho}$$

and

$$Y_{WKB}(\rho) = \rho^{-1/2} \exp(\pm i\rho). \tag{7.30}$$

This is in agreement with the leading terms of (6.63) and (6.64), found with a little more effort.

The WKB solutions sometimes even help us to obtain the exact solution of an equation.

Problem for the Reader

Solve in closed form

$$y'' + \left(x^2 + \frac{2}{9x^2}\right) y = 0. \tag{7.31}$$

Solution

The WKB solutions of (7.31), valid for $|x| \gg 1$, are easily found to be

$$y_{WKB}(x) = x^{-1/2} \exp\left(\pm \frac{x^2 i}{2}\right). \tag{7.32}$$

The exponent in (7.32) is equal to a constant times x^2; thus infinity is a singular point of the ODE of rank two. (We see once again that using the WKB method is an easy way to obtain the rank of a singular point.)

Let us look into the possibility of y being a Bessel function. Comparing the exponent of the solution of (7.30) with that of (7.32) suggests to us that

$$\rho = \frac{x^2}{2}. \tag{7.33}$$

With this identification of the independent variables, the exponential function of the solution of (7.30) is now equal to that of (7.32). Yet the factor multiplying the exponential function of (7.30) is $\rho^{-1/4}$, which differs from that of (7.32) by a factor of $\rho^{-1/4}$. Let

$$Y = \rho^{-1/4} y; \tag{7.34}$$

then the asymptotic forms of Y are exactly the same as those given by (7.30). Therefore, we make the change of the independent variable (7.33) and the change of the dependent variable (7.34) for equation (7.31). Then it is straightforward to show that eq. (7.31) becomes the Bessel equation with order p equal to $1/12$. Thus the general solution of (7.31) is

$$y(x) = x^{1/2}\left[aJ_{1/12}(x^2/2) + bJ_{-1/12}(x^2/2)\right], \tag{7.35}$$

where a and b are constants.

Problem for the Reader

Show that the parabolic function $D_\nu(x)$ satisfying the equation

$$y'' + (\nu + 1/2 - x^2/4)y = 0$$

is not directly related to the Bessel functions unless $\nu = -1/2$.

Solution

Let

$$\eta = \sqrt{x^2/4 - \nu - 1/2}.$$

When x is large and positive, we have

$$\eta \simeq x/2 - (\nu + 1/2)x^{-1}.$$

Thus

$$\int \eta dx \simeq x^2/4 - (\nu + 1/2)\ln x.$$

Note that while a term in η that is equal to a constant times x^{-1} is small when x is large, it generates a term in $\int \eta dx$ that is equal to a constant times $\ln x$. Such a term is large when x is large, and cannot be ignored. The WKB solutions for this equation are

$$y^{+}_{WKB}(x) = x^{\nu} e^{-x^2/4}$$

and

$$y^{-}_{WKB}(x) = x^{-\nu-1} e^{x^2/4}.$$

The power functions for these two solutions are different unless $\nu = -1/2$, while those for the Bessel functions are the same. Thus, unless $\nu = -1/2$, the parabolic cylinder function $D_\nu(x)$ is not equal to a power function times a Bessel function $Z_p(\rho)$, where p is any number and ρ is any function of x.

Finally, we mention that while one may get the notion that (7.4) is likely to hold in the region of x where $p(x)$ is very large, this is not always the case. Consider

$$p(x) = 1/x, \tag{7.36}$$

which is very large when x is small. Yet

$$\frac{d}{dx}\frac{1}{p(x)} = 1,$$

which is not small when x is small.

C. Higher-Order WKB Approximation

We shall in this section find the higher-order terms of the WKB approximation. For this purpose let us return to eq. (7.11). Since this equation has a small parameter ϵ and is linear, it is straightforward to use it to derive successive corrections to the WKB approximations. We put

$$v = v_0 + \epsilon v_1 + \epsilon^2 v_2 + \cdots, \tag{7.37}$$

where v_n, $n = 0, 1 \cdots$, are independent of ϵ. The series in (7.37) is called a perturbation series, which is expected to be useful when ϵ is small. We substitute (7.37) into (7.11) and get

$$\begin{aligned}(v_0 + \epsilon v_1 + \epsilon^2 v_2 + \cdots)' + \frac{P'}{2P}(v_0 + \epsilon v_1 + \epsilon^2 v_2 + \cdots) \\ = \pm \frac{i\epsilon}{2P}(v_0 + \epsilon v_1 + \epsilon^2 v_2 + \cdots)''.\end{aligned} \tag{7.38}$$

In the lowest-order approximation we set ϵ in (7.38) to zero and get

$$v_0{}' + \frac{P'}{2P} v_0 = 0.$$

This equation gives

$$v_0(x) = \frac{1}{\sqrt{P(x)}},$$

which is, aside from an immaterial constant multiple, (7.13).

Setting to zero the sum of terms in (7.38) that are proportional to ϵ, we get

$$v_1' + \frac{P'(x)}{2P(x)} v_1 = \pm \frac{i}{2P(x)} \left(\frac{1}{\sqrt{P(x)}} \right)''. \tag{7.39}$$

Solving this first-order linear equation, we find that

$$v_1(x) = \pm \frac{i}{2\sqrt{P(x)}} \int \frac{1}{\sqrt{P(t)}} \left(\frac{1}{\sqrt{P(t)}} \right)'' dt. \tag{7.40}$$

Now we are ready to give a justification of the WKB method, which is approximating the solution of (7.1) by truncating the series of (7.37). Strictly speaking, truncating a series is justified if we succeed in proving that the sum of terms neglected is much less than the sum of terms kept. But proving this is sometimes difficult to do. We shall be content with proving that the $(n+1)^{\text{th}}$ term in the series is much less than the nth term if ϵ is sufficiently small. Thus we will accept that the WKB solutions (7.8) are good first-order approximations if

$$|\epsilon v_1| \ll |v_0|. \tag{7.41}$$

Since ϵ is small, (7.41) is satisfied provided that $v_1(x)/v_0(x)$ does not blow up, which is true unless $P(x)$ happens to vanish.

If $P(x)$ vanishes at x_0, the differential equation (7.1) is said to have a turning point at x_0. At a turning point of the differential equation, we may prove from (7.40) that the ratio v_1/v_0 blows up, and the WKB approximation fails.

How far away from the turning point must it be in order for the WKB approximation to work? If when x is near x_0, $P(x)$ goes to zero like $(x-x_0)^n$, then $v_1(x)$ blows up like

$$(x-x_0)^{-1-3n/2}, \tag{7.42}$$

while $v_0(x)$ blows up like $(x-x_0)^{-n/2}$. Thus (7.41) requires

$$|x-x_0| \gg \frac{1}{\lambda^{1/(1+n)}}. \tag{7.43}$$

Aside from a multiplicative constant, (7.43) is the same condition as (7.15).

We may find all higher-order terms of the solution from (7.38). This is done by gathering all the terms in (7.38) proportional to ϵ^m and setting the sum to zero. We get

$$v'_m + \frac{P'}{2P}v_m = \pm\frac{i}{2P}v''_{m-1}. \tag{7.44}$$

Thus

$$v_m(x) = \pm\frac{i}{2\sqrt{P(x)}}\int \frac{dx}{\sqrt{P(x)}}\frac{d^2}{dx^2}v_{m-1}(x). \tag{7.45}$$

From (7.45), we obtain the mth-order term of the perturbation series once the $(m-1)^{\text{th}}$-order term of the perturbation series has been found. Thus we obtain all v_m by successive iteration.

If $P(x)$ has no zero, all v_m are finite. When ϵ is sufficiently small, we have

$$|\epsilon v_m| \ll |v_{m-1}|. \tag{7.46}$$

Thus the WKB approximation is justified to higher orders. Here we like to give the reader a reminder: The WKB approximation has been justified to higher orders only if $p(x)$ is of the form (7.5) or $\eta(x)$ is of the form of (7.19), and neither of them vanishes. (I feel obligated to say it as I have seen the WKB approximations being too liberally applied.)

We may show that if $P(x)$ vanishes at x_0 and is given by (7.14) near x_0, the condition (7.46) is satisfied provided that x is sufficiently far away from x_0 so that (7.15) is satisfied. (See homework problem 3.)

The high-order WKB approximations for (7.16) can be obtained in a similar way. We put

$$y \equiv e^{\pm \int \eta(x)dx} v. \tag{7.47}$$

Then we have

$$v' + \frac{\eta'}{2\eta} v = \mp \frac{1}{2\eta} v''. \tag{7.48}$$

We may use (7.48) to obtain successive approximations of the WKB solutions of (7.16).

D. Turning Points

As we have mentioned, the WKB approximation is useful in problems of wave propagation. In this section we demonstrate this by applying it to the wave equation that governs the quantum mechanical behavior of a particle.

Consider the time-independent Schrödinger equation of one spatial dimension discussed at the end of Chapter 5. We will write this equation in the form

$$\frac{d^2\phi}{dx^2} + \lambda^2 \left[E - V(x)\right] \phi = 0, \tag{7.49}$$

where λ is equal to $\sqrt{2m}$ divided by the Planck constant, with m the mass of the particle. We shall consider λ as a very large number.

As we have mentioned, $|\phi|^2$ is the probability density of the particle, E is the energy of the particle, and $V(x)$ is the potential. We note that $E - V$ is equal to the kinetic energy of the particle.

In the region where

$$E > V(x),$$

the kinetic energy of the particle is positive. Comparing with (7.1), we identify $p(x)$ with

$$\lambda\sqrt{E - V(x)}.$$

The WKB solutions are

$$p^{-1/2}e^{i\int pdx} \text{ and } p^{-1/2}e^{-i\int pdx}. \tag{7.50}$$

Thus the wavefunctions ϕ in the classically accessible region are oscillatory as a function of x.

In the region where

$$E < V(x),$$

the kinetic energy of the particle is negative, and the momentum of the particle is imaginary. In classical mechanics, the momentum is not allowed to be imaginary. Thus this region is inaccessible to the classical particle. In quantum mechanics, the wavefunction ϕ is exponentially small in this region. The WKB solutions of the wave equation are

$$\eta^{-1/2}e^{\int \eta dx} \text{ and } \eta^{-1/2}e^{-\int \eta dx}, \tag{7.51}$$

where

$$\eta = \lambda\sqrt{V(x) - E}.$$

The first solution above is exponentially increasing as x increases, and the second solution above is exponentially decreasing as x increases. Since a particle is rarely observed in the classically inaccessible region, we require the wavefunction ϕ in this region to be the solution that decreases to smaller and smaller values as x goes deeper and deeper into the classically inaccessible region.

Let there be a point x_0 at which

$$E - V(x_0) = 0.$$

Note that x_0 is the point at which the momentum of the particle vanishes. By (7.4), the WKB approximation fails at x_0.

Let us study the behavior of the wavefunction near x_0. We shall assume that $V'(x_0)$ is different from zero; hence $E - V(x)$ is negative at one side of x_0 and positive at the other side of x_0. The point x_0 is the dividing point between a classically accessible region and a classically inaccessible region. In classical mechanics, the particle cannot move into the region where its kinetic energy is negative, and must turn back as it arrives at x_0. That is why x_0 is called a turning point of the wave equation. As we may expect, the qualitative behavior of the quantum wavefunction goes through a transition near x_0. More precisely, the WKB solution changes from an oscillatory behavior from one side of the turning point to an exponential behavior at the other side of the turning point. However, we cannot continue the solution from one side of the turning point to the other side of the turning point with the WKB solutions alone. This is because the WKB approximation fails in a small neighborhood around the turning point x_0.

Fortunately, another approximation for the solution is available when x is close to x_0. In order to make the following discussions appear as simple as possible, we shall choose $x_0 = 0$.

Near the turning point that is chosen to be the origin, we have by assumption

$$[E - V(x)] \approx -\alpha x,$$

with α equal to $V'(0)$. This linear approximation is valid for

$$|x| \ll 1. \tag{7.52}$$

We will call the region given by (7.52) the turning region. In the turning region, the wavefunction ϕ is approximately described by

$$\frac{d^2\phi}{dx^2} - \alpha\lambda^2 x\phi = 0, \tag{7.53}$$

the general solution of which is a linear superposition of Airy functions.

We expect the WKB solutions to be good when x is sufficiently far away from the turning point. How far away from the turning point should x be in order for the WKB approximation to be valid? Not very far, as it turns out. Indeed, by (7.4), the region where the WKB approximation is valid is determined to be

$$|x| \gg \Lambda^{-2/3}, \tag{7.54}$$

where

$$\Lambda^2 = |\alpha|\lambda^2.$$

Since Λ is very large compared to unity, $\Lambda^{-2/3}$ is very small compared to unity. Thus the WKB solutions are good even when x is close, although not too close, to the turning point.

This means that there are values of x for which both (7.52) and (7.54) are fulfilled. These are the values of x satisfying the inequality

$$1 \gg |x| \gg \Lambda^{-2/3}.$$

The regions of x satisfying the inequality above will be called the overlapping regions, in which both the WKB solutions and the Airy function solutions are good approximations of the wavefunction. Because x may be either positive or negative, there are two disjoint overlapping regions. They are

$$1 \gg x \gg \Lambda^{-2/3} \tag{7.55a}$$

and

$$1 \gg -x \gg \Lambda^{-2/3} \tag{7.55b}$$

Visually, these two overlapping regions form the two fringes of the turning region.

The existence of these two overlapping regions is crucial. The Airy functions approximate the wavefunction well throughout the turning region, which overlaps with the region in which the WKB solutions are good. This fact enables us to join the WKB solution from one fringe to the other, using the Airy functions to interpolate the wavefunction through the central part of the turning region where the WKB approximation fails.

We now demonstrate specifically how this is done. First we consider the case in which (7.49) has only one turning point. Without loss of generality we shall let α be positive. (For if α is negative, we may make a change of variable, referring to x as $-x$.) If α is positive, the kinetic energy of the particle is negative in the region $x > 0$. In classical mechanics, this is the region inaccessible to the particle. Thus we require the wavefunction ϕ to vanish rapidly as x increases in this region. The WKB solution in the region $x > 0$ satisfying this requirement is

$$\phi_{WKB}(x) = \frac{e^{-\int_0^x \eta(x')dx'}}{\sqrt{\eta(x)}}. \tag{7.56}$$

Incidentally, we may choose the solution to be a constant times the right side of (7.56), but this affects only the overall normalization of the wavefunction and we will leave it the way it is.

We use the solution of (7.56) to describe the wavefunction in the region $x > 0$. This WKB solution fails as x gets into the interior of the turning region. To obtain an approximation of the solution that is good throughout the turning region, we note that when x is in the turning region, we may approximate (7.49) by (7.53). By changing the independent variable, we transform eq. (7.53) into the Airy equation

$$\frac{d^2}{d\rho}y - \rho y = 0,$$

where

$$\rho = \Lambda^{2/3}x. \tag{7.57}$$

Thus the two independent solutions of (7.53) are the Airy functions denoted by $Ai(\rho)$ and $Bi(\rho)$.

Note that ρ is related to x by a change of scale. Therefore, a region of unit width in the x variable corresponds to a region of width $\Lambda^{2/3}$ in the ρ variable. Since $\Lambda^{2/3}$ is a very large number, the region of unit width in the x variable corresponds to a region of very large width in the ρ variable. When we study the solution as a function of ρ, it is like studying the function with a magnifier, under which the scale is enlarged.

In terms of the variable ρ, the turning region is given by

$$|\rho| \ll \Lambda^{2/3}, \tag{7.58}$$

which is a very large region, and the overlapping regions are given by

$$1 \ll \rho \ll \Lambda^{2/3} \tag{7.59a}$$

and

$$1 \ll -\rho \ll \Lambda^{2/3}. \tag{7.59b}$$

As we will discuss in the next chapter, the asymptotic forms of $Ai(\rho)$ when the magnitude of ρ is large are given by

$$Ai(\rho) \sim \frac{e^{-2\rho^{3/2}/3}}{2\sqrt{\pi}\rho^{1/4}}, \quad \rho \to \infty, \tag{7.60}$$

and

$$Ai(\rho) \sim \frac{\sin\left[\frac{2}{3}(-\rho)^{3/2} + \frac{\pi}{4}\right]}{\sqrt{\pi}(-\rho)^{1/4}}, \quad \rho \to -\infty. \tag{7.61}$$

Since $Ai(\rho)$ vanishes exponentially when ρ is large and positive, the Airy function solution that matches the WKB solution (7.56) is

$$\phi(x) = cAi(\rho), \tag{7.62}$$

where c is a constant.

To determine c, we match the solution (7.62) with the solution (7.56) in the first overlapping region (7.59a) or, equivalently, (7.55), where both approximations are good. Now when x is small and positive, we have

$$\eta(x) \approx \Lambda x^{1/2} = \Lambda^{2/3}\rho^{1/2}$$

and

$$-\int_0^x \eta(x')dx' \approx -\frac{2\rho^{3/2}}{3}.$$

Thus the solution (7.62) and the solution (7.56) have the same functional form in the overlapping region. They match exactly if we choose

$$c = \frac{2\sqrt{\pi}}{\Lambda^{1/3}}.$$

Therefore, as we continue the wavefunction (7.56) into the turning region, the wavefunction becomes

$$\phi(x) = \frac{2\sqrt{\pi}}{\Lambda^{1/3}} Ai(\rho). \tag{7.63}$$

Next we continue the wavefunction further into the region $x < 0$. By (7.61), the Airy function solution (7.63) as ρ is negative and large is

$$\phi(x) \approx \frac{2}{\Lambda^{1/3}} \frac{\sin\left[\frac{2}{3}(-\rho)^{3/2} + \frac{\pi}{4}\right]}{(-\rho)^{1/4}}. \tag{7.64}$$

In the second overlapping region (7.55b) or, equivalently, (7.60b), we have

$$p(x) \approx \Lambda(-x)^{1/2} = \Lambda^{2/3}(-\rho)^{1/2}$$

and

$$\int_x^0 p(x')dx' \approx \frac{2}{3}(-\rho)^{3/2}.$$

Therefore, the WKB solution for $x < 0$ that matches (7.64) in the second overlapping region is

$$\frac{2\sin\left[\int_x^0 p(x')dx' + \pi/4\right]}{\sqrt{p(x)}}. \tag{7.65}$$

We point out that as it is being continued into the region $x < 0$, the WKB solution (7.56) for $x > 0$ turns into neither a multiple of the WKB solution

$$p^{-1/2}e^{i\int pdx}$$

nor a multiple of the WKB solution

$$p^{-1/2}e^{-i\int pdx}.$$

Instead, it turns into a linear superposition of equal magnitude of these two solutions.

With (7.56), (7.63), and (7.65), we have, for the wave equation of one turning point, an approximate solution covering all values of x. It is easy to generalize the results above to the case in which the turning point is located at x_0 rather than at the origin. We simply call $X = x - x_0$, and the point $x = x_0$ corresponds to the point $X = 0$. Thus we only need to replace x in the results above by X. I remind the reader that if we re-express the results so obtained in terms of the variable x, the limits of integration given in the formulae above should be changed accordingly. For example, $X = 0$ corresponds to $x = x_0$.

We are now ready to treat the wave equation with two turning points. Let x_0 and x_1 be the two turning points, with $x_0 > x_1$. Let $E - V(x)$ be positive inside the region $x_1 < x < x_0$, and be negative outside the region. In classical physics, the region $x_1 < x < x_0$ is the region to which the particle is confined. Thus we must require the wavefunction to decrease rapidly as x leaves the region $x_1 < x < x_0$. In particular, we require that the wavefunction vanish as x approaches plus infinity or minus infinity.

This is a boundary-value problem with the trivial solution

$$\phi = 0.$$

This trivial solution satisfies both (7.49) and the conditions of vanishing at plus infinity and minus infinity.

A nontrivial solution exists only if E takes some special values called eigenvalues. These eigenvalues can be found approximately with the WKB method, as we shall presently show.

Since it is required to vanish as $x \to -\infty$, the WKB solution in the region $x < x_1$ is chosen to be

$$\frac{e^{-\int_x^{x_1} \eta(x')dx'}}{\sqrt{\eta(x)}}. \tag{7.66}$$

Since it is required to vanish as $x \to \infty$, the WKB solution in the region $x > x_0$ is of the form

$$\phi_{WKB}(x) = a \frac{e^{-\int_{x_0}^{x} \eta(x')dx'}}{\sqrt{\eta(x)}}, \tag{7.67}$$

where a is a constant. Since the normalization of the wavefunction has been set by choosing the wavefunction in the region $x < x_1$ to be precisely the expression in (7.66), we no longer have any freedom to choose a for the wavefunction in the region $x > x_0$. The constant a will be determined with a calculation.

We shall give the wavefunction only in the region that is inside the interval $x_0 > x > x_1$ but is sufficiently far away from the turning points x_0 and x_1. The WKB approximation holds throughout this region. Since the matching with the Airy functions has already been performed, we shall be spared the chores of doing it once again if we are not interested in the wavefunction near the turning points.

We find with the use of (7.65) that the WKB solution in the region $x_0 > x > x_1$ that matches with (7.66) is

$$\frac{2\sin\left[\int_{x_1}^{x} p(x')dx' + \pi/4\right]}{\sqrt{p(x)}}.$$

We also find with (7.65) that the WKB solution in the region $x_0 > x > x_1$ that matches with (7.67) is

$$\frac{2a\sin\left[\int_{x}^{x_0} p(x')dx' + \pi/4\right]}{\sqrt{p(x)}},$$

where the factor a is the same as in (7.67). Since these two solutions are valid in the same region $x_1 < x < x_0$, they are required to be the same. Hence we have

$$\sin\left[\int_{x_1}^{x} p(x')dx' + \pi/4\right] = a\sin\left[\int_{x}^{x_0} p(x')dx' + \pi/4\right]. \tag{7.68}$$

Setting $x = x_0$, we get from (7.68)

$$\sin(I + \pi/4) = a/\sqrt{2}, \tag{7.69}$$

where

$$I \equiv \int_{x_1}^{x_0} p(x')dx'.$$

Differentiating (7.68) with respect to x and setting $x = x_0$, we get

$$\cos(I + \pi/4) = -a/\sqrt{2}. \tag{7.70}$$

Taking the ratio of (7.69) and (7.70), we get

$$\tan(I + \pi/4) = -1.$$

Therefore,

$$I = \int_{x_1}^{x_0} p(x)dx = (n + 1/2)\pi, n = 0, 1, 2, \cdots, \tag{7.71}$$

which is known as the Bohr quantization rule giving the approximate energy eigenvalues. Substituting this value of I into (7.68), we get

$$a = (-1)^n.$$

Problem for the Reader

With the use of the WKB method, find the quantum energy eigenvalues of the harmonic oscillator for which

$$V(x) = \frac{1}{2}\kappa x^2,$$

where κ is a constant.

Solution

The turning points are obtained by setting

$$E - \frac{1}{2}\kappa x^2 = 0,$$

which gives

$$x_1 = -L,\ x_0 = L,$$

where

$$L = \sqrt{\frac{2E}{\kappa}}.$$

We have

$$I = \lambda \int_{-L}^{L} \sqrt{E - \kappa x^2/2} dx.$$

To evaluate the integral above, it is best to scale the variable of integration so that the limits of integration are -1 and 1. Thus we put

$$X = x/L,$$

and get

$$I = \lambda E \sqrt{\frac{2}{\kappa}} \int_{-1}^{1} \sqrt{1 - X^2} dX.$$

Since

$$\int_{-1}^{1} \sqrt{1 - X^2} dX = \pi/2,$$

we find

$$I = \lambda E \pi \sqrt{\frac{1}{2\kappa}}.$$

Therefore, Bohr's quantization rule says that

$$E_n = \sqrt{2\kappa}\lambda^{-1}(n + 1/2), n = 0, 1, 2 \cdots,$$

which turns out to be the exact answer.

We note that in classical physics, the energy of a harmonic oscillator can take any value from zero to infinity. But in quantum mechanics, the energy can only take the discrete values given above. The lowest of the energy eigenvalues is

$$\frac{1}{2}\sqrt{2\kappa}/\lambda,$$

while the other energy eigenvalues are evenly spaced with distance $\sqrt{2\kappa}/\lambda$ apart. We say that the energy of a harmonic oscillator is quantized. (A more in-depth discussion of the solution near a turning point will be given in Section E of Chapter 9.)

Homework Problems for This Chapter

Solutions to the Homework Problems can be found at www.lubanpress.com.

1. Show that the Wronskian of y^+_{WKB} and y^-_{WKB} given by (7.4) is a constant.

Hint: To calculate the Wronskian of y^+_{WKB} and y^-_{WKB}, it is efficient to use the formula (1.7).

2. The WKB solutions (7.8) can also be derived by putting $y = e^{i\lambda S}$ and substituting this expression for y into

$$\frac{d^2y}{dx^2} + \lambda^2 P^2(x)y = 0.$$

a. Show that S satisfies the nonlinear second-order differential equation

$$i\lambda S'' - \lambda^2 (S')^2 + \lambda^2 P^2 = 0.$$

b. Explain why we may drop the term iS'' in the equation above and obtain

$$(S')^2 - P^2 = 0.$$

This equation is known as the Hamilton-Jacobi equation. Show that this approximation is justified as long as

$$|\lambda S''| \ll (\lambda S')^2.$$

c. Show that the solutions of the Hamilton-Jacobi equation are $S = \pm \int P dx$, which will yield the zeroth-order WKB approximation. Show also that, with S given by $\pm \int P dx$, the inequality given in (b) is the same as (7.7).

d. Obtain the additional factor $1/\sqrt{P(x)}$ in the WKB solutions by expressing S as the perturbation series

$$S = S_0 + \epsilon S_1 + \cdots$$

and solving for S_1, where $\epsilon = \lambda^{-1}$ is a small number.

e. Discuss how to obtain the second-order WKB approximation with this approach. Compare this method of getting the second-order WKB approximation with the method given in Section C of this Chapter.

3. Show that when $P(x)$ has a zero of order n at x_0, (7.46) is true if (7.15) is satisfied, where n can be any positive number. What if $P(x)$ vanishes at $x = 0$ like e^{-1/x^2}?

4. Find the zeroth-order and the first-order terms for the solutions of

$$\frac{d^4y}{dx^4} + \lambda^4 U(x)y = 0, \quad \lambda \gg 1,$$

where $U(x)$ does not vanish. Consider both the cases $U(x) > 0$ and $U(x) < 0$. How do you find the higher-order terms of the solutions?

5. Solve the following equations in closed forms:

a. $y'' + x^m y = 0$,

b. $y'' + (x^2 + 3x^{-2}/16)y = 0$,

c. $y'' - (x^4 - 3x^{-2}/16)y = 0$.

6. Obtain the WKB solutions for the equations below. For what positive values of t are these approximations good? Can you solve them in closed forms?

a. $\dfrac{d^2y}{dt^2} + (1 + e^{-\epsilon t})y = 0$, where $\epsilon \ll 1$.

b. $\dfrac{d^2y}{dt^2} + e^{-\epsilon t}y = 0$, where $\epsilon \ll 1$.

Hint: To obtain the solutions of these equations in closed forms, put

$$\tau = \frac{2}{\epsilon}e^{-\epsilon t/2}.$$

The solutions for the equation of (a) are $J_{\pm 2i/\epsilon}(\tau)$, whereas those of (b) are $J_0(\tau)$ and $Y_0(\tau)$.

7. For the example in (6b), make the transformation $T = \epsilon t$ and cast the equation into the form

$$\frac{d^2y}{dT^2} + \lambda^2 P^2(T)y = 0.$$

What is λ and what is $P(T)$? Show that $P(T)$ has only one zero. What is the order of this zero of $P(T)$?

8. Consider $y'' + x^{-2}y = 0$ for x of the order of ϵ, $\epsilon \ll 1$. Let $x \equiv \epsilon X$ and show that there is no large parameter in the resulting equation. What do you conclude from this example?

9. Find the WKB solutions of the confluent hypergeometric equation

$$xy'' + (c - x)y' - ay = 0,$$

where a and c are constants. Determine the values of x for which the WKB solutions are good approximations.

10. With the use of the WKB method, find the approximate quantum energy eigenvalues of a particle moving in the potential

$$V(x) = \frac{1}{2}\kappa x^4,$$

where κ is a constant.

Hint: The integral

$$\int_0^1 \sqrt{1 - X^4}dX$$

can be expressed by a Beta function

$$B(p, q) = \int_0^1 t^{p-1}(1-t)^{q-1}dt \quad (p, q > 0),$$

which is equal to

$$\frac{\Gamma(p)\Gamma(q)}{\Gamma(p+q)}.$$

Chapter 8

Asymptotic Expansions of Integrals

A. Integral Representation

In the preceding two chapters, we have shown how to find approximate solutions of linear homogeneous ordinary differential equations. For example, in the case of the Airy equation, we have obtained explicitly the two independent solutions $y_1(x)$ and $y_2(x)$, given by (6.15) and (6.16) respectively, which are Maclaurin series. While the Maclaurin series solution is useful when x is small, it is not very convenient to use when x is large. First of all, such series solutions of differential equations may not converge for large values of x. And even if the series converges, one has to take, when x is large, a large number of terms of the series, which reveals little of the qualitative behavior of the function. The convenient formulae for computing the solutions of the Airy equation when x is large are those given by (7.28), which are asymptotic series.

While we have found the solutions that are useful in different regions of x, it remains for us to tie them together. Take again the example of the Airy equation. We know that the asymptotic form of each of its Maclaurin series solutions when x is large is a linear superposition of its WKB solutions. But what we have done so far tells us nothing about the constants of superposition. Nor does it tell us the constants of superposition with which the WKB solutions of the Airy equation valid for x negative and large are related to those valid for x positive and large.

In problems such as scattering problems in quantum mechanics, it is important to know the constants of superposition relating the solutions of the equation valid in different regions. For such purposes, it will be helpful if we are able to express solutions of differential equations in a closed form that is valid both for x small and for x large.

The integral representation of functions is one such form. In this representation, a solution of a differential equation is expressed in the form of an integral, which holds for all values of x. One may derive from it series of expansion around a finite point, or its asymptotic series around an irregular singular point of the differential equation the function satisfies. Thus an integral representation for a function often serves as a bridge between the solutions for different regions of the independent variable x.

Let us demonstrate as an example how to derive the integral representation of the Airy functions that satisfy the Airy equation

$$y'' - xy = 0.$$

The method we use is due to Laplace. In this method, we express the solution of the Airy equation in the form

$$y(x) = \int_c e^{-xt} Y(t) dt, \tag{8.1}$$

where c is a contour in the complex t-plane to be specified later.

If c in (8.1) is chosen to be the positive real axis, then $y(x)$ is the Laplace transform of $Y(t)$. However, the contour c is not always chosen to be the positive real axis. It will be specified differently from case to case.

From (8.1), we have

$$\frac{d}{dx} y(x) = \int_c -e^{-xt} tY(t) dt \tag{8.2}$$

and

$$xy(x) = \int_c - \left(\frac{\partial e^{-xt}}{\partial t} \right) Y(t) dt = e^{-xt} Y(t)|_a^b + \int_c e^{-xt} \frac{d}{dt} Y(t) dt,$$

where b and a are the endpoints of the contour c. We will choose the endpoints of the contour c so that

$$e^{-xt} Y(t)|_a^b = 0. \tag{8.3}$$

This is the only requirement we impose on the contour c. With this requirement satisfied, we have

$$xy(x) = \int_c e^{-xt} \frac{d}{dt} Y(t) dt. \tag{8.4}$$

Equation (8.2) states that if the transform of $Y(t)$ is $y(x)$, then the transform of $-tY(t)$ is dy/dx. We express this statement as

$$\frac{d}{dx} \to -t. \tag{8.5}$$

Similarly, (8.4) states that

$$x \to \frac{d}{dt}. \tag{8.6}$$

Substituting (8.1) into the Airy equation and making use of (8.5) and (8.6), we find that the Airy equation is satisfied provided that

$$t^2 Y(t) - Y'(t) = 0,$$

and that the contour c is so chosen that (8.3) is fulfilled.

The equation for Y is first-order, which makes it possible for us to solve it in a closed form. We find that the solution of this equation is

$$Y(t) = e^{t^3/3},$$

and hence

$$y(x) = \int_c e^{-xt+t^3/3} dt. \tag{8.7}$$

It remains to choose the contour c so that (8.3) is satisfied. Let us express the complex variable t in polar form $t = \rho e^{i\phi}$, where ρ and ϕ are the magnitude and the phase of t, respectively. Then

$$t^3 = \rho^3 e^{3i\phi},$$

and (8.3) is satisfied if the endpoints of the contour are at infinity ($\rho = \infty$) in the region satisfying $\cos 3\phi < 0$. This is because

$$|e^{t^3/3}| = e^{\rho^3 \cos 3\phi/3},$$

which vanishes at the infinity of such a region. There are three such regions:

$$\begin{aligned}&(a)\ \pi/6<\phi<\pi/2,\\&(b)\ 5\pi/6<\phi<7\pi/6,\\&(c)\ 3\pi/2<\phi<11\pi/6.\end{aligned}$$

These regions are illustrated as unshaded regions in the figure below.

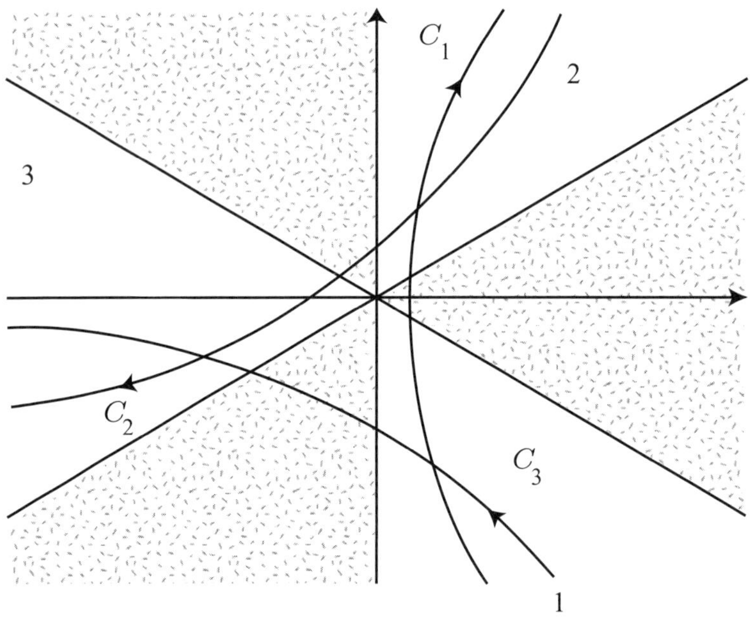

Figure 8.1

Thus both

$$y_1(x)=\int_{c_1}e^{-xt+t^3/3}dt$$

and

$$y_2(x)=\int_{c_2}e^{-xt+t^3/3}$$

satisfy the Airy equation, where c_1 and c_2 are drawn in the figure above. We may think of the contour c_1 as one going from the infinity of region 1 to that of region 2, and the contour c_2 as that going from the infinity of region 2 to that of region 3. The solutions y_1 and y_2 are independent. On the other hand,

$$y_3(x) = \int_{c_3} e^{-xt+t^3/3} dt$$

is equal to the sum $y_1 + y_2$. This is because the contour $c_1 + c_2$ is a contour starting at the infinity of region 1 passing through the infinity of region 2 and ending at the infinity of region 3. This contour can be deformed into c_3. Thus y_3 is not independent of y_1 and y_2.

When x is small, we may obtain the Maclaurin series of the Airy functions by expressing the factor e^{-xt} in the integrand of (8.7) as the series $\sum(-xt)^n/n!$. But when x is large, we need to evaluate asymptotically the integral (8.7) when x is large. We'll show how this is done later in this chapter.

This method of obtaining the solution of the Airy equation is due to Laplace; they can be used to solve other linear and homogeneous differential equations of any order, provided that the coefficients of this differential equation are linear functions of x. This is because

$$x \to \frac{d}{ds},$$

so the equation becomes first-order after the Laplace transform. Some differential equations not of this form can be cast into this form with a change of variable.

B. The Laplace Method

We begin with integrals of the form

$$I(\lambda) = \int_a^b e^{-\lambda v(x)} h(x) dx, \tag{8.8}$$

where $v(x)$ and $h(x)$ are independent of the parameter λ. The variable of integration x is real. We shall show how to find the asymptotic form of $I(\lambda)$ as $\lambda \to \infty$.

We note that the integrand in the integral above is maximum when v is minimum. Indeed, let x_0 be the point inside the interval $[a, b]$ at which $v(x)$ is minimum. Then the ratio of the integrand at the point x_0 and the integrand at any other point x inside $[a, b]$ is

$$\frac{e^{-\lambda v(x_0)} h(x_0)}{e^{-\lambda v(x)} h(x)} = e^{\lambda[v(x)-v(x_0]} \frac{h(x_0)}{h(x)}.$$

Since $v(x) - v(x_0)$ is positive, this ratio is exponentially large as λ goes to infinity. As a result, the magnitude of integrand at x_0 is by an exponential function of λ larger than that at any other point inside $[a, b]$. Therefore, to evaluate the integral of (8.8) approximately, it is crucial to approximate $v(x)$ and $h(x)$ accurately near x_0. This is because we need not be concerned with the integrand away from x_0, as the dominant contribution to the integral comes from a very small region near x_0.

In the above, we assume that $h(x_0)$ does not vanish. (We will show a little later that in the case $h(x_0) = 0$, the answer depends on how fast $h(x)$ approaches zero as x approaches x_0.) In a practical calculation, we often replace $v(x)$ and $h(x)$ by the first few terms of their respective Taylor series expanded around x_0, if such expansion series exist.

Problem for the Reader

Find the leading asymptotic term of the integral

$$I(\lambda) = \int_{-\infty}^{\infty} \exp(-\lambda \sinh^2 x) dx, \lambda \gg 1. \tag{8.9}$$

Solution

Comparing (8.9) with (8.8), we identify

$$v(x) = \sinh^2 x$$

and

$$h(x) = 1.$$

The function $\sinh^2 x$ is minimum at $x = 0$. As a result, the integrand is sharply peaked around the point $x = 0$. Thus, to obtain the asymptotic behavior of the integral, it suffices to approximate $\sinh^2 x$ and $h(x)$ accurately near the origin.

We have

$$\sinh^2 x \approx x^2,$$

while $h(x) = 1$ is already in the form of a Taylor series expanded around $x = 0$. We therefore have

$$I(\lambda) \approx \int_{-\infty}^{\infty} e^{-\lambda x^2} dx.$$

Let

$$\rho \equiv \sqrt{\lambda} x$$

and

$$dx = d\rho/\sqrt{\lambda}.$$

With this change of variable the exponent is independent of λ. Then we have

$$I(\lambda) \approx \frac{1}{\sqrt{\lambda}} \int_{-\infty}^{\infty} e^{-\rho^2} d\rho = \sqrt{\frac{\pi}{\lambda}}. \tag{8.10}$$

In the formulae above, we have made use of the fact that the Gaussian integral

$$\int_{-\infty}^{\infty} e^{-\rho^2} d\rho$$

is equal to $\sqrt{\pi}$.

Not all problems have closed-form solutions, and we often have to resort to finding an approximate solution for the problem we face. However, if one makes a mistake in an approximation, the answer one gets can be far off the mark. Thus it is often useful to have a quick answer before we embark on a long and detailed calculation. We shall show how to get a simple and rough estimate of the asymptotic form of the integral $I(\lambda)$ of (8.8).

To be specific, we shall use the example of the integral in (8.9). The integrand of $I(\lambda)$ falls off sharply when we move away from the point $x = 0$, and hence there is a sharp peak around the origin. Indeed, the peak becomes steeper and steeper, or the width of the peak becomes smaller and smaller, as λ becomes larger and larger. What is the width of the peak? We observe that the integrand drops by a factor of e when

$$\lambda x^2 = 1,$$

or

$$x = 1/\sqrt{\lambda}.$$

Thus $1/\sqrt{\lambda}$ is roughly the width of the region that gives the dominant contributions to the integral. Indeed, the width $1/\sqrt{\lambda}$ of the region of dominant contribution is the same factor relating dx with $d\rho$. We shall say that the scale of x is $1/\sqrt{\lambda}$.

The integral $I(\lambda)$ is roughly equal to its integrand at $x = 0$ times the width of the region of dominant contributions. Since the integrand at $x = 0$ is equal to unity and the width of the region of dominant contributions is of the order of $1/\sqrt{\lambda}$, $I(\lambda)$ is of the order of $1/\sqrt{\lambda}$. This produces the answer (8.10) up to a multiplicative constant. We express this estimate as

$$I(\lambda) = O\left(1/\sqrt{\lambda}\right).$$

For certain problems, such an estimate is already adequate for the purpose. If so, we are spared the chore of a tedious calculation. The factor missing in the estimate is the multiplicative constant $\sqrt{\pi}$, which is the value of the Gaussian integral.

Let us apply the Laplace method to another problem. As we know, the Laplace transform of a function $f(x)$ is

$$L(s) = \int_0^\infty f(x)e^{-sx}\,dx.$$

The integral above is in the form of the integral of (8.8) with λ equal to s, $v(x)$ equal to x, and $h(x)$ equal to $f(x)$. Therefore, the Laplace method enables us to obtain the asymptotic form of $L(s)$ when s is large.

Problem for the Reader

Find the asymptotic form of the Laplace transform of e^{-x^3}, which can be expressed as

$$I(\lambda) = \int_0^\infty e^{-\lambda x - x^3}\,dx, \lambda \gg 1. \tag{8.11}$$

Solution

Comparing (8.11) with (8.8), we find that

$$v(x) = x \text{ and } h(x) = e^{-x^3}.$$

The minimum of $v(x)$ occurs at $x = 0$. Thus we expand $v(x)$ and $h(x)$ near the origin. There is nothing we need to do with $v(x) = x$, which is already in the form of a Taylor series expanded around $x = 0$. Also, the Taylor series of $h(x)$ expanded around $x = 0$ is

$$h(x) = 1 - x^3 + \cdots.$$

Approximating $v(x)$ and $h(x)$ by the first terms of their Taylor series, we get

$$I(\lambda) \approx \int_0^\infty e^{-\lambda x} dx = 1/\lambda. \tag{8.12}$$

A rough estimate of this result can be obtained by noting that the integrand at $x = 0$ is equal to unity. Since the integrand drops by a factor of e as $\lambda x = 1$, the width of the region of dominant contributions is of the order of $1/\lambda$. Hence we have

$$I(\lambda) = O(1/\lambda), \tag{8.13}$$

which agrees with (8.12).

Comparing (8.10) with (8.12), we see that the powers of λ are different in these two cases. This is because for the integral of (8.9), we have $v'(x_0) = 0$. Thus

$$v(x) - v(x_0) = \frac{v''(x_0)}{2}(x - x_0)^2 + \cdots, \tag{8.14}$$

which is approximately a quadratic function of $(x - x_0)$. Thus the width of the region of dominant contributions for this integral is given by $\lambda(x - x_0)^2 = 1$, and the width of the region of dominant contributions is of the order of $\lambda^{-1/2}$. But for the integral of (8.11), $v'(x_0) \neq 0$, and we have

$$v(x) - v(x_0) = v'(x_0)(x - x_0) + \cdots,$$

which is approximately a linear function of $(x - x_0)$. Thus the width of the region of dominant contributions for this integral is determined by $\lambda|x - x_0| = 1$, and the width of the region of dominant contributions is of the order of λ^{-1}. We shall illustrate with more examples.

Problem for the Reader

Find the leading asymptotic term for the integral

$$I(\lambda) = \int_{-1}^{2} (1 + x^2) \exp(-\lambda x^6) dx. \tag{8.15}$$

Solution

The limits of integration of $I(\lambda)$ above are finite. We shall show that for the purpose of obtaining the asymptotic form of the integral above, we may replace the limits of integration above by $-\infty$ and ∞.

We put

$$I(\lambda) = I_1(\lambda) + I_2(\lambda) + I_3(\lambda),$$

where

$$I_1(\lambda) = \int_{-\infty}^{\infty} (1 + x^2) \exp(-\lambda x^6) dx,$$

$$I_2(\lambda) = -\int_{2}^{\infty} (1 + x^2) \exp(-\lambda x^6) dx,$$

and

$$I_3(\lambda) = -\int_{-\infty}^{-1} (1 + x^2) \exp(-\lambda x^6) dx.$$

For all three integrals, $v(x) = x^6$ and $h(x) = 1 + x^2$. They differ by their limits of integration.

Inside the interval $[2, \infty]$, $v(x)$ is minimum at $x = 2$. Thus the magnitude of the integrand of $I_2(\lambda)$ is maximum at $x = 2$, where it is equal to $5e^{-64\lambda}$. Since $v'(2) \neq 0$, the integral $I_2(\lambda)$ is of the order of $e^{-64\lambda}/\lambda$, which is exponentially small and negligible.

Inside the interval $[-\infty, -1]$, $v(x)$ is minimum at $x = -1$. Thus the magnitude of the integrand of $I_3(\lambda)$ is largest at $x = -1$. Since the integrand at $x = -1$ is of the order of $e^{-\lambda}$ and since $v'(-1) \neq 0$, the integral $I_3(\lambda)$ is of the order of $e^{-\lambda}/\lambda$, which is exponentially small and negligible.

Thus we may approximate $I(\lambda)$ by $I_1(\lambda)$. As we shall see, this makes it easy to calculate the asymptotic form of the integral.

Let us first make a rough estimate of the value of $I_1(\lambda)$. The function $v(x)$ in the region $[-\infty, \infty]$ is minimum at $x = 0$. At the origin, the integrand is equal to 1. The integrand drops by a factor e when

$$\lambda x^6 = 1,$$

or

$$x = \lambda^{-1/6}.$$

Thus the width of the dominant region of contributions is $\lambda^{-1/6}$. Therefore, the integral is of the order of $\lambda^{-1/6}$. We also have an alternative way to understand why we may replace the limits of integration by $-\infty$ and ∞. The dominant contributions come from a small neighborhood of the origin, and all other regions do not matter.

To determine precisely the asymptotic form of the integral, we note that the functions $v(x)$ and $h(x)$ are already in the form of Taylor series expanded around the origin. Let us express the integral as

$$I_1(\lambda) = \int_{-\infty}^{\infty} (1 + x^2) e^{-\lambda x^6} dx = 2 \int_0^{\infty} (1 + x^2) e^{-\lambda x^6} dx,$$

the last step being justified as the integrand is an even function of x. To evaluate this integral, we simplify the exponent by calling it t:

$$\lambda x^6 = t,$$

or

$$x = \lambda^{-1/6} t^{1/6}.$$

We get

$$I_1(\lambda) = \frac{1}{3\lambda^{1/6}} \int_0^{\infty} (1 + \lambda^{-1/3} t^{1/3}) e^{-t} t^{-5/6} dt,$$

which can be easily integrated to yield

$$I(\lambda) \approx I_1(\lambda) = \frac{\Gamma(1/6)}{3\lambda^{1/6}} + \frac{\Gamma(1/2)}{3\lambda^{1/2}}. \tag{8.16}$$

Note that we have obtained not only the leading term but also the next leading term, the latter being smaller than the former by a factor of $\lambda^{1/3}$. The terms neglected are by an exponential factor of λ smaller than the terms in (8.16).

We remark that if $v(x)$ has several local minima, only the absolute minimum counts. This is because the integrands at other local minima are exponentially smaller than that at the absolute minimum. Thus we may neglect the contributions from such local minima if we are not interested in terms exponentially smaller than the leading term.

We also note that the derivative of $v(x)$ does not necessarily vanish at the point of absolute minimum. An example is the integral of (8.11). In this example, $v'(x)$ is always positive, and hence $v(x)$ is monotonically increasing. Then the minimum point is the lower endpoint. At this point of minimum, $v'(x)$ does not vanish.

Thus we classify the integral $I(\lambda)$ of (8.8) into three different categories. Each of them is handled differently.

a. $v(x)$ is monotonically increasing inside $[a, b]$, as shown in the figure below.

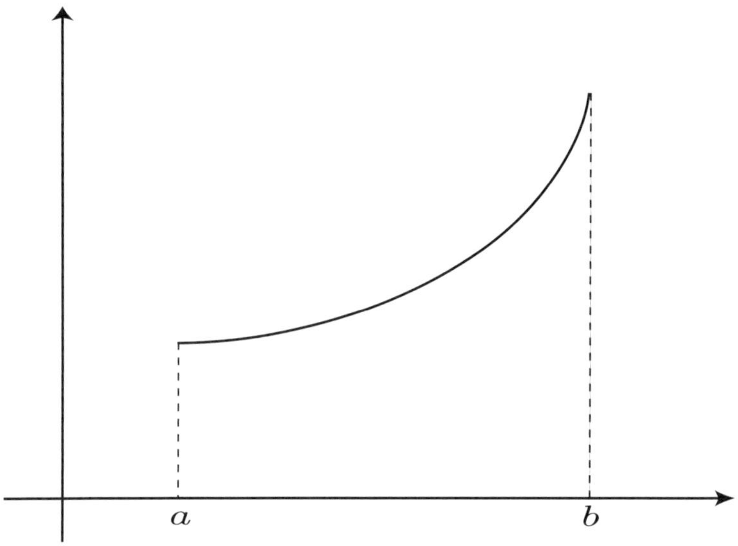

Figure 8.2

We see that $v(x)$ is minimum at the lower endpoint of the interval. In this case, we make the approximations

$$v(x) \approx v(a) + v'(a)(x - a) \tag{8.17a}$$

and

$$h(x) \approx h(a). \tag{8.17b}$$

By making the change of variable

$$t = \lambda v'(a)(x - a),$$

we get

$$I(\lambda) \approx \frac{e^{-\lambda v(a)} h(a)}{\lambda v'(a)} \int_0^\infty e^{-t} dt.$$

In the above, we replace the upper limit of integration by infinity. This does not affect the asymptotic form of the integral, as the dominant contribution comes from a small neighborhood near $t = 0$. Thus we have

$$I(\lambda) \approx \frac{e^{-\lambda v(a)} h(a)}{\lambda v'(a)}. \tag{8.18}$$

We see that the leading asymptotic term is in the form of an exponential function of λ divided by λ. Note that

$$e^{-\lambda v(a)} h(a)$$

is the value of the integrand at $x = a$ and the factor $1/\lambda$ is the width of the region of dominant contributions. We also note that $v'(a)$ is positive as $v(x)$ is an increasing function of x.

It may happen that $v'(x)$ vanishes at $x = a$. In this case one must represent $v(x) - v(a)$ by its first nonvanishing term other than the right side of (8.17a). Similarly $h(a)$ may vanish, and in this case we represent $h(x)$ by its lowest nonvanishing term in its expansion near a.

b. $v(x)$ is monotonically decreasing inside $[a, b]$; then $x_0 = b$. This is illustrated in the figure below.

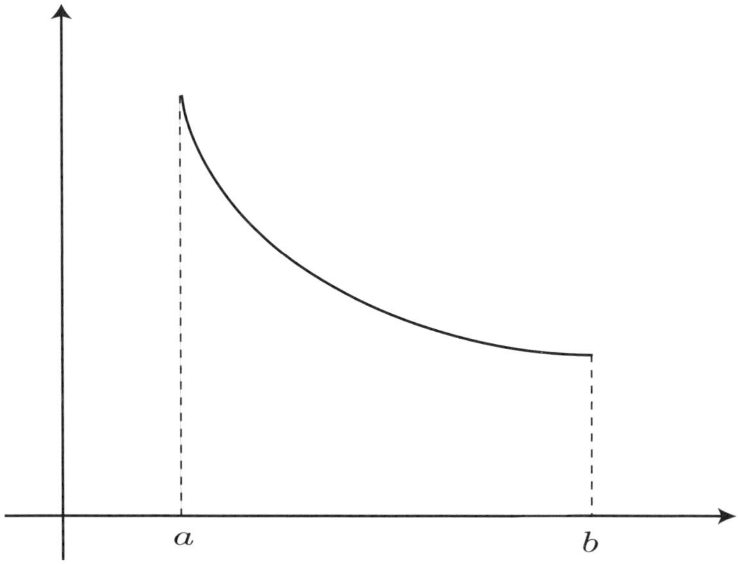

Figure 8.3

In this case, we make the approximations

$$v(x) - v(b) \approx v'(b)(x - b) \tag{8.19a}$$

and

$$h(x) \approx h(b). \tag{8.19b}$$

We then make the change of variable

$$t = -\lambda v'(b)(x - b)$$

(note that $v'(b)$ is negative) and get

$$I(\lambda) \approx -\frac{e^{-\lambda v(b)}h(b)}{\lambda v'(b)} \int_{-\infty}^{0} e^t dt = -\frac{e^{-\lambda v(b)}h(b)}{\lambda v'(b)}. \tag{8.20}$$

Again,

$$e^{-\lambda v(b)}h(b)$$

is the value of the integrand at $x = a$, and the factor $1/\lambda$ is the width of the region of dominant contributions. If $v'(b)$ or $h(b)$ or both vanish, we replace the right side of (8.19a) and (8.19b) by the respective lowest-order nonvanishing terms.

c. $v(x)$ is minimum at an interior point x_0 of the interval $[a, b]$, as shown in the figure below.

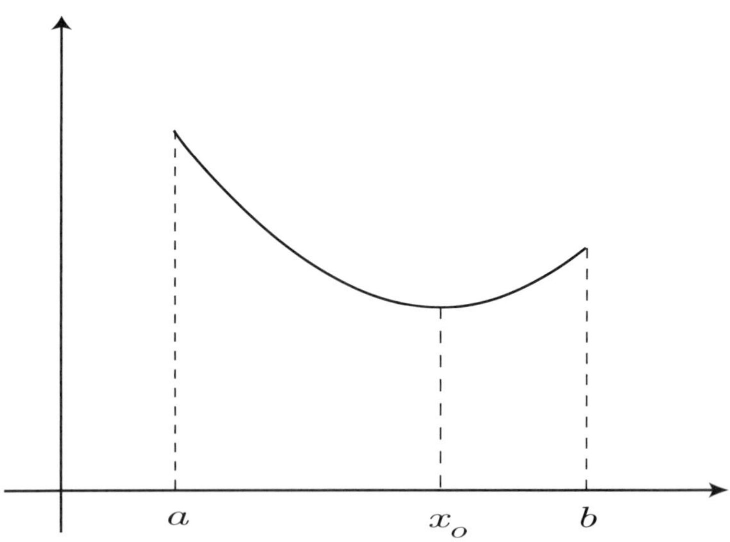

Figure 8.4

If

$$v'(x_0) = 0$$

and

$$v''(x_0) > 0,$$

we make the approximations

$$v(x) \approx v(x_0) + \frac{v''(x_0)}{2!}(x - x_0)^2 + \cdots \tag{8.21a}$$

and

$$h(x) \approx h(x_0). \tag{8.21b}$$

Calling

$$t^2 = \frac{\lambda}{2} v''(x_0)(x - x_0)^2,$$

we get

$$I(\lambda) \approx \frac{e^{-\lambda v(x_0)} h(x_0)}{\sqrt{\frac{\lambda}{2} v''(x_0)}} \int_{-\infty}^{\infty} e^{-t^2} dt = \sqrt{\frac{2\pi}{\lambda v''(x_0)}} e^{-\lambda v(x_0)} h(x_0). \tag{8.22}$$

Note that the right side of (8.22) is proportional to the value of the integrand at x_0 times $1/\sqrt{\lambda}$, which is the width of the dominant region of contributions. If $v''(x_0)$ or $h(x_0)$ vanishes, we represent $v(x) - v(x_0)$ or $h(x)$ by its first nonvanishing term, respectively, before we carry out the integration. The result is then different from (8.22).

Problem for the Reader

Give an estimate of each of the leading terms of the following integrals:

1. $\displaystyle\int_{-1}^{4} e^{-\lambda x^3}(1 + x^4)dx.$

2. $\displaystyle\int_{1}^{\infty} \sqrt{x - 1}e^{-\lambda \cosh x} dx.$

3. $\displaystyle\int_{0}^{2} e^{\lambda x(1-x)} dx.$

Solution

For the first integral,

$$v(x) = x^3$$

is minimum at the lower endpoint -1. The value of the integrand at this point is $2e^{\lambda}$. As x_0 is an endpoint minimum with $v'(x_o) \neq 0$, the width of the dominant region of integration is of the order of $1/\lambda$.
Thus the integral is of the order of

$$2e^{\lambda}/\lambda.$$

Note that the integrand at $x = -1$ blows up exponentially as $\lambda \to \infty$. For this reason, the integral is exponentially large when λ is large.

For the second integral, $x_0 = 1$. As the leading term comes from an endpoint contribution with $v'(x_o) \neq 0$, the width of the dominant region of integration is again $1/\lambda$. This means $(x-1)$ is of the order of $1/\lambda$ in the region of dominant contribution. As a result, the factor $\sqrt{x-1}$ in the integrand is of the order of $\sqrt{1/\lambda}$. Thus the leading term of this integral is of the order of $e^{-\lambda}\lambda^{-3/2}$, which is exponentially small. We note that since $h(x) = \sqrt{x-1}$ vanishes at $x = 1$, it contributes a vanishing factor $1/\sqrt{\lambda}$ to the integral.

More generally, consider the case in which $h(x)$ vanishes like $(x - x_0)^n$ near x_0. If $v'(x)$ does not vanish at x_0, then $h(x)$ contributes a factor λ^{-n} to the integral. But if $v'(x_0) = 0$ and $v''(x_0) \neq 0$, then $(x - x_0)$ is of the order of $\lambda^{-1/2}$. Thus $h(x)$ contributes a factor $\lambda^{-n/2}$ to the integral. Note that in both cases, the vanishing of $h(x)$ at the minimum of $v(x)$ makes the asymptotic forms of $I(\lambda)$ smaller, as we may expect.

For the third integral, the dominant contribution comes from the interior point $x_0 = 1/2$ with $v'(1/2) = 0$ and $v''(1/2) > 0$. Thus the integral is of the order of

$$e^{\lambda/4}/\sqrt{\lambda},$$

which is exponentially large when λ is large because the integrand is exponentially large at $x = 1/2$.

Next we discuss how to find the next-order term, and indeed the entire asymptotic series of the integral. The principle to do this is simple: We just make successive approximations of the integrand.

Problem for the Reader

Find the entire asymptotic series for the integral given by (8.11).

Solution

The width of the region of dominant contributions near $x = 0$ is of the order of $1/\lambda$. We shall say that the scale of x is $1/\lambda$. We put

$$\lambda x = t,$$

and we have

$$I(\lambda) = \frac{1}{\lambda}\int_0^\infty e^{-t}e^{-t^3/\lambda^3}dt.$$

Note that the scale of t is unity. This makes it easy to see that t^3/λ^3 is of the order of $1/\lambda^3$. Therefore, we may expand e^{-t^3/λ^3} into a Taylor series. We have

$$I(\lambda) = \frac{1}{\lambda}\int_0^\infty e^{-t}\left(1 - \frac{t^3}{\lambda^3} + \cdots + \frac{(-1)^n t^{3n}}{\lambda^{3n} n!} + \cdots\right)dt.$$

Strictly speaking, we should sum up the infinite series before we carry out the integration. This is because changing the order of summation and integration is sometimes illegitimate, which is the case for the present problem. As a result, the series we shall obtain is a divergent asymptotic series. With this in mind we shall carry out the integration first and sum up the series later. We get

$$I(\lambda) = \frac{1}{\lambda}\left(1 - \frac{3!}{\lambda^3} + \cdots + \frac{(-1)^n(3n)!}{\lambda^{3n}n!} + \cdots\right). \tag{8.23}$$

While the sum in (8.23) is divergent for all λ, it can be proved that it is useful as an asymptotic series. We also note that each term in this series is smaller than the preceding term by a factor $1/\lambda^3$. Thus when λ is very large the nth term is smaller than the $(n-1)$th term by a factor $1/\lambda^3$. For example, if $\lambda = 10$, approximating $I(\lambda)$ by $1/\lambda$ is accurate to about 0.6 %.

Problem for the Reader

Find the entire asymptotic series for the integral of (8.9).

Solution

We may express $\sinh^2 x$ as a Taylor series expanded around the origin, but since $\sinh^2 x$ is the exponent of the integrand, we then need to expand the exponential function of an infinite series. Such an expansion is difficult to handle.

Instead, let us simplify the exponent and define

$$\lambda \sinh^2 x = t^2.$$

Then

$$\sinh x = \frac{t}{\sqrt{\lambda}}.$$

We have

$$dx = \frac{dt}{\sqrt{\lambda}\cosh x} = \frac{dt}{\sqrt{\lambda(1+t^2/\lambda)}}.$$

The integrand $e^{-\lambda \sinh^2 x}$ is now simply equal to e^{-t^2}, which drops by a factor e as t changes by one unit. This means that the scale of t is unity. Now we merely need to expand $(1+t^2/\lambda)^{-1/2}$ into a binomial series. We have

$$\begin{aligned} I(\lambda) &= \int_{-\infty}^{\infty} e^{-t^2} \frac{dt}{\sqrt{\lambda(1+t^2/\lambda)}} \\ &= \frac{1}{\sqrt{\lambda}} \int_{-\infty}^{\infty} e^{-t^2} \sum_{n=0}^{\infty} \frac{\Gamma(n+1/2)}{\Gamma(1/2)n!} \left(-\frac{t^2}{\lambda}\right)^n . \end{aligned} \tag{8.24}$$

Again, it is actually not legitimate to integrate term by term. But we will do it and get

$$I(\lambda) = \frac{1}{\sqrt{\lambda\pi}} \sum_{n=0}^{\infty} \frac{\Gamma^2(n+1/2)}{n!} \left(-\frac{1}{\lambda}\right)^n . \tag{8.25}$$

The series of (8.25) diverges for all λ, although $I(\lambda)$ is a convergent integral. This divergence is due to the illegitimacy of changing the order of summation and integration. Still, the series of (8.25) is useful as an asymptotic series. (See homework problem 4.)

In some cases, the function $h(x)$ in (8.8) vanishes exponentially at the point where $e^{-\lambda v(x)}$ is maximum. In such cases, a somewhat modified approach is called for.

Problem for the Reader

Find the first leading term for the Gamma function

$$\Gamma(\lambda+1)=\int_0^\infty e^{-x}x^\lambda dx, \lambda \gg 1. \tag{8.26}$$

Solution

Let us rewrite (8.26) as

$$\Gamma(\lambda+1)=\int_0^\infty e^{\lambda \ln x - x}dx.$$

Comparing the integrand above with the integrand in (8.8), we identify

$$v(x)=-\ln x$$

and

$$h(x)=e^{-x}.$$

Now $v(x)$ is minimum at $x=\infty$ with

$$v(\infty)=-\infty;$$

hence $e^{-\lambda v(\infty)}$ is infinite. Thus the maximum value of the factor $e^{-\lambda v(x)}$ in the integrand is infinity. But there is an additional factor e^{-x} in the integrand. This factor vanishes at $x=\infty$. Thus at $x=\infty$, the integrand is equal to the product of infinity times zero. As we know, the product of e^{-x} and x^λ actually goes to zero as x goes to infinity, no matter how large λ is. Thus the integrand is equal to zero at $x=\infty$, and is not maximum there.

This means that we must take care of both factors in the integrand in order to determine the point at which the integrand is maximum. Let us put

$$f(x)\equiv x-\lambda \ln x;$$

then the integrand is equal to $e^{-f(x)}$, and is maximum when $f(x)$ is minimum. The minimum of $f(x)$ occurs at the point

$$f'(x)=1-\lambda/x=0,$$

or

$$x=\lambda.$$

We note that unlike in the cases we have considered so far, x_0, the point of minimum of v in the present problem, depends on λ. While x_0 does go to infinity as λ goes to infinity, the asymptotic form of $\Gamma(\lambda+1)$ cannot be obtained by simply setting x_0 to infinity.

Let us make a change of variable

$$x = \lambda t;$$

then the maximum of the integrand occurs at

$$t = 1,$$

independent of λ. After making the change of variable, we have

$$\Gamma(\lambda+1) = \lambda^{\lambda+1} \int_0^\infty \exp\left[-\lambda(t - \ln t)\right] dt. \tag{8.27}$$

The integral above is in the form of (8.8), with

$$v(t) = t - \ln t, h(t) = 1.$$

We have

$$v'(t) = 1 - 1/t$$

and

$$v'(1) = 0.$$

We also have

$$v''(t) = 1/t^2$$

and

$$v''(1) = 1.$$

Using (8.22), we get

$$\Gamma(\lambda+1) \approx \sqrt{2\pi}\lambda^{\lambda+1/2}e^{-\lambda}, \tag{8.28}$$

which is known as the Stirling formula.

The problem above demonstrates that the designation of $v(x)$ and $h(x)$ sometimes requires a bit more care. Generally, when $h(x)$ vanishes exponentially at x_0, the minimum of $v(x)$, we cannot ignore $h(x)$ in our search of the point of maximum for the integrand. In such a case, the location of this point usually depends on λ. While it is not necessary, it is often convenient to make a change of the variable of integration to scale out this dependence on λ.

Take the example of

$$I(\lambda) = \int_0^\infty e^{-\lambda x - 1/x^2}\,dx.$$

If we identify $v(x)$ as x and $h(x)$ as e^{-1/x^2}, the minimum of $v(x)$ is at the lower endpoint $x = 0$. But $h(x)$ vanishes exponentially at $x = 0$. Thus the integrand vanishes at $x = 0$, and the magnitude of the integrand is certainly not maximum at this point. Indeed, by differentiating the exponent of the integrand, we find that the maximum of the integrand is at

$$\lambda - 2/x^3 = 0,$$

or

$$x_0 = (2/\lambda)^{1/3},$$

not at $x = 0$. When λ is very large, x_0 is indeed very small. Still, we must not replace x_0 by zero.

We make the change of variable

$$x = (2/\lambda)^{1/3} t.$$

Then we have

$$I(\lambda) = (2/\lambda)^{1/3} \int_0^\infty e^{-\Lambda\left(t + \frac{1}{2}t^{-2}\right)}\,dt,$$

where

$$\Lambda = 2^{1/3}\lambda^{2/3}.$$

The point where the integrand is maximum is at

$$t = 1,$$

independent of λ. The answer is easily found with the use of (8.22) . We get

$$I(\lambda) \approx 2^{1/6}\lambda^{-2/3}\sqrt{\pi}e^{-3(\lambda/2)^{2/3}}.$$

There are also integrals with a large parameter that are not of the form (8.8). Some of them can be transformed into such a form by a change of variable. We give one example below.

Problem for the Reader

Find the entire asymptotic series for

$$I(\lambda) = \int_{\lambda}^{\infty} e^{-t^2} dt, \lambda \gg 1. \tag{8.29}$$

Solution

In (8.29), the large parameter λ appears in the lower limit of integration rather than the integrand. We note that the integrand is maximum at the lower limit of integration $t = \lambda$, where it is equal to $e^{-\lambda^2}$, a very small number. Also, the integrand drops exponentially when we move away from $t = \lambda$. Thus we expect that the dominant region of integration is in a small neighborhood near $t = \lambda$. Expanding the exponent t^2 around the lower endpoint $t = \lambda$ in a Taylor series, we see that the width of this neighborhood is inversely proportional to the derivative of t^2 at λ. Therefore, we estimate the integral given by (8.29) to be of the order of $e^{-\lambda^2}/\lambda$.

Let us determine the leading term of the integral above more precisely. One way to do this is to use integration by parts. We write

$$I(\lambda) = -\int_{\lambda}^{\infty} \frac{1}{2t} d(e^{-t^2}).$$

Performing integration by parts, we get

$$I(\lambda) = \frac{e^{-\lambda^2}}{2\lambda} - \int_{\lambda}^{\infty} \frac{1}{2t^2} e^{-t^2} dt. \tag{8.30}$$

What we have gained by performing an integration by parts is that since t is at least as large as λ, the integrand in (8.30) is smaller than that in (8.29) by at least a factor of λ^2. Indeed, we easily estimate the integral in (8.30) to be of the order of $e^{-\lambda^2}/\lambda^3$. Thus the leading term of $I(\lambda)$ is the first term on the right side of (8.30), and we have

$$I(\lambda) \approx \frac{e^{-\lambda^2}}{2\lambda}.$$

By applying integration by parts again to the integral of (8.30), we obtain the next leading term of the integral in (8.29). Indeed the entire asymptotic series for the integral of (8.29) can be obtained by succesive applications of integration by parts.

Alternately, we may define

$$x = t/\lambda.$$

Then the integral in (8.29) becomes

$$I(\lambda) = \lambda \int_1^\infty e^{-\Lambda x^2} dx,$$

where

$$\Lambda = \lambda^2.$$

We may apply the Laplace method to obtain the asymptotic series of $I(\lambda)$. In the interval $[1, \infty]$, x^2 is minimum at $x = 1$. At $x = 1$, the exponent of the integrand is equal to Λ. Thus we put

$$\Lambda x^2 - \Lambda = \rho,$$

or

$$x = \sqrt{1 + \rho/\Lambda}.$$

With such a change of variable, the exponent is simplified and the maximum of the integrand occurs at $\rho = 0$. We get

$$I(\lambda) = \frac{e^{-\lambda^2}}{2\lambda} \int_0^\infty e^{-\rho}(1 + \rho/\Lambda)^{-1/2} d\rho.$$

Since the scale of ρ is unity, ρ/λ is small and we may express $(1 + \rho/\Lambda)^{-1/2}$ by its binomial expansion. It is then easy to find that

$$\begin{aligned} I(\lambda) &= \frac{e^{-\lambda^2}}{2\lambda} \int_0^\infty e^{-\rho} \sum_{n=0}^{\infty} (-1)^n \frac{\Gamma(n + 1/2)}{\Gamma(1/2) n!} (\frac{\rho}{\Lambda})^n d\rho \\ &= \frac{e^{-\lambda^2}}{2\lambda\sqrt{\pi}} \sum_{n=0}^{\infty} \Gamma(n + 1/2) \frac{(-1)^n}{\lambda^{2n}}. \end{aligned}$$

Sometimes it is not possible to obtain the entire asymptotic series. But we may find successive terms in the asymptotic series by successively approximating the integrand. Let's look at an example.

Problem for the Reader

Find the next leading term for the Gamma function $\Gamma(\lambda + 1)$ for $\lambda \gg 1$.

Solution

We use the expression in eq. (8.27) for the Gamma function. The maximum of the integrand is at $t = 1$. We simplify the algebra somewhat by making the change of variable

$$s = t - 1.$$

The maximum of the integrand is at $s = 0$. Thus we expand the exponent of the integrand in (8.27) around $s = 0$ and get

$$v(s) = s - \ln(1 + s) = s^2/2 - s^3/3 + s^4/4 + \cdots .$$

The Gamma function is

$$\Gamma(\lambda + 1) = \lambda^{\lambda+1} e^{-\lambda} \int_{-1}^{\infty} \exp\left[-\lambda(s^2/2 - s^3/3 + s^4/4 + \cdots)\right] ds.$$

We estimate the scale of the region of dominant contributions by setting

$$\lambda s^2/2 = 1.$$

Thus the scale of s is $(2/\lambda)^{1/2}$. We shall call

$$u = \sqrt{\frac{\lambda}{2}} s.$$

We get

$$\Gamma(\lambda + 1) \approx \sqrt{2} \lambda^{\lambda+1/2} e^{-\lambda} \int_{-\infty}^{\infty} e^{-u^2} \exp\left(\frac{2}{3}\sqrt{\frac{2}{\lambda}} u^3 - \frac{u^4}{\lambda} - \cdots\right) du.$$

The only approximation we have made so far is replacing the lower limit of integration $-\sqrt{\lambda/2}$ by $-\infty$. This means that we have neglected terms that are exponentially smaller than the leading terms.

Since $u = O(1)$, it is easy to estimate the order of magnitude of each of the terms in the exponent. For example, the term

$$\frac{2}{3}\sqrt{\frac{2}{\lambda}} u^3$$

in the exponent is $O(1/\sqrt{\lambda})$, while the term

$$-\frac{u^4}{\lambda}$$

in the exponent is $O(1/\lambda)$. We may therefore make the expansion

$$\exp\left(\frac{2}{3}\sqrt{\frac{2}{\lambda}}u^3 - \frac{u^4}{\lambda} - \cdots\right) = 1 + \frac{2}{3}\sqrt{\frac{2}{\lambda}}u^3 - \frac{1}{\lambda}\left(u^4 - \frac{4u^6}{9}\right) + \cdots, \quad (8.31)$$

where the terms not exhibited are at least as small as $1/\lambda^{3/2}$. Substituting (8.31) into the integral above for $\Gamma(\lambda+1)$, we find that since

$$\int_{-\infty}^{\infty} e^{-u^2} u^3 du = 0,$$

we have

$$\Gamma(\lambda+1) \approx \sqrt{2\pi}\lambda^{\lambda+1/2}e^{-\lambda}\left(1 + \frac{1}{12\lambda} + \cdots\right). \quad (8.32)$$

We remark that while $1/(12\lambda)$ is small compared to unity when λ is large, the first-order correction of the Gamma function, being $\sqrt{2\pi}\lambda^{\lambda+1/2}e^{-\lambda}/(12\lambda)$, is itself a very large number.

C. Method of Stationary Phase

In 1911, E. Rutherford performed an experiment in which α-particles were made to pass through a thin metal foil. He measured the α-particles emerging from the foil at an angle, and found that the rate for such events, while very low, was much higher than what had been expected. More precise experiments performed later showed that with the scattering angle at 90 degrees, the rate of such events is 1 event out of 8000 events. While such an event is rare, it is at least by a factor 10^{10} more frequent than what had been predicted by J. J. Thomson's atomic model. Rutherford concluded that this means that an atom has a nucleus that is much smaller than even the tiny atom. His discovery ushered in the nuclear age. Scientists have since found that if we imagine an atom to be of the size of the moon, then in proportion a moderate-size nucleus would have about the volume of a school gymnasium. Rutherford's experiment illustrates that it is not enough to know that an event is rare. We must also know how rare it is.

While the scattering of α-particles from the atoms in the metal foil obeys the laws of quantum mechanics, the phenomenon Rutherford observed falls within a general pattern of wave interference. Consider, for example, a monochromatic wave with the amplitude

$$e^{i\vec{k}\cdot\vec{x}}$$

passing through two slits in a wall, arriving at a point on a screen behind the wall. We illustrate this in the figure below.

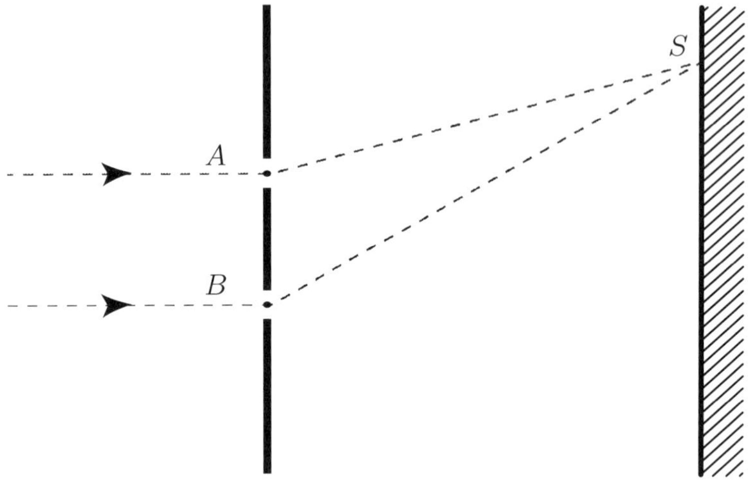

Figure 8.5

The phase of the ray arriving at the screen passing through the first slit is different from that passing through the second slit, as the lengths of the two ray paths designated by AS and BS are different. The amplitude of the wave at the wall is equal to the sum of the amplitudes of the two rays, which interfere with each other.

If there are more than two slits, then there will be more rays interfering with one another. When we add up a large number of waves with the same frequency but different phases, the interference of the waves tends to cancel out. This is especially so as the frequency becomes very high.

To see this, let us consider the example of the integral

$$\int_0^b e^{i\lambda x} dx.$$

This integral is a superposition of waves of the same wave number λ but with a phase dependent on x. The magnitude of a wave is equal to unity when $0 < x < b$, and zero otherwise.

We easily carry out the integration and get

$$\int_0^b e^{i\lambda x} dx = \frac{e^{i\lambda b} - 1}{i\lambda}. \tag{8.33}$$

As we see from (8.33), this integral vanishes like $1/\lambda$ when λ is large. Since the absolute value of the integrand for $0 < x < b$ is unity, there are cancellations among the waves.

To understand how the cancellations come about, we observe that $e^{i\lambda x}$ is a rapidly varying function of x when λ is large. Indeed, as x varies by the small amount $1/\lambda$, the phase of $e^{i\lambda x}$ varies by the amount unity. Let us change the scale of the integration variable and define

$$t = \lambda x.$$

As x varies by the small amount $1/\lambda$, t varies by the amount unity. The integral in (8.33) becomes

$$I(\lambda) = \frac{1}{\lambda}\int_0^{b\lambda} e^{it}dt. \tag{8.34}$$

While the integrand of (8.33), considered as a function of x, is of the scale of $1/\lambda$, the integrand of (8.34), considered as a function of t, is of the scale of unity, and is not a rapidly varying function. This is like looking at a photograph with tiny structures through a magnifier, and the small-scale variations become variations of normal scales. The factor $1/\lambda$ multiplying the integral in (8.34) is due to the fact that dx is smaller than dt by a multiplicative factor λ. If we use t as the integration variable, the region of integration is from 0 to $b\lambda$, which is a very large interval. Yet as we integrate from $t = 0$ to $t = 2\pi n$, where n is any integer, we get exactly zero. This is because both the real part and the imaginary part of the function e^{it} are positive half of the time and negative half of the time, and the contributions of the waves integrated over a complete cycle of width 2π exactly cancel one another. Thus the region of uncancelled contributions in the t-variable is the region near the endpoint $t = \lambda b$ with a width less than 2π. If we put down the magnifier, we see that the uncancelled region in the x-variable is a region near the endpoint $x = b$ with a width less than $2\pi/\lambda$. That is why the integral of (8.33) is of the order of $1/\lambda$.

Now let us change the lower limit of integration from zero to a and consider the integral

$$\int_a^b e^{i\lambda x}dx.$$

It is clear that uncancelled contributions to the integral come from two small regions with widths of the order of $1/\lambda$ near the endpoints. The integral above can be exactly calculated. We have

$$\int_a^b e^{i\lambda x}dx = \frac{e^{i\lambda b} - e^{i\lambda a}}{i\lambda}. \tag{8.35}$$

Note that $e^{i\lambda b}$ and $e^{i\lambda a}$ are the values of the integrand at the endpoints b and a, respectively.

Now consider more generally the integral

$$I(\lambda) = \int_a^b e^{i\lambda u(x)} h(x) dx. \tag{8.36}$$

In (8.36), the phase of the integrand is $\lambda u(x)$ rather than λx, and the integrand has an additional factor $h(x)$.

Since

$$|e^{i\lambda u(x)}| = 1,$$

the absolute value of the integrand of (8.36) is of the order of unity, no matter how big or how small $u(x)$ is. This is in contrast to the integrand in (8.8), the absolute value of which is an exponential function of λ, and is the largest when $v(x)$ is the smallest.

From the example given above, we expect that the uncancelled contributions come from the endpoints a and b only. This will be proven more rigorously in the next section. We will therefore make the following approximations for the integrand in (8.36). We put, near the endpoint $x = b$,

$$u(x) - u(b) \approx u'(b)(x - b) \tag{8.37a}$$

and

$$h(x) \approx h(b). \tag{8.37b}$$

Similarly, we put, near the endpoint $x = a$,

$$u(x) - u(a) \approx u'(a)(x - a) \tag{8.38a}$$

and

$$h(x) \approx h(a). \tag{8.38b}$$

As we have seen, the dominant region of integration near an endpoint is of a width of the order of $1/\lambda$, provided that $v'(x)$ does not vanish at this endpoint. Thus the approximations made above for u and h are accurate provided that they are good approximations at the distance of $O(1/\lambda)$ from the points of expansion. This is to say that the functions $u(x)$ and $h(x)$ must not vary appreciably as x varies by an amount $1/\lambda$. In other words, for the cancellations to take place, the scale of the functions $u(x)$ and $h(x)$ must be much larger than $1/\lambda$.

Let us illustrate this point with the physical example of α-particle scattering. The size of an atom is of the order of 10^{-8} cm. The wavelength of the α-particles is inversely proportional to the wave number λ. In Rutherford's experiment, the wavelength of the α-particle is in the range of 10^{-12} cm. Since the scale of the atom is about 10^4 times larger than the scale of the wavelength of the α-particles, there is a large number of cancellations among rays that are scattered from various parts of the atom. On the other hand, the radius of the nucleus of the hydrogen atom is of the order of 10^{-13} cm, which is smaller than 10^{-12} cm. Thus Rutherford would not be able to see the inner structure of a nucleus. Today, billions of dollars are being spent to build accelerators producing particles with wavelengths as small as 10^{-17} cm, so that scientists may probe into the world of very small dimensions and get a look at the fundamental physical laws that reign at such small distances.

If λ is so large that (8.37) and (8.38) are good approximations, then, referring to (8.35), we conclude that the contributions to the integral of (8.36) by the endpoints are given by

$$\frac{e^{i\lambda u(b)}h(b)}{i\lambda u'(b)} - \frac{e^{i\lambda u(a)}h(a)}{i\lambda u'(a)}, \tag{8.39}$$

which vanishes as fast as λ^{-1} when λ goes to infinity. This result is consistent with the Riemann-Lebesgue lemma, which states that the integral $\int_a^b e^{i\lambda x}h(x)dx$ vanishes as λ goes to infinity provided that $h(x)$ satisfies some condition of integrability. The Riemann-Lebesgue lemma is a rigorous mathematical result; a rough outline of a proof of this lemma is as follows. We divide the interval of integration into a number of very fine subintervals. Over each of the subintervals we will approximate $h(x)$ by a constant, the errors in doing this being arbitrarily small as each of the subdivisions is made arbitrarily fine. Since, by (8.35), the integral over each subinterval vanishes as λ goes to infinity, so does the sum over all subintervals.

While it is comforting that there is a rigorous lemma to confirm the vanishing of integrals of this kind, this lemma does not tell us how fast the integral goes to zero as λ becomes very large. The Riemann-Lebesgue lemma does not give us the definite answer because there is no definite answer until the functions u and h are specifically given. Indeed, the asymptotic form of an integral is not always given by (8.39). Take the example of the integral

$$I(\lambda) = \int_{-\infty}^{\infty} e^{i\lambda x^2}dx, \lambda \gg 1. \tag{8.40}$$

In this example, $u(x) = x^2$ and $h(x) = 1$, and the endpoints are $\pm\infty$, i.e., there are no finite endpoints. As we vary the x-coordinate by Δx, the phase varies by the amount

$$2\lambda x\Delta x. \tag{8.41}$$

Thus, if Δx is equal to the small amount $1/\lambda$, the phase λx^2 changes by the amount $2x$.This is to say that the phase of the integrand can change appreciably as x changes by as little as $1/\lambda$. A sharp reader will probably detect that this statement is not true at $x = 0$. This is because $u'(0) = 0$ and at $x = 0$ the phase does not change at all. We say that the phase is stationary at the origin. Actually, the phase λx^2 does vary somewhat. Indeed, it goes from zero to unity as x varies from zero to $1/\sqrt{\lambda}$. Thus the scale of the function λx^2 near the origin is of the order of $1/\sqrt{\lambda}$, somewhat bigger than the scale $1/\lambda$ at $x \neq 0$.

Let us put $t = \sqrt{\lambda}x$. Then the integral in (8.40) becomes

$$I(\lambda) = \frac{1}{\sqrt{\lambda}} \int_{-\infty}^{\infty} e^{it^2} dt. \tag{8.42}$$

Just like the integrand of (8.34), the integrand of (8.42) is not rapidly varying. But unlike with the integrand of (8.34), there is no complete cycle the integration over which gives zero. Still, the integral is convergent, hence the region of integration contributing to the integral covers all finite values of t. Such values are $t = 0(1)$, or eqivalently, $x = 0(\frac{1}{\sqrt{\lambda}})$. To illustrate this, we plot below the function $\sin(\pi\rho^2)$, which is the imaginary part of the integrand in (8.42) with $t = \sqrt{\pi}\rho$.

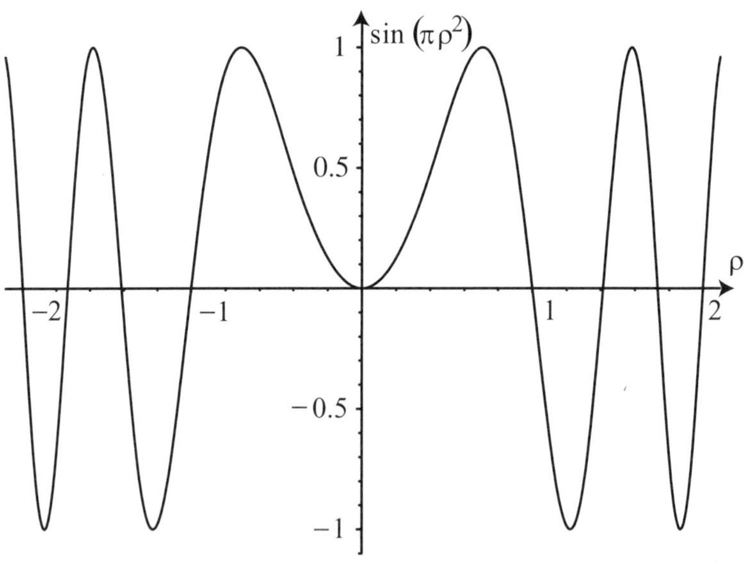

Figure 8.6

We see from the figure above that the function $\sin(\pi\rho^2)$ varies between the values of -1 and 1. However, the area under the curve in the interval from $\rho = 0$ to $\rho = 1$, where $\sin(\pi\rho^2)$ is positive, is the largest as the width of the interval is the largest. The area under the curve in the interval from $\rho = 1$ to $\rho = 1.414\cdots$, where $\sin(\pi\rho^2)$ is negative, is smaller as the width of the interval is smaller. As ρ increases, the function $\sin(\pi\rho^2)$ keeps on alternating between -1 and 1, but each of the intervals in which $\sin(\pi\rho^2)$ is of the same sign keeps on shrinking. Thus the integral is an infinite sum of terms with alternating signs and with magnitudes decreasing to zero. The sum of these terms is finite, and the integral is convergent.

The total sum can be calculated exactly. This is because the integral in (8.42) is related to the Gaussian integral by a deformation of the contour of integration. To see this, we may, instead of taking the contour from $-L$ to L (where L will be taken to be infinity at the end) along the real axis, take the contour with two vertical lines and a straight line making an angle $\pi/4$ with respect to the real axis, as illustrated in the figure below.

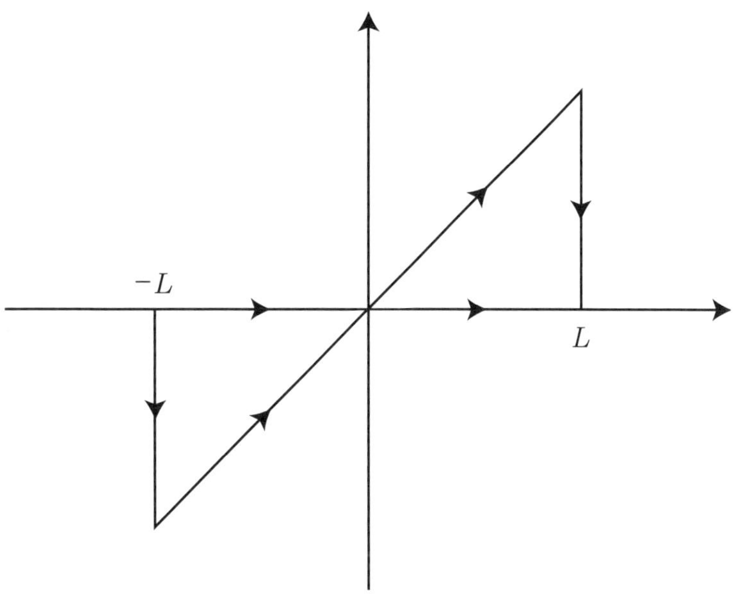

Figure 8.7

When t is complex, we express it in the polar form $t = re^{i\theta}$. Hence

$$t^2 = r^2 e^{i2\theta} = r^2(\cos 2\theta + i \sin 2\theta).$$

Therefore, the integrand is

$$e^{it^2} = e^{ir^2 \cos 2\theta} e^{-r^2 \sin 2\theta},$$

which vanishes as $r \to \infty$ provided that $\sin 2\theta$ is positive. As a result, the contributions from the two vertical lines vanish as L goes to infinity. Thus we have justified using the line $x = y$ instead of the real axis as the contour of integration. (See Appendix B of this chapter for more details.) On the line $x = y$, we have

$$t = \rho e^{i\pi/4};$$

therefore

$$\int_{-\infty}^{\infty} e^{it^2}\, dt = e^{i\pi/4} \int_{-\infty}^{\infty} e^{-\rho^2}\, d\rho = e^{i\pi/4}\sqrt{\pi}.$$

Note that $\pi/4$, the phase of the value of the integral $\int_{-\infty}^{\infty} e^{it^2}\, dt$, is the angle the deformed contour $x = y$ makes with the real axis. Mathematically, it comes about as $t = \rho e^{i\pi/4}$ on this deformed contour.

From the above, we easily find that

$$\int_{-\infty}^{\infty} e^{i\lambda x^2}\, dx = e^{i\pi/4}\sqrt{\frac{\pi}{\lambda}}. \tag{8.43a}$$

By taking the complex conjugate of (8.43a), we get

$$\int_{-\infty}^{\infty} e^{-i\lambda x^2}\, dx = e^{-i\pi/4}\sqrt{\frac{\pi}{\lambda}}. \tag{8.43b}$$

Next we consider the slightly more general integral

$$\int_{-\infty}^{\infty} e^{-ax^2}\, dx,$$

where a is a complex number. In order for the integral to be convergent, we must have $\text{Re}a \geq 0$. While the phase of the complex number a is of multiple values, we choose it to be the one in the interval $-\pi/2 \leq \arg(a) \leq \pi/2$. We deform the contour of integration into the straight line making the angle $-\frac{1}{2}\arg(a)$ with the positive x axis, plus the two vertical lines joining the real axis and the straight line we just mentioned. Since we choose the phase of a to be between $-\pi/2$ and $\pi/2$, the integrand vanishes at the vertical lines. Thus we have

$$\int_{-\infty}^{\infty} e^{-ax^2}\, dx = \int_c e^{-az^2}\, dz,$$

where c is the straight line making the angle $-\frac{1}{2}\arg(a)$ with the positive x axis. On this straight line, az^2 is real and positive, and we may put $az^2 = t^2$, or $z = t/\sqrt{a}$. Hence we have

$$\int_c e^{-az^2}\, dz = a^{-1/2} \int_{-\infty}^{\infty} e^{-t^2}\, dt = e^{-i\arg(a)/2}\sqrt{\frac{\pi}{|a|}}. \tag{8.44}$$

Note that the phase of the integral is the angle the deformed contour makes with the real axis.

Let us extend this result to the integral of (8.36). From the considerations we make above, we expect that there is a contribution from a small region around the point of stationary phase x_0, i.e., the point at which

$$u'(x_0) = 0.$$

Near the point x_0, we make the approximation

$$u(x) - u(x_0) \approx \frac{u''(x_0)}{2!}(x - x_0)^2.$$

Thus the width of the region of dominant contributions is of the order of $\sqrt{\lambda}$.

If $u''(x_0)$ is positive, we make use of (8.43a) and find that the contribution from the neighborhood of x_0 is asymptotically equal to

$$e^{i\pi/4}\sqrt{\frac{2\pi}{\lambda u''(x_0)}}e^{i\lambda u(x_0)}h(x_0). \tag{8.45}$$

If $u''(x_0)$ is negative, we make use of (8.43b) and find that the contribution from the neighborhood of x_0 is asymptotically equal to

$$e^{-i\pi/4}\sqrt{\frac{2\pi}{\lambda|u''(x_0)|}}e^{i\lambda u(x_0)}h(x_0). \tag{8.46}$$

We note that the magnitudes of the right sides of both (8.45) and (8.46) are slightly larger than that of (8.39), as the former are of the order of $1/\sqrt{\lambda}$ while the latter is of the order of $1/\lambda$.

If $u''(x_0)$ turns out to be zero, the right sides of (8.45) and (8.46) are infinity. This means that these formulae do not apply. In such a case, we have

$$u(x) - u(x_0) \approx u'''(x_0)(x - x_0)^3/6,$$

provided that $u'''(x_0) \neq 0$. The contribution from the point of stationary phase is then of the order of $(\lambda)^{-1/3}$. This is somewhat larger than the contribution from a point of stationary phase with $u''(x_0) \neq 0$. More generally, if the order of the lowest nonvanishing derivative at x_0 is n, the contribution of the point x_0 is of the order of $\lambda^{-1/n}$. This method of computing the asymptotic form of the integral (8.36) is known as the method of stationary phase.

Problem for the Reader

Find the leading asymptotic terms for the integral

$$I(\lambda) = \int_{-1}^{1} e^{i\lambda x^2}\,dx. \tag{8.47}$$

Solution

We have

$$u(x) = x^2.$$

The point $x = 0$ is a point of stationary phase, and $x = \pm 1$ are the endpoints.

The values of u and u' at the endpoints are

$$u(-1) = u(1) = 1$$

and

$$u'(1) = -u'(-1) = 2.$$

Therefore, we have from (8.39) that the lower endpoint and the upper endpoint both give the contrbution

$$\frac{e^{i\lambda}}{2i\lambda}.$$

Also, we have from (8.45) that the contribution for the point of stationary phase $x_0 = 0$ is $e^{i\pi/4}\sqrt{\pi/\lambda}$. Thus we have

$$I(\lambda) \approx \frac{e^{i\lambda}}{i\lambda} + e^{i\pi/4}\sqrt{\pi/\lambda}. \tag{8.48}$$

Problem for the Reader

Find the leading term of the Fourier coefficient

$$a_n = \frac{1}{2\pi}\int_0^{2\pi} f(\theta)e^{-in\theta}d\theta, n \gg 1.$$

Solution

There is no point of stationary phase for this integral. Thus the leading terms of this integral come from the endpoints.

From (8.39), we have

$$I(\lambda) \approx \frac{f(2\pi) - f(0)}{-2\pi in}. \tag{8.49}$$

Thus the Fourier ceofficient a_n vanishes like $1/n$ as $n \to \infty$.

Since the coefficient of $1/n$ in (8.49) is proportional to $f(2\pi) - f(0)$, we see that if $f(\theta)$ is periodic, i.e.,

$$f(2\pi) = f(0),$$

the expression in (8.49) vanishes and the Fourier coefficient a_n must vanish faster than $1/n$. In such a case, we carry out an integration by parts and get

$$a_n = \frac{1}{-2\pi in}\int_0^{2\pi} f(\theta)d(e^{-in\theta}) = \frac{1}{i2\pi n}\int_0^{2\pi} f'(\theta)e^{-in\theta}d\theta. \tag{8.50}$$

Applying (8.39) to the integral above, we have

$$a_n \approx \frac{f'(2\pi) - f'(0)}{2\pi n^2}. \tag{8.51}$$

We see that a_n vanishes like $1/n^2$ as n goes to infinity.

Needless to say, if we have, in addition,

$$f'(0) = f'(2\pi),$$

then a_n vanishes like $1/n^3$. Indeed, if

$$f^{(m)}(0) = f^{(m)}(2\pi)$$

for all m, then a_n vanishes faster than any inverse power of n as n goes to infinity.

So far we have implicitly assumed that the function $f(\theta)$ and all its derivatives are continuous at all interior points. But if $f(\theta)$ is discontinuous at θ_0, we should break up the integral for a_n into two integrals, one with limits of integration 0 and θ_0, and the other with limits of integration θ_0 and 2π. Applying (8.39) to each of these two integrals, we see that there is an endpoint contribution equal to $ie^{-in\theta_0}/n$ times the discontinuity of $f(\theta)$ at θ_0. And if $f(\theta)$ is continuous at θ_0, but $f'(\theta)$ is discontinuous at θ_0, then a_n has a term proportional to $e^{-in\theta_0}/n^2$. Therefore, the asymptotic form of a_n tells us how "smooth" $f(\theta)$ is as a function of θ.

As another example, consider the integral

$$I(\lambda) = \int_{-1}^{1} dxe^{-in\pi x}/(1+x^2),$$

where n is a very large positive integer. The integrand does not have any point of stationary phase. Let

$$h(x) = 1/(1+x^2);$$

then

$$h(1) = h(-1).$$

Consequently, the leading terms from the two endpoints cancel. By performing an integration by parts, we get

$$I(\lambda) = \frac{2i}{n\pi} \int_{-1}^{1} \frac{x}{(1+x^2)^2} e^{-in\pi x} dx \approx \frac{(-1)^{n+1}}{n^2\pi^2},$$

where we have made use of (8.39).

In summary, the asymptotic form of the integral of (8.36) is derived from two kinds of contributions: The first kind are the endpoint contributions and the second kind are the contributions from points of stationary phase. The first kind of contributions are given by (8.39), and the second kind of contributions are given by (8.45) and (8.46). Note that if there are two points of stationary phase in the region of integration, and assuming that at one of them $u(x)$ is maximum and at the other $u(x)$ is minimum, we have to take into account the contributions from both points. This is because the contributions of these two points are of the same order of magnitude. In contrast, for the integral of (8.8), the contribution from the point of maximum of $v(x)$ is negligible. I also remind the reader that if $u'(x)$ vanishes at an endpoint, or if $u''(x_0)$ vanishes at a point of stationary phase, we must expand $u(x)$ to the next nonvanishing order before we carry out the integration.

If the integral (8.36) has neither points of stationary phase nor finite endpoints, the method of stationary phase fails. As an example, consider the integral

$$I(\lambda) = \int_{-\infty}^{\infty} \frac{e^{-i\lambda x}}{\sqrt{1+x^2}} dx, \lambda \gg 1.$$

The function $I(\lambda)$ is the Fourier transform of $(1+x^2)^{-1/2}$. Identifying $u(x)$ with $-x$, we see that, since $u'(x) = -1 \neq 0$, this integral has no points of stationary phase. Nor does this integral have finite endpoints. Now the contribution from a point x_0, where x_0 is either an endpoint or a point of stationary phase, is of the order of an inverse power of λ times the value of the integrand at x_0. Since there is no such x_0 on the contour of integration, and since the magnitude of the integrand on the contour of integration is of the order of unity, we conclude that $I(\lambda)$ is smaller than any inverse power of λ when λ is large. However, the method outlined in this section is incapable of telling us exactly how small $I(\lambda)$ is. The asymptotic form of the integral above can be found with the use of the saddle point method, which we shall discuss in the next section.

Before we close this section, let us apply the method of stationary phase to a general theory of wave propagation. The simplest theory of wave propagation is given by the wave equation discussed in Section B of Chapter 4. If the velocity of the wave is equal to c, this wave equation reads

$$c^2\phi_{xx} - \phi_{tt} = 0,$$

where $\phi(x,t)$ is the amplitude of the wave. The general solution of this equation is

$$\phi(x,t) = f(x+ct) + g(x-ct),$$

where f and g are arbitrary functions to be determined from the initial conditions. If we set $c = 1$, this solution is equal to the one given by (4.26).

Let the initial conditions be

$$\phi(x,0) = p(x), \phi_t(x,0) = 0,$$

where $p(x)$ is nonvanishing only in a neighborhood of the origin $x = 0$. Then we have, for $t > 0$,

$$\phi(x,t) = \frac{p(x+ct) + p(x-ct)}{2},$$

which is nonvanishing only in neighborhoods around $x = -ct$ or $x = ct$. Such a solution describes two wave packets; one travels to the right with velocity c, and the other travels to the left with velocity c. If we view this solution in the xt-plane, the function $\phi(x,t)$ is nonvanishing only around the straight line $x = ct$ and the straight line $x = -ct$.

We shall express this solution as the sum of two Fourier integrals. Let $P(k)$ be the Fourier transform of $p(x)$. Then

$$\phi(x,t) = \frac{1}{2}\int_{-\infty}^{\infty} P(k)e^{ik(x-ct)}\frac{dk}{2\pi} + \frac{1}{2}\int_{-\infty}^{\infty} P(k)e^{ik(x+ct)}\frac{dk}{2\pi}.$$

Not all wave theories are so simple. In other wave theories, the amplitude of a wave with the same initial conditions is given by

$$\phi(x,t) = \int_{-\infty}^{\infty} e^{ikx-i\omega(k)t}P(k)\frac{dk}{2\pi} + \int_{-\infty}^{\infty} e^{ikx+i\omega(k)t}P(k)\frac{dk}{2\pi}.$$

In the above, the angular frequency $\omega(k)$ may depend on the wave number k in a nonlinear way. This relation between ω and k is called the dispersion relation.

We consider

$$I(x,t) = \int_{-\infty}^{\infty} e^{ikx-i\omega(k)t}P(k)\frac{dk}{2\pi},$$

which is the first integral of $\phi(x,t)$. The behavior of the second integral of $\phi(x,t)$ can be obtained in the same way with $t \to -t$.

Problem for the Reader

Find $I(x, t)$ when x and t are very large.

Solution

When x and t are very large, the dominant contributions to $I(x, t)$ come from the point (or points) of stationary phase. Let us denote a point of stationary phase by k_0. The value of k_0 is obtained by setting to zero the derivative of the exponent $kx - \omega(k)t$ taken with respect to the integration variable k. Thus we have

$$x - \omega'(k_0)t = 0,$$

or

$$\frac{x}{t} = \omega'(k_0).$$

The value $\omega'(k_0)$ is called the group velocity. As a point of reference, the group velocity for the solutions of the wave equation above is simply $\pm c$.

If $\omega''(k_0)$ is positive, the contribution to $I(x, t)$ from the point k_0 is

$$e^{-i\pi/4}\sqrt{\frac{1}{2\pi\omega''(k_0)t}}e^{ik_0x - i\omega(k_0)t}P(k_0).$$

And if $\omega''(k_0)$ is negative, the contribution to $I(x, t)$ from the point k_0 is

$$e^{i\pi/4}\sqrt{\frac{1}{2\pi\left|\omega''(k_0)\right|t}}e^{ik_0x - i\omega(k_0)t}P(k_0).$$

Problem for the Reader

We see from the expression above that when x and t are large, the magnitude of $I(x, t)$ is inversely proportional to $\sqrt{t}$. Can you explain this qualitatively?

Solution

Consider the integral $I(x, t)$ on the straight line $x = Ct$, where C is a constant. For all points on this straight line, the dominant contributions come from the same

k_0 that is the root of the equation $\omega'(k_0) = C$. We note that for a wave theory in which the range of $\omega'(k)$ is $[a, b]$, i.e.,

$$a < \omega'(k) < b,$$

there is no solution for the equation $x/t = \omega'(k_0)$ if the value of x/t is outside this range. Therefore, $I(x, t)$ vanishes rapidly unless x/t is inside $[a, b]$:

$$a < x/t < b.$$

To compare this solution with the solution of the idealized wave equation (4.24) we just discussed, we note that the latter describes two wave packets that are nonzero only around the lines $x = \pm t$. On the other hand, the integral $I(x, t)$ vanishes rapidly everywhere except in the whole region bounded by the straight lines $x = at$ and $x = bt$. This means that at a fixed t, the wave does not vanish rapidly in the region

$$at < x < bt.$$

Thus the size of the wave packet is asymptotically $(b - a)t$, which grows linearly with t. As the magnitude of the wave decreases like $t^{-1/2}$,

$$\int_{-\infty}^{\infty} |I(x,t)|^2 \, dx$$

is a constant.

Let the wave be nearly monochromatic or, more precisely, $P(k)$ be nonzero only in a small neighborhood of k_0, vanishing quickly but smoothly outside this region. The value of $\omega'(k)$ for all k in this small neighborhood is then approximately equal to $\omega'(k_0)$. In such a case, $I(x, t)$ is nonzero only near $x/t = \omega'(k_0)$, and the wave travels with the group velocity $\omega'(\kappa_0)$, not with the phase velocity $\omega(k_0)/k_0$. The size of this wave is ϵt, where ϵ is the width of the neighborhood of $\omega(k_0)$ in which $P(k)$ is nonzero.

We have stated that the method of stationary phase can be used to obtain the behavior of the solution of a wave theory when x and t are "large". But how large is "large"?

Question for the Reader

How do you estimate how large x and t must be in order for the method of stationary phase to be applicable to the integral $I(x, t)$?

Solution

In our calculation, we replace the exponent $kx - \omega(k)t$ by the first few terms of its Taylor series expanded around k_0. More precisely, we make the approximation

$$kx - \omega(k)t \approx k_0 x + (k - k_0)x - \omega(k_0)t - \omega'(k_0)(k - k_0)t \\ - \frac{\omega''(k_0)}{2}(k - k_0)^2 t.$$

Note that, k_0 being the point of stationary phase, the sum of terms linear in $(k - k_0)$ vanishes. Thus we have

$$kx - \omega(k)t \approx k_0 x - \omega(k_0)t - \frac{\omega''(k_0)}{2}(k - k_0)^2 t.$$

The region of dominant contributions is

$$\left|\frac{\omega''(k_0)}{2}(k - k_0)^2 t\right| = O(1),$$

or

$$|k - k_0| = O\left(\sqrt{\frac{2}{|\omega''(k_0)t|}}\right).$$

For the approximation to be valid, a necessary condition is that the first term we have dropped from the Taylor series be small in the region of dominant contributions. Thus we require that

$$\left|\frac{\omega'''(k_0)(k - k_0)^3 t}{6}\right| \ll 1$$

in the dominant region of contributions. A little calculation gives

$$t \gg \frac{2}{9}\frac{[\omega'''(k_0)]^2}{[\omega''(k_0)]^3}.$$

Also,

$$|x| = |\omega'(k_0)t| \gg \frac{2}{9}\frac{|\omega'(k_0)|\,[\omega'''(k_0)]^2}{[\omega''(k_0)]^3}.$$

D. The Saddle Point Method

The Laplace method enables us to deal with the integral of the form of (8.8), where the coefficient of λ in the exponent is real. The method of stationary phase enables us to deal with the integral of the form of (8.36), where the coefficient of λ in the exponent is purely imaginary.

In this section, we shall deal with integrals of the form

$$I(\lambda) = \int_c e^{i\lambda f(z)} h(z) dz, \tag{8.52}$$

where c is a contour in the complex z-plane and where $f(z)$ and $h(z)$ are analytic functions of z. As we shall see, finding the solution of this problem will also help us to calculate integrals of the form (8.36) in which there are neither finite endpoints nor points of staionary phase.

Let

$$f(z) = u(x, y) + iv(x, y);$$

then the exponential function in (8.52) is equal to

$$e^{i\lambda f(z)} = e^{i\lambda u(x,y)} e^{-\lambda v(x,y)}.$$

We see that the integral of (8.8) and that of (8.36) are special cases of the integral of (8.52).

If the value of v on the contour c is minimum at z_0, then the magnitude of the integrand is maximum at z_0. However, z_0 is not necessarily the point that gives the dominant contribution to the integral. This is because while the magnitude of the integrand along c is sharply peaked around z_0, the integrand also oscillates rapidly in the neighborhood of z_0 as long as $u'(z_0)$ does not vanish. If such is the case, the oscillations of the integrand around z_0 wipe out entirely the contribution of z_0. But if we are able to deform the contour c into a contour c' on which u is a constant, then we have

$$I(\lambda) = e^{i\lambda a} \int_{c'} e^{-\lambda v(z)} h(z) dz,$$

where a is the value of u on c'. Now the integrand does not oscillate along c'. The dominant contribution to the integral then comes from the point z_0' on c' at which v is minimum. The asymptotic form of the integral can therefore be obtained with the use of Laplace's method.

The point z_0' may be an endpoint of c'. In such a case, we may obtain the asymptotic forms of the integral by expanding the integrand around the endpoint and carrying out the integration.

The point z_0' may also be an interior point of c'. We shall argue that if z_0' is an interior point of c', all first-order derivatives of u and v must vanish at z_0'. This is because, since v is minimum at z_0' along the curve c', the variation of v in the tangential direction of c' must vanish at z_0'. Also, since u is constant on c', the variation of u in the tangential direction of c' is zero. By the Cauchy-Riemann equations, the vanishing of the variation of u in the tangential direction of c' implies the vanishing of the variation of v in the direction of the normal of c'. Thus all first-order derivatives of v vanish at z_0'. By the Cauchy-Riemann equations, so do all first-order derivatives of u at z_0'.

The point z_0' is known as a saddle point for the integral (8.52). Since by assumption v is minimum at z_0' along c', the magnitude of the integrand is maximum at z_0' on the contour of integration c'. In addition, the phase of the integrand is a constant along the contour c'. Thus there is no cancellation and the dominant contributions to the integral come from the saddle point z_0'. The asymptotic form of the integral can be obtained by expanding the integrand around z_0' and carrying out the Gaussian integrals.

A curve on which u is a constant is called a level curve of u. As we try to deform the contour into a level curve of u passing through one or more saddle points on which the magnitude of the integrand is maximum, we may encounter singularities that keep us from deforming the contour any further. As a result, the singularity may give dominant contributions to the integral. Therefore, the possible points of contribution for the integral of the general form (8.52) are the singular points, the endpoints, and the saddle points.

The integrand of the integral may have more than one saddle point and singular point. Not all of them contribute to the asymptotic form of the integral. We must decide which of these points are relevant as well as what contour c' to deform to. We shall show how to do this below. But such an exposition will require a little understanding of complex analysis. Here we'll add to our previous discussions on this topic in Chapter 2.

A curve on which u is a constant is called a level curve of u. A curve on which v is a constant is called a level curve of v. We shall show that, with certain notable exceptions, the level curves of u intersect orthogonally with the level curves of v. The exceptions occur at the points where $f'(z)$ vanishes.

Problem for the Reader

Consider in the complex plane the function

$$f(z) = z^2 = (x + iy)^2.$$

Draw the level curves of u as well as those of v for this function.

Solution

We have

$$u(x, y) = x^2 - y^2$$

and

$$v(x, y) = 2xy.$$

We draw the level curves of u and the level curves of v in the figure below.

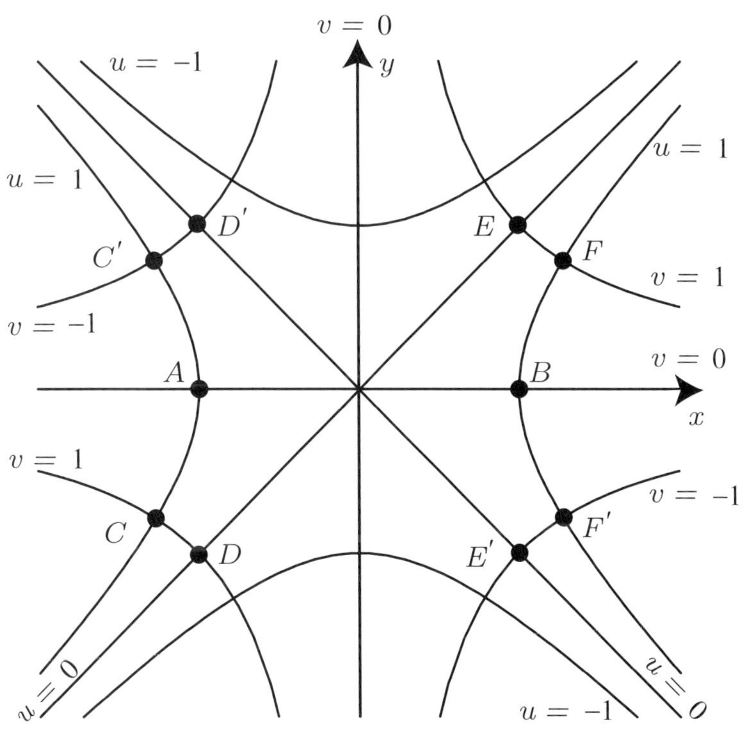

Figure 8.8

Question for the Reader

Do the level curves of u and those of v intersect orthogonally?

Solution

As we can see from Figure 8.8, the level curves of u generally intersect orthogonally with those of v. Also, level curves of u generally do not intersect with one another. Similarly for the level curves of v.

But there is one notable exception. There are two level curves of u and two level curves of v intersecting at the origin. Furthermore, the level curves of u and those of v do not intersect orthogonally at the origin. Instead, the two level curves of u intersect orthogonally at the origin. So do the two level curves of v. We note also that the derivative of $f(z)$ vanishes at the origin.

As a general rule, if u and v are harmonic conjugates of each other, the level curves of u and those of v intersect orthogonally. To see this, we express the infinitesimal variation of $u(x, y)$ as

$$du = u_x dx + u_y dy.$$

Now on a level curve of u, the value of u does not vary. Therefore we have

$$u_x dx + u_y dy = 0,$$

where dx and dy are the variations of x and y between two neighboring points on the level curve of u. This says that $\vec{\nabla} u$ is perpendicular to the tangent vector of the level curve of u. Thus $\vec{\nabla} u$ is in the direction of the normal of the level curve of u. Similarly, $\vec{\nabla} v$ is perpendicular to the tangent vector of the level curve of v, and is hence in the direction of the normal of the level curve of v.

The scalar product of the vectors $\vec{\nabla} u$ and $\vec{\nabla} v$ is $u_x v_x + u_y v_y$. If u and v are harmonic conjugates of each other, we get, using the Cauchy-Riemann equations (2.9),

$$u_x v_x + u_y v_y = 0.$$

Therefore, at the point a level curve of u and a level curve of v intersect, the normals of these two curves are orthogonal to each other, and hence these two curves intersect orthogonally.

But there are exceptions. The points of exception are the points at which the derivative of $f(z)$ vanishes. At each of such points, all of the first-order derivatives of u and v vanish. Thus $\vec{\nabla} u$ is a null vector at such a point and has no direction, and has nothing to do with any of the normals of the level curves of u at such a point.

Similarly for $\vec{\nabla} v$. While the scalar product of $\vec{\nabla} u$ and $\vec{\nabla} v$ vanishes at a zero of f', it does not imply that the level curve of u intersects with that of v orthogonally.

Note also that two level curves of u generally do not intersect. Let a level curve be given by

$$u(x, y) = c.$$

And let (x_0, y_0) be a point on this curve. Now by the implicit function theorem, the equation above has a unique solution in a neighborhood of (x_0, y_0) in the form

$$y = y(x),$$

provided that $u_y(x_0, y_0) \neq 0$. Similarly, it has a unique solution in a neighborhood of (x_0, y_0) in the form

$$x = x(y),$$

provided that $u_x(x_0, y_0) \neq 0$. Thus, unless both u_x and u_y vanish, there can only be one level curve passing through the point (x_0, y_0). But at the point z_0 at which $f'(z_0) = 0$, both u_x and u_y vanish, and there is no reason why two level curves of u cannot intersect at z_0. Indeed, we see in the example above that two level curves of u do intersect at the origin. Two level curves of v in this example also intersect at the origin.

We have learned in calculus that if the first derivative of a function of a real variable x vanishes at x_0, then it has an extremum at x_0 provided that its second derivative does not vanish. This is quite different for $u(x, y)$ and $v(x, y)$, which are functions of two variables. Near the point z_0 at which the derivative of $f(z)$ vanishes, we have

$$f(z) - f(z_0) \simeq f''(z_0)(z - z_0)^2/2.$$

Therefore, if $f''(z_0) \neq 0$, the behavior of $f(z)$ near z_0 is the same as the function z^2 near the origin.

Problem for the Reader

Let $v(x, y) = 2xy$, the imaginary part of z^2. What is the behavior of v on the line $x = y$ and the line $x = -y$? What does v look like near $x = y = 0$?

Solution

The line $x = y$ is a level curve of u. On this line $v = 2x^2$, and is minimum at the origin. The line $x = -y$ is also a level curve of u. On this line $v = -2x^2$, and is maximum at the origin.

If we consider $v(x, y)$ as a function of x and y in the three-dimensional xyv-space, it resembles the surface of a saddle near the point $x = y = 0$, where the line $x = y$ is in the front-back direction of the saddle, and the line $x = -y$ is in the sidewise direction of the saddle. The function $v(x, y)$ is neither absolutely maximum nor absolutely minimum at the origin. Thus we call the origin a saddle point of v. Similarly, the origin is a saddle point of $u = x^2 - y^2$.

More generally, if u and v are the real part and the imaginary part of an analytic function $f(z)$ the derivative of which vanishes at z_0, then z_0 is a saddle point of u and that of v provided that $f''(z_0) \neq 0$. If $f''(z_0) = 0$ but $f'''(z_0) \neq 0$, the behavior of u and v are similar to those of their counterparts of the function z^3.

By definition, the value of a function $v(x, y)$ does not vary on any of its level curves. Thus it varies most rapidly in the direction orthogonal to its level curve. Since the level curves of u intersect those of v orthogonally, v varies most rapidly on a level curve of u, and u varies most rapidly on a level curve of v.

As a consequence, the function $e^{-\lambda v}$ varies most rapidly on a level curve of u. Now we have seen that there are two level curves of u passing through the point z_0. We have also seen that the function $v(x, y)$ is minimum on one of these curves and maximum on the other curve. Thus the function $e^{-\lambda v(z)}$ decreases most rapidly as one leaves z_0 on the former curve. This curve is called the path of steepest descent. Similarly, $e^{-\lambda v(z)}$ increases most rapidly as one leaves z_0 on the latter curve, which is called the path of steepest ascent.

Consider now the integral in the general form of (8.52). If c is a level curve of $u(x, y)$, the phase of the exponential function of (8.52) is a constant and can be taken outside the integral. Also, the magnitude $e^{-\lambda v(x,y)}$ varies most rapidly on c. Thus the integral is of the form of (8.8), and the dominant contributions come from the point at which v is minimum. It is therefore a good idea to deform the contour of the integral of (8.52) into a path of steepest descent, for then the asymptotic form of the integral can be obtained with the Laplace method.

Take the example of (8.47), in which $f(z) = z^2$. We cannot easily go from -1 to 1 on a curve of constant u. Let us therefore take the path ACDEFB in Figure 8.8. Along the path AC, the phase $u(x, y)$ is a constant and $v(x, y)$ is minimum at the endpoint A. Thus the dominant contribution to the integral along AC is given by a small neighborhood of width $1/\lambda$ near the endpoint A, and can be calculated by the Laplace method. The result is the same as we have obtained by using the formula (8.39) and keeping only the contribution from the endpoint A. The contour CD is not a level curve of u. Instead, it is a level curve of v, on which $v = 1$. The integrand on CD is as small as $e^{-\lambda}$; hence the integral over CD will be neglected.

Next we consider the contour DE. This contour is a path of steepest descent, with the saddle point at the origin. Thus we may again use the Laplace method to evaluate the contribution of the origin, and the answer is the same as that calculated with the use of (8.45). The contribution of the contour EF, which is not a level curve of u, will be neglected, as the integrand is exponentially small on this contour. The dominant contribution of the contour FB, which is a level curve of u, comes from the point B. Applying the Laplace method gives us the same result obtained with the use of (8.39). Thus, by deforming the contour of integration, we succeed in verifying (8.48) with the application of the Laplace method.

Note that if we apply the method of stationary phase to the integral (8.47), with the contour of integration joining -1 to 1 on the real axis, the origin is considered as a point of stationary phase. But if we deform the contour as we demonstrated above, the origin is considered the saddle point. How you classify the origin depends on the direction of the contour of integration you choose.

We note that it will not be fruitful to deform the contour from the real axis to the curve AC′D′E′F′B. For if we do this, we will find that on the curve AC′, v is minimum not at A but at C′. Thus the dominant contributions to the integral over AC′ come from point C′, where the integrand is of the order of e^{λ}. This large term is cancelled by an equally large term with the opposite sign coming from the integral over C′D′. Also, while the contour D′E′ passes through the saddle point $z = 0$, this contour is one of steepest ascent and the contributions of this contour come not from the saddle point $z = 0$ but from the endpoints D′ and E′. Indeed, the integral over each of the segments of the curve AC′D′E′F′B has a very large leading term. These terms cancel one another and it is extremely difficult to determine what remains after such extensive cancellations.

Therefore, as we start from an endpoint of the contour and try to deform the contour to a level curve of u, be sure to go in the direction v increases. We mention that it is possible to deform the contour into a sum of curves all of which are level curves of u. Instead of taking the contour ACDEFB to go from A to B, we may go from A to C on the level curve of u as before, but instead of making a turn at C, we follow this level curve all the way to infinity. We then change over at infinity to the line $x = y$, and go from one infinite end of this straight line to the other infinite end of this straight line. Then we change over at infinity to the level curve on which F and B lie to arrive at B. In this way u is a constant on each of the segments of the contours.

Applying the Laplace method to each of the contours, we may obtain not only the leading term but also the next leading terms. To do this, we merely have to keep more terms in the expansions of the integrand around the points of contribution and carry out the integration.

Problem for the Reader

Find the entire asymptotic series for the integral

$$I(\lambda) = \int_0^\infty e^{i\lambda \sinh^2 x} dx. \tag{8.53}$$

Solution

There is only one finite endpoint, which is the origin. This point is also the only point of stationary phase on the contour of integration. Thus this point gives the dominant contribution to the integral.

Let

$$t = \lambda \sinh^2 x;$$

then

$$I(\lambda) = \frac{1}{2\sqrt{\lambda}} \int_0^\infty \frac{e^{it}}{\sqrt{t(1+t/\lambda)}} dt. \tag{8.54}$$

We make the binomial expansion

$$(1+t/\lambda)^{-1/2} = \sum_{n=0}^{\infty} \frac{(-1)^n \Gamma(n+1/2)}{n!\Gamma(1/2)} \left(\frac{t}{\lambda}\right)^n \tag{8.55}$$

and substitute it into (8.54). We find that the nth term in (8.55) is proportional to the integral

$$\int_0^\infty e^{it} t^n dt, \tag{8.56}$$

which is divergent.

Let us deform the contour in the t-plane to the positive imaginary t axis. This is possible since the integrand in (8.54) vanishes exponentially at the infinity of the first quadrant of the complex t-plane. (See Appendix B in this chapter for more discussion of the deformation of contours.) The positive imaginary axis is indeed the path of steepest descent, on which $t = i\tau$, with τ running from zero to infinity. After deforming the contour, (8.54) becomes

$$I(\lambda) = \frac{i}{2\sqrt{\lambda}} \int_0^\infty \frac{e^{-\tau}}{\sqrt{e^{i\pi/2}\tau(1 + i\tau/\lambda)}} d\tau.$$

Making the binomial expansion for $(1 + i\tau/\lambda)^{-1/2}$, we get

$$I(\lambda) = \frac{e^{i\pi/4}}{2} \sum_{n=0}^{\infty} \frac{(-i)^n \Gamma^2(n + 1/2)}{\Gamma(1/2) n! \lambda^{n+1/2}}. \tag{8.57}$$

Next we consider integrals of the general form

$$I(\lambda) = \int_{-\infty}^{\infty} e^{i\lambda u(x)} h(x) dx.$$

If $u'(x)$ does not vanish for $-\infty < x < \infty$, the integrand has no points of stationary phase. Since neither does this integral have finite endpoints, the method of stationary phase cannot be applied. We note that in the special case of $u = -x$, $I(\lambda)$ is the Fourier transform of $h(x)$. We shall show that the dominant contributions to such an integral come from the singularities or the saddle points of the integrand in the complex plane.

As an example of a saddle point contribution, consider

$$I(\lambda) = \int_{-\infty}^{\infty} e^{i\lambda x} e^{-x^2} dx. \tag{8.58}$$

This integral can be calculated in a closed form, but we shall use it as another example to demonstrate how to change the contour of integration to our advantage.

This integral is of the form (8.36), with $u(x) = x$ and $h(x) = e^{-x^2}$. There are no finite endpoints. Also, since $u'(x)$ does not vanish anywhere, neither are there points of stationary phase. We conclude that the integral must vanish faster than any inverse power of λ, as the integrand of $I(\lambda)$ is of the order of unity.

The integral does have a saddle point in the complex plane. Let

$$f(z) = i\lambda z - z^2;$$

then

$$f'(z) = i\lambda - 2z.$$

Thus the saddle point is at $z = i\lambda/2$, which is very far away from the real axis when λ is large. Indeed, we have

$$f(z) = -(z - i\lambda/2)^2 - \lambda^2/4.$$

Therefore, on the horizontal line

$$z = x + i\lambda/2,$$

which passes through the saddle point,

$$f(z) = -x^2 - \lambda^2/4$$

is real. Thus, if the contour of integration is on the horizontal line passing through the saddle point, we may use the Laplace method to calculate the asymptotic form of this integral.

While the contour of integration is on the real axis, we will deform it into the horizontal line passing through the saddle point $i\lambda/2$. We will think of the real axis as the horizontal line joining $-L$ to L, where L is a very large number that eventually will be taken to be infinite. Let us deform the contour to the line $z = x + i\lambda/2$, as illustrated in the figure below, which we will call c.

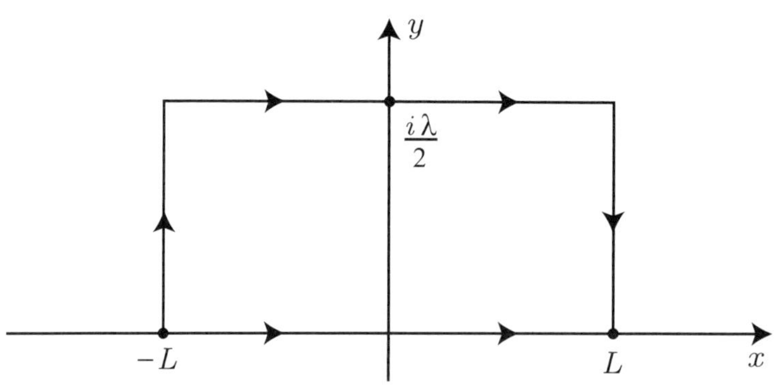

Figure 8.9

Since the integrals over the two vertical lines vanish as L goes to infinity, we are left with the integral over the horizontal line $z = x + i\lambda/2$. Thus we have

$$I(\lambda) = e^{-\lambda^2/4} \int_c e^{-(z-i\lambda/2)^2} dz = e^{-\lambda^2/4} \int_{-\infty}^{\infty} e^{-x^2} dx = \sqrt{\pi} e^{-\lambda^2/4}. \quad (8.59)$$

Note that the Gaussian $e^{-\lambda^2/4}$ above vanishes very rapidly when λ is very large, indeed faster than $e^{-\lambda a}$ for any a. As a matter of fact, $e^{-\lambda^2/4}$ is the value of the integrand $e^{i\lambda z} e^{-z^2}$ at the saddle point $z = i\lambda/2$.

That $I(\lambda)$ is so small may appear surprising, as the magnitude of the integrand with the contour of integration on the real axis is equal to e^{-x^2}, which is finite as $\lambda \to \infty$. Let us think of the integral in (8.58) as a summation of waves each of which is of the same wave number λ. The weighting factor is e^{-x^2}, and the summation is over all possible values of x. The integral vanishes rapidly because these waves cancel one another to a very high degree. The degree of cancellation depends on how "smooth" the weighting factor is. For example, we have considered in the preceding section the example in which the weighting factor is unity in the interval $[a, b]$ and zero outside this interval. The resulting integral is the one given by (8.35), which vanishes only as fast as the inverse power of λ. This is because the weighting factor of (8.35) has discontinuities at $x = a$ and at $x = b$. In contrast, the waves in the integral (8.58) havc the weighting factor e^{-x^2}. Not only is this weighting factor continuous for all values of x, all of its mth-order derivatives are continuous for all values of x. Moreover, this function is analytic in the entire complex plane. The superposition of waves with such a smooth weighting factor cancels to an extremely large degree, and the integral vanishes like a Gaussian as λ goes to infinity.

This is why it is important to know, in some physical problems, the rate at which an integral of this form vanishes. For example, in Rutherford's time, people knew that inside an atom are electrons, which carry negative electric charges. Since an atom is neutral electrically, it must also possess other stuff that carries positive charges. In J. J. Thomson's atomic model, this stuff of positive charges, which carries most of the mass of the atom, is smeared out smoothly throughout the atom. According to this model, the superposition of α-particle waves scattered from these smoothly distributed positive charges must vanish very rapidly. Since this was not observed, Rutherford concluded that the positive charges are confined to a tiny volume, almost pointlike, which is now known as the atomic nucleus.

Let us give another mathematical example to demonstrate how the form of the weighting factor affects the asymptotic form of $I(\lambda)$. Consider the integral

$$I(\lambda) = \int_{-\infty}^{\infty} e^{i\lambda x} \frac{1}{1+x^2} dx. \tag{8.60}$$

For this problem, the weighting factor is smooth as a function of the real variable x. Indeed, this weighting factor, being equal to $1/(1+x^2)$, has continuous derivatives to all orders for all values of x. Since there are no finite endpoints nor points of stationary phase, we expect this integral to vanish faster than any inverse power of λ.

To find out exactly how fast the integral vanishes as $\lambda \to \infty$, we replace the real variable x in the integrand by the complex variable z. Considered as a function of z, the integrand is no longer smooth. In fact, $1/(1 + z^2)$ blows up at $z = \pm i$. The value of $e^{i\lambda z}$ at $z = -i$ is e^{λ}, which is exponentially large. Since we expect the integral to vanish as $\lambda \to \infty$, the point $z = -i$ must not be relevant to the asymptotic form of the integral. On the other hand, the value of $e^{i\lambda z}$ at $z = i$ is $e^{-\lambda}$, which vanishes as $\lambda \to \infty$. Therefore, we surmise that this singularity of the integrand is relevant to the asymptotic form of the integral. This also means that the integral vanishes as fast as $e^{-\lambda}$, which is exponentially fast, but not as fast as the Gaussian.

To confirm this, we note that considered as a function of z, the integrand vanishes at the infinity of the upper half-plane (see Appendix B in this chapter for more discussion of this point). Thus we close the contour of integration by adding an infinite semicircle in the upper half-plane to the contour. We may then use the Cauchy residue theorem to find the value of this integral. Since the residue at i is equal to

$$\lim_{z \to i}(z - i)\frac{e^{i\lambda z}}{1 + z^2} = \frac{e^{-\lambda}}{2i},$$

we have

$$I(\lambda) = \pi e^{-\lambda}. \tag{8.61}$$

This confirms our guess.

As a general rule, a singularity of the integrand at z_0 in the complex plane may contribute a term of the order of $e^{i\lambda z_0}$ to the integral. But the integrand may have more than one singularity. Which one of them contributes?

Let us consider the Fourier transform of $h(x)$ given by

$$I(\lambda) = \int_{-\infty}^{\infty} dx e^{i\lambda x} h(x). \tag{8.62}$$

This integral has neither finite endpoints nor points of stationary phase on the real axis, which is the contour of integration. If $h(x)$ is analytic at all points of the real axis, then $I(\lambda)$ vanishes faster then any inverse power of λ. Let us deform the contour to the horizontal line $z = x + ia$, where a is positive. We then have, on this deformed contour, $e^{i\lambda z} = e^{i\lambda x}e^{-\lambda a}$, the magnitude of which is $e^{-\lambda a}$. Since the integrand has the phase factor $e^{i\lambda x}$, which is a rapidly oscillatory function of x, there are great cancellations and we estimate that $I(\lambda)$ must be much smaller than $e^{-\lambda a}$. To find a better estimate, we move the horizontal contour upward to the horizontal line $z = x + ib$, with $b > a$. We then estimate that $I(\lambda)$ is much smaller than even $e^{-\lambda b}$, which is a better estimate. Since we can continue to do this until we encounter a singularity z_0 of $h(z)$ in the upper half complex plane, we conclude that the integral vanishes roughly like $e^{i\lambda z_0}$, where z_0 is a singularity of $h(z)$ in the upper half-plane.

We mention that it is not fruitful to deform the contour to one in the lower half-plane. For if we deformed the contour to a straight horizontal line $z = x - ia$, where a is positive, we would conclude that, due to the oscillations of the integrand on this line, the integral is smaller than $e^{a\lambda}$. This conclusion, while correct, is not very informative, as we already know before we deform the contour that the integral must vanish as λ goes to infinity.

If the integrand has no singularity in the upper half-plane, then the integral vanishes faster than $e^{-\lambda a}$ for any a. In this case we may continue to deform the contour upward, until we encounter a saddle point of the integrand. In the example of (8.58), $h(z) = e^{-z^2}$, which has no singularities in the finite plane. The asymptotic form of the integral (8.58) is then determined by the location of the saddle point located at $z = i\lambda/2$, where the derivative of the exponent vanishes. At this point the integrand of the integral is equal to $e^{-\lambda^2/4}$, which is consistent with (8.59). We'll demonstrate this with five more examples.

Problem for the Reader

Find the asymptotic series of the integral

$$I(\lambda) = \int_{-\infty}^{\infty} \frac{e^{i\lambda x}}{\sqrt{1+x^2}} dx, \lambda \gg 1.$$

Solution

There are neither finite endpoints nor points of stationary phase. Thus we expect that dominant contributions come from the singularities of the integrand at $z = i$.

The singularities at $\pm i$ are branch points of $(1 + z^2)^{-1/2}$. (See Chapter 2 for a discussion of branch points.) We draw branch cuts from these branch points to make the integrand a single-valued function. There is some freedom in choosing the branch cuts as long as none of the cuts crosses the real axis, as the integrand is by assumption continuous on the real axis. We draw a set of branch cuts as illustrated in Figure 8.10. This is discussed in more detail in Chapter 2.

Let us deform the contour upstairs in the way illustrated in the figure. Since the integral along the infinite semi-circle in the upper half-plane vanishes, we get

$$I(\lambda) = \int_{c_1} \frac{e^{i\lambda z} dz}{\sqrt{z^2+1}} + \int_{c_2} \frac{e^{i\lambda z} dz}{\sqrt{z^2+1}}.$$

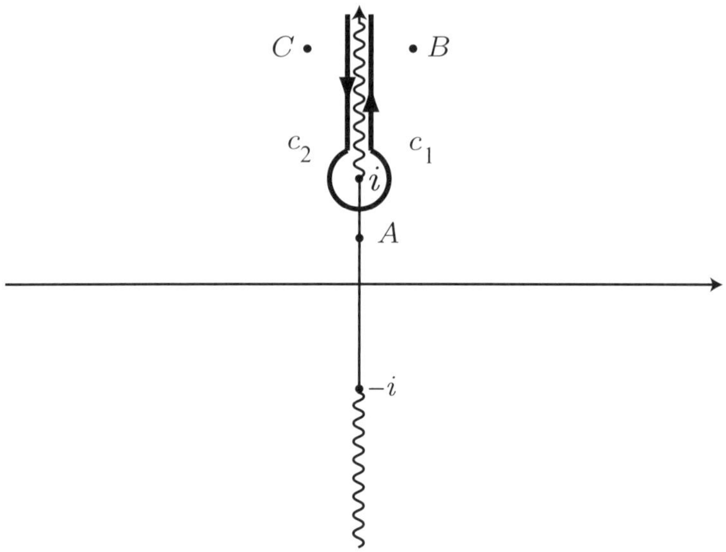

Figure 8.10

Next we determine the values of the function

$$(1+z^2)^{-1/2} = (z+i)^{-1/2}(z-i)^{-1/2}$$

on c_1 and c_2. We note that this function is real on the real axis, and at the origin it is equal to 1. As we move up the imaginary axis from the origin, by continuity this function is equal to the positive real number $(1-y^2)^{-1/2}$ provided that $y < 1$. When $y > 1$, which is the case when z gets to either c_1 or c_2, the value of the function can be either of $\pm i(y^2-1)^{-1/2}$. To determine which sign it is, we start from point A, which is immediately below the branch point $z = i$. The function is real and positive at A. As we go from A to B, we must travel along a counterclockwise curve and the phase of $(z-i)$ increases by π. Hence the phase of $(z-i)^{-1/2}$ decreases by $-\pi/2$. Since $e^{-i\pi/2} = -i$, this function at B must be equal to $-i\sqrt{y^2-1}$. By a similar argument, we find that this function at C is equal to $i\sqrt{y^2-1}$. Thus we have

$$I(\lambda) = 2\int_1^\infty \frac{e^{-\lambda y}}{\sqrt{y^2-1}}dy. \tag{8.63}$$

We may apply the Laplace method to obtain the asymptotic series of the integral in (8.63). Let us scale the integration variable by

$$t = \lambda(y-1).$$

We then get

$$I(\lambda) = \sqrt{2}e^{-\lambda}\lambda^{-1/2}\int_0^\infty \frac{e^{-t}}{\sqrt{t(1+t/2\lambda)}}dt.$$

By making the binomial expansion for the factor $(1+t/2\lambda)^{-1/2}$ in the integrand, we get

$$I(\lambda) = \sqrt{2}e^{-\lambda}\lambda^{-1/2}\sum_{n=1}^{\infty}\frac{\Gamma^2(n+1/2)}{\Gamma(1/2)n!}\left(-\frac{1}{2\lambda}\right)^n. \tag{8.64}$$

In particular, the leading term is

$$\sqrt{2\pi}e^{-\lambda}\lambda^{-1/2}.$$

The exponential function above is the value of $e^{i\lambda z_0}$ with $z_0 = i$, the singularity of the integrand in the upper half-plane. Note that the contours c_1 and c_2 in the problem are level curves of u. This is why we can apply the Laplace method to find the asymptotic form of the integral.

Next we consider the solution for the Airy function given by the integral (8.7) with the contour c chosen to be c_1 in Figure 8.1. We may make c_1 approach the imaginary axis and the solution becomes

$$I(x) = \int_{-\infty}^{\infty} e^{i(x\tau+\tau^3/3)}d\tau, \tag{8.65}$$

where t in the integral is replaced by $-i\tau$. If we expand $e^{ix\tau}$ into a Taylor series, we obtain $I(x)$ as a linear superposition of the two series solutions of the Airy equation given by (6.15) and (6.16). (See homework problem 13.) These series are useful when x is not large. We shall calculate the asymptotic form of this integral when x is large.

Problem for the Reader

Find the asymptotic form of the integral for the Airy function given by (8.65) when the magnitude of x is large.

Solution

First of all, we observe that the integral is of the form (8.36) and the integrand is of the magnitude unity on the contour of integration. As $|x|$ is large, the integrand is a rapidly varying function of τ and we expect the integral to be much smaller than its integrand. Thus we expect the integral to be much smaller than unity. We also observe that the integral has no finite endpoints and the integrand has no singularities.

If we identify $u(\tau)$ as τ, then $u'(\tau) = 1$ and there are no points of stationary phase on the contour of integration. However, when τ is large, τ^3 is much larger than τ and it is not justified to ignore τ^3 in our search for points of stationary phase. Let

$$f(\tau) = x\tau + \tau^3/3;$$

then

$$f'(\tau) = x + \tau^2.$$

Thus $\tau = \pm\sqrt{-x}$ are the saddle points of the integrand. Note that these saddle points are very far away from the origin.

If $x < 0$, the saddle points are located at the contour of integration of the integral of (8.65), which is the real axis of τ. The problem can then be handled with the method of stationary phase. If $x > 0$, the saddle points are located in the complex τ-plane, and one must deform the contour of integration to a level curve of u passing through a saddle point.

Case A: $x < 0$

When x is negative, the two saddle points $\tau = \pm\sqrt{-x}$ are located on the contour of integration, which is the entire real axis. In such a case, these saddle points are also called points of stationary phase. It is straightforward to apply (8.45) and (8.46) to obtain the asymptotic form of the integral. But since the point of stationary phase depends on the large parameter of the integral, it is a little cleaner to scale out this parameter. Let $s = \tau/\sqrt{-x}$; then the points of stationary phase are $s = \pm 1$, independent of x. We have

$$x\tau + \tau^3/3 = \lambda(-s + s^3/3),$$

where $\lambda = (-x)^{3/2}$. The integral of (8.65) is equal to

$$I(x) = \sqrt{-x}\int_{-\infty}^{\infty} ds e^{i\lambda(-s+s^3/3)}. \tag{8.66}$$

The integral above is in the form of (8.36), where

$$u(s) = -s + s^3/3, h(s) = 1.$$

By (8.45) and (8.46) we get

$$I(x) \approx e^{-i\pi/4}\pi^{1/2}(-x)^{-1/4}\exp\left[\frac{2(-x)^{3/2}i}{3}\right] + c.c., \tag{8.67}$$

where *c.c.* means complex conjugation. The right side of (8.67) is a linear superposition of the two WKB solutions of the Airy equation.

Case B: $x > 0$

When x is positive, the saddle points are located at $\tau = \pm i\sqrt{x}$, which are on the imaginary axis of the complex τ-plane. Since the integrand has no singularities, the dominant contributions are expected to come from one or both of the saddle points $\tau = \pm i\sqrt{x}$. Let us scale the variable of integration by setting $s = \tau/\sqrt{x}$. We then have

$$I(x) = \sqrt{x}\int_{-\infty}^{\infty} ds e^{i\lambda f(s)},$$

where

$$f(s) = (s + s^3/3)$$

and where $\lambda = x^{3/2}$. The saddle points are now at $s = \pm i$, which are independent of x.

The integrand at $\pm i$ is given by

$$e^{i\lambda f(s)}|_{s=\pm i} = e^{\mp 2\lambda/3}.$$

Thus the integrand is exponentially large at $s = -i$. Since we know that the integral is exponentially small, the point $s = -i$ cannot possibly contribute to the integral. Thus we try to deform the contour into a level curve of u passing through the saddle point $s = i$. To make sure that the level curve we choose to deform to is a path of steepest descent, we study the integrand in the complex plane. We write out explicitly the real part and the imaginary part of the complex variable s as $s = t + iw$; then

$$f(s) = u + iv,$$

where

$$u = t(1 + t^2/3 - w^2)$$

and

$$v = w(1 + t^2 - w^2/3).$$

Since, at the saddle point $s = i$, we have $f(i) = 2i/3$, which has no real part, the level curves of u passing through this saddle point is

$$t(1 + t^2/3 - w^2) = 0.$$

The equation is satisfied if

$$t = 0 \tag{8.68}$$

or

$$1 + t^2/3 - w^2 = 0. \tag{8.69}$$

Each of the two equations above represents a level curve of u. As we have discussed, one of the level curves of u passing through the saddle point is a path of steepest descent and the other is a path of steepest ascent. We illustrate these two curves in the figure below.

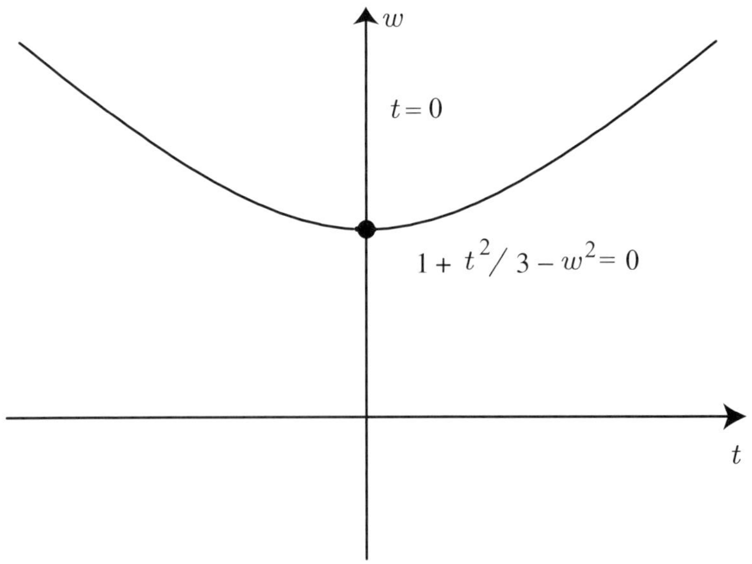

Figure 8.11

Now we have

$$f(i) = 2i/3, f'(i) = 0, f''(i) = 2i.$$

Thus near the point $s = i$, $f(s)$ is approximately given by

$$i\lambda f(s) = -2\lambda/3 - \lambda(t + iw)^2 + \cdots .$$

On the level curve of u given by (8.68), the integrand is approximately equal to $e^{-2\lambda/3+\lambda w^2}$, which increases rapidly as one leaves the point $s = i$. This shows that the path of (8.68) is a path of steepest ascent. Next we turn to the curve of (8.69). Near the point $s = i$, this curve is approximately horizontal. Therefore, on this level curve of u, we may set ω to zero and the integrand is approximately equal to $e^{-2\lambda/3-\lambda t^2}$. This integrand decreases rapidly as one leaves the point $s = i$ along this level curve. Thus this level curve is the path of steepest descent.

It is straightforward to prove that we may deform the contour of integration from the real axis to the curve $1 + t^2/3 - \omega^2 = 0$ illustrated in Figure 8.11. Let us give the justification in some detail. The original contour of integration is the real axis, which we shall take to be the part of the real axis from $-L$ to L, where L will be taken to be infinity at the end. We draw a circle with radius L and with the center at the origin. This circle intersects the curve $1 + t^2/3 - \omega^2 = 0$ in Figure 8.11 at two points. The point of intersection in the second quadrant will be called A and the point of intersection in the first quadrant will be called B. Since the integrand has no singularities, we may deform the contour of integration into any contour as long as the endpoints of this contour are $z = -L$ and $z = L$. We shall deform the contour into the one that consists of three parts. The first part is the circular arc joining the point $z = -L$ with A. The second part is the portion of the curve $1 + t^2/3 - \omega^2 = 0$ illustrated in Figure 8.11 between A and B. The third part is the circular arc joining B with $z = L$. Referring to the first diagram of Figure 8.16 in Appendix B of this chapter, we see that the two arcs of this deformed contour lie in the unshaded areas of the complex plane. (We may use the diagram in Figure 8.16 as $s + s^3/3$ is approximately equal to $s^3/3$ when the magnitude of s is large.) Therefore, the integrand of $I(x)$ at any interior point of these two arcs vanishes exponentially as $L \to \infty$. We may then prove that these two arcs can be ignored and the deformed contour of integration is the contour $1 + t^2/3 - \omega^2 = 0$ illustrated in Figure 8.11.

With the contour deformed into the path of steepest descent passing the saddle point $s = i$, we expand the exponent around the saddle point $s = i$ and get

$$I(x) \approx \sqrt{x}e^{-2\lambda/3} \int_{-\infty}^{\infty} dse^{-\lambda(s-i)^2} = \sqrt{\pi x/\lambda}e^{-2\lambda/3}$$
$$= \sqrt{\pi}x^{-1/4}e^{-2x^{3/2}/3}. \tag{8.70}$$

We recognize that this is one of the solutions given by (7.21) obtained with the WKB method.

In retrospect, there is no need to determine the entire path of steepest descent. This is because the dominant contributions come from a small neighborhood near the saddle point, where the contour of (8.69) is approximately horizontal. Thus we may do just as well by deforming the contour to the horizontal straight line $s = t + i$.

We will illustrate this simpler way of calculation with two more examples. But before we do this, let me make an observation. We note that the asymptotic form of $I(x)$ given by (8.65) depends on the sign of x. More precisely, it vanishes exponentially if $x \to \infty$, and is oscillatory if $x \to -\infty$.

Let us consider more generally the integral

$$I(z) = \int_{-\infty}^{\infty} e^{i(z\tau+\tau^3/3)}d\tau,$$

where z is complex. This integral is obtained from the integral of (8.65) by replacing x with z.

The saddle points of this integral are

$$\tau = \pm i\sqrt{z}.$$

It is clear that as long as z is not on the negative real axis, neither saddle point is on the real axis. Indeed, let

$$z = re^{i\theta}.$$

When z is not on the negative real axis, we may choose

$$-\pi < \theta < \pi.$$

The saddle point $\pm i\sqrt{z}$ is given by

$$\pm i\sqrt{z} = \pm i\sqrt{r}\left(\cos\frac{\theta}{2} + i\sin\frac{\theta}{2}\right),$$

and both have nonzero imaginary parts if z is not on the negative real axis. Indeed, the saddle point $i\sqrt{z}$ has a positive imaginary part if z is not on the negative real axis. Thus the saddle point $i\sqrt{z}$ is above the real axis. As before, it is easy to prove that we are always allowed to deform the contour upward to the horizontal line passing through the saddle point $i\sqrt{z}$. Thus we get, just as we have done in case B,

$$I(z) \approx \pi^{1/2}(z)^{-1/4}e^{-2z^{3/2}/3}, |z| \to \infty,$$

which holds for $-\pi < \arg z < \pi$. On the other hand, the asymptotic form of $I(z)$ is given by a different expression, namely (8.67), at $\arg z = \pi$ or $-\pi$, as was shown in case A.

That the asymptotic form of a function is different in different directions is called the Stokes phenomenon, and the line at which the transition takes place is called a Stokes line. For example, the negative real axis is a Stokes line for the Airy function.

Problem for the Reader

Find the leading term of the integral

$$I(\lambda) = \int_{-\infty}^{\infty} e^{i\lambda x}e^{-x^4/4}dx. \tag{8.71}$$

Note that this integral is the Fourier transform of $e^{-x^4/4}$.

Solution

The integral of (8.71) is in the form of the integral of (8.36), with $u(x)$ and $h(x)$ equal to x and $e^{-x^4/4}$, respectively. The integral has neither points of stationary phase nor finite endpoints. Since the magnitude of the integrand is always less than one, the integral vanishes faster than any inverse power of λ as λ goes to infinity.

To find out precisely how fast this integral vanishes, we scale the variable of integration by the transformation $x = \lambda^{1/3}t$, and the integral of (8.71) becomes

$$I(\lambda) = \lambda^{1/3}\int_{-\infty}^{\infty} dte^{i\Lambda f(t)}, \tag{8.72}$$

where

$$f(t) = t + it^4/4, \tag{8.73}$$

with $\Lambda = \lambda^{4/3}$. We see that the integral is in the form (8.52), with the large parameter equal to Λ and with $h(z)$ equal to unity. From (8.73) we get

$$f'(t) = 1 + it^3. \tag{8.74}$$

There are three roots for

$$f'(t_0) = 0. \tag{8.75}$$

They are $t_0 = e^{i\pi/6}$, $e^{5i\pi/6}$, and $-i$. To find out which ones of them contribute to the asymptotic form of the integral, we evaluate the value of $f(t)$ at each of the roots. To facilitate the evaluation, we make use of (8.75), which says that $t_0^3 = i$. Thus we have

$$f(t_0) = 3t_0/4.$$

The value of the integrand at $-i$ is exponentially large. Hence $-i$ is an irrelevant saddle point. The values of the integrand at the other two saddle points are both exponentially small and are of the same order of magnitude. Thus these two saddle points give the dominant contributions to the integral. Both of these saddle points are in the upper half t-plane, as shown in the figure below.

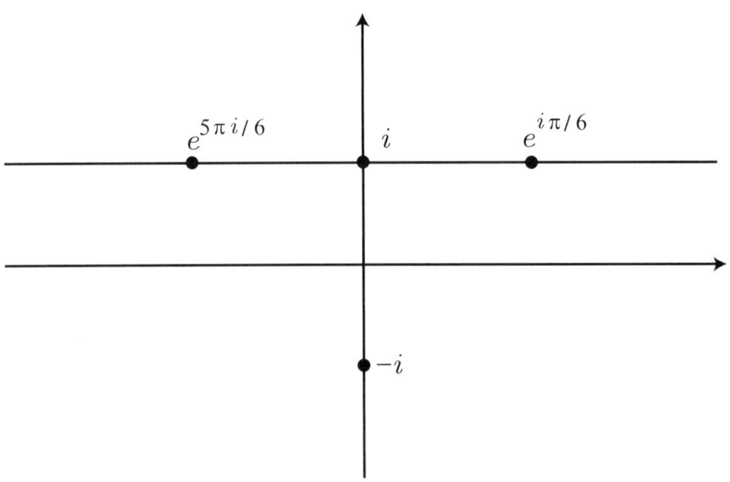

Figure 8.12

We have

$$f(e^{i\pi/6}) = 3e^{i\pi/6}/4,$$

and

$$f''(e^{i\pi/6}) = -3e^{-i\pi/6}.$$

Also

$$f(e^{5i\pi/6}) = -3e^{-i\pi/6}/4,$$

and

$$f''(e^{5i\pi/6}) = 3e^{i\pi/6}.$$

It turns out that the level curves of u passing through the two saddle points do not connect with each other. While it is possible to draw them, we will simply deform the contour of integration to the horizontal straight line $\mathrm{Im}t = 1/2$, which passes through the two saddle points $e^{i\pi/6}$ and $e^{5i\pi/6}$. Near the point $e^{i\pi/6}$, we have the Taylor series expansion

$$i\Lambda f(t) \approx \frac{3}{8}\Lambda\left(\sqrt{3}i - 1\right) - \frac{3}{2}\Lambda e^{i\pi/3}\left(t - e^{i\pi/6}\right)^2.$$

Thus the contribution from the point $e^{i\pi/6}$ is easily obtained to be

$$\lambda^{1/3}\sqrt{\frac{\pi}{3\Lambda e^{i\pi/3}/2}}\exp\frac{3}{8}\Lambda\left(\sqrt{3}i - 1\right).$$

The contribution from the point $e^{5i\pi/6}$ is given by the complex conjugation of the above. The sum of these two terms gives

$$I(\lambda) \approx \sqrt{\frac{8\pi}{3}}\lambda^{-1/3}\exp\left(-\frac{3}{8}\lambda^{4/3}\right)\cos\left(3\sqrt{3}/8\lambda^{4/3} - \pi/6\right). \tag{8.76}$$

Since the contour of integration we use is not a level curve of u, let us justify our procedure. On the horizontal line $t = T + i/2$, which is the contour of integration we have deformed to, the function $f(z)$ is given by

$$f(T + i/2) = u(T) + iv(T),$$

where

$$u(T) = 9T/8 - T^3/2$$

and

$$v(T) = T^4/4 - 3T^2/8 + 3/4.$$

It is easy to verify that $v'(T)$ vanishes at the two saddle points on this line. Also, $v''(T)$ is positive at these two points. Thus, on the contour of $t = T + i/2$, $v(T)$

is minimum at these two saddle points. Consequently, along this contour of integration, the magnitude of the integrand is maximum at the two saddle points. In addition, since $f'(t)$ vanishes at the saddle points, the phase of $e^{i\Lambda f(t)}$ is stationary at these points. Thus the contributions of the saddle points on the contour dominate compared to those from other points of the contour and can be calculated as we have done before. Indeed, since the magnitude of the integrand decreases as one leaves the saddle points, the contour is a path of descent, although not of steepest descent. Thus we may even obtain the next leading term for the integral by keeping more terms in the expansions of the exponent around the two saddle points.

Problem for the Reader

Find the leading term for the integral

$$I(\lambda) = \int_{-\infty}^{\infty} e^{i\lambda x} \left(e^{-1/x^2} - 1\right) dx.$$

Solution

The integral is of the form (8.36) with u and h equal to x and $(e^{-1/x^2} - 1)$, respectively. But since there are neither points of stationary phase nor finite endpoints, the method of stationary phase cannot be applied directly.

Due to the rapid oscillation of the integrand, the integral must vanish faster than any inverse power of λ as $\lambda \to \infty$. Thus we need to deform the contour to the complex plane. The integrand has no singularity in the complex plane. Let us deform the contour to the contour c, the endpoints of which are at the infinities of the upper half-plane. This is shown in the figure 8.13.

Since

$$\int e^{i\lambda z} dz = \frac{e^{i\lambda z}}{i\lambda},$$

$\int_c e^{i\lambda z} dz$ is equal to zero as $e^{i\lambda z}$ vanishes at the endpoints of c. Thus we have

$$I(\lambda) = \int_c e^{i\lambda z - 1/z^2} dz. \tag{8.77}$$

The integral above is convergent as $e^{i\lambda z}$ vanishes at the infinities of c.

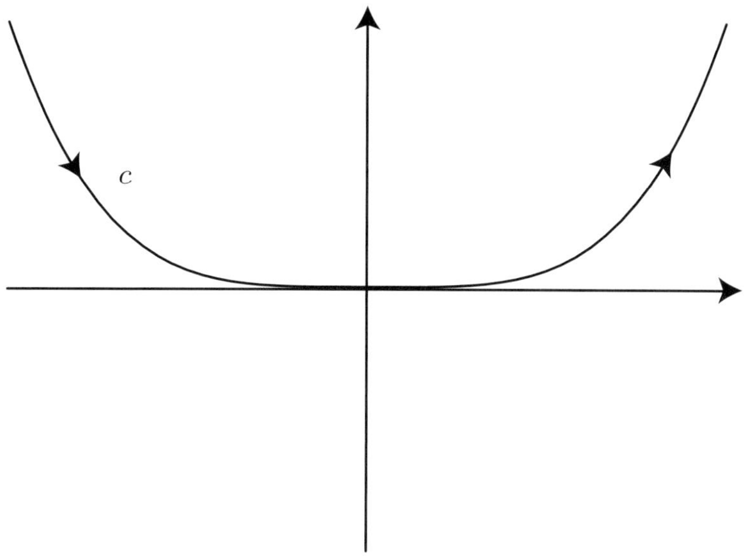

Figure 8.13

Let us scale the variable of integration by putting

$$z = 2^{1/3}\lambda^{-1/3}t;$$

then

$$I(\lambda) = 2^{1/3}\lambda^{-1/3}\int_c e^{i\Lambda f(t)}dt, \tag{8.78}$$

where

$$\Lambda = (\lambda/2)^{2/3}$$

and

$$f(t) = 2t + i/t^2. \tag{8.79}$$

We get

$$f'(t) = 2 - 2i/t^3 \tag{8.80}$$

and

$$f''(t) = 6i/t^4. \tag{8.81}$$

From (8.80), we find the relevant saddle points to be

$$t_0 = e^{i\pi/6}$$

and

$$t_0' = e^{5i\pi/6}.$$

The third saddle point at $t = -i$ cannot be relevant as

$$f(i) = -3i,$$

which makes the integrand exponentially large at $t = -i$.

We have

$$i\Lambda f(e^{i\pi/6}) = 3\Lambda e^{2i\pi/3}$$

and

$$i\Lambda f''(e^{i\pi/6}) = -6\Lambda e^{-2i\pi/3}.$$

Also,

$$i\Lambda f(e^{5i\pi/6}) = 3\Lambda e^{-2i\pi/3}$$

and

$$i\Lambda f''(e^{5i\pi/6}) = -6\Lambda e^{2i\pi/3}.$$

The contribution for the point $e^{i\pi/6}$ is

$$2^{1/3}\lambda^{-1/3}\sqrt{\frac{2\pi}{6\Lambda e^{-2i\pi/3}}}\exp\left(3\Lambda e^{2i\pi/3}\right).$$

The contribution from the point $e^{5i\pi/6}$ is the complex conjugation of the above. Thus we get

$$I(\lambda) \approx 2\left(\frac{2}{\lambda}\right)^{2/3}\sqrt{\frac{\pi}{3}}e^{-\frac{3}{2}(\lambda/2)^{2/3}}\cos\left[\frac{3\sqrt{3}}{2}(\lambda/2)^{2/3} + \frac{\pi}{3}\right]. \tag{8.82}$$

Finally, we give an example in which there are infinitely many saddle points.

Problem for the Reader

Find the leading term for the integral

$$I(\lambda) = \int_{-\infty}^{\infty} e^{i\lambda \sinh x} dx, \ \lambda \gg 1. \tag{8.83}$$

Solution

The endpoints of this integrand are $\pm\infty$. In addition, since

$$\frac{d}{dx}\sinh x = \cosh x \neq 0,$$

there is no point of stationary phase on the real axis. Thus the dominant contributions to this integral come from point or points in the complex plane.

Since the integrand has no singularities in the complex plane, any points of contribution must be saddle points in the complex plane. And since the integrand is of the order of unity on the real axis, we expect that the integral vanishes exponentially as $\lambda \to \infty$.

Let

$$f(z) = \sinh z.$$

The saddle points are the zeroes of

$$f'(z) = \cosh z.$$

Thus a saddle point z_0 satisfies

$$e^{z_0} + e^{-z_0} = 0,$$

which can be written as

$$e^{2z_0} = -1.$$

Taking the log of both sides, and remembering that

$$\log(-1) = i\pi + 2in\pi,$$

where n is any integer, we get

$$z_0 = \pm i\pi/2, \pm 3i\pi/2, \pm 5i\pi/2, \cdots \tag{8.84}$$

There are an infinite number of saddle points.

To see which of the saddle points are the contributing saddle points, we evaluate the integrand at each of the saddle points. We have

$$\sinh(x + iy) = \sinh x \cos y + i \cosh x \sin y.$$

Thus

$$\sinh\left(\pm i\pi \frac{2n+1}{2}\right) = \pm i(-1)^n. \tag{8.85}$$

The integrand of (8.83) is hence exponentially large at

$$z_0 = -i\pi/2, 3i\pi/2, -5i\pi/2, 7i\pi/2 \cdots,$$

which are eliminated as points of contribution.

Thus we deform the contour of integration to the upper half-plane. The original contour of integration can be considered to be the portion of the real axis from $-L$ to L, with L taken to be infinite at the end. Since the integrand of the integral in (8.83) is an entire function of the variable of integration, we may deform the contour of integration into any other contour as long as the endpoints of the deformed contour are $-L$ and L. The goal is to make the deformed contour pass through one or more of the saddle points, with the magnitude of the integrand largest at the saddle points.

Since

$$|e^{i\lambda \sinh(x+iy)}| = e^{-\lambda \cosh x \sin y},$$

the integrand vanishes as $x \to \pm\infty$ provided that $\sin y > 0$. Let us deform the contour into the one illustrated in Figure 8.9 with λ replaced by π. This contour consists of three parts: the vertical line joining $-L$ to $-L + i\pi/2$, the vertical line joining $L + i\pi/2$ to L, and the horizontal line $y = \pi/2$. In the limit as L goes to infinity, the integrand vanishes at the first two parts of the contour. Thus the two vertical lines can be ignored and the deformed contour of integration is just the horizontal line $y = \pi/2$.

To add a few words of clarification, we may not, for example, move the contour of integration to the horizontal line $y = -\pi/2$. This is because the integrand blows up at all points of the vertical line joining $-L$ to $-L - i\pi/2$ and the vertical line joining L to $L - i\pi/2$.

On the line $y = \pi/2$,

$$\sinh z = \sinh(x + i\pi/2) = i \cosh x.$$

Hence we have

$$I(\lambda) = \int_{-\infty}^{\infty} e^{-\lambda \cosh x} dx.$$

It is easy to see that the function $\cosh x$ is smallest at the point $x_0 = 0$. Note that since $y = \pi/2$, this is the saddle point $z_0 = i\pi/2$. The integrand is sharply peaked around this point. Applying the Laplace method yields

$$I(\lambda) \simeq \int_{-\infty}^{\infty} e^{-\lambda(1+x^2/2)} dx = e^{-\lambda}\sqrt{\frac{2\pi}{\lambda}},$$

which vanishes exponentially as λ goes to zero, as expected.

Appendix A: Gaussian Integrals

We give the values of some Gaussian integrals here.

They are

$$\int_{-\infty}^{\infty} e^{-x^2} dx = \Gamma(1/2) = \sqrt{\pi}, \tag{A.1}$$

$$\int_{-\infty}^{\infty} e^{-x^2} x^{2n} dx = \Gamma(n + 1/2), \tag{A.2}$$

and

$$\int_{-\infty}^{\infty} e^{-x^2} x^{2n+1} dx = 0. \tag{A.3}$$

Appendix B: Infinite Contours

Next we'll briefly discuss how to close the contour at infinity when the integrand involves exponential functions.

Consider the function $e^{iz} = e^{ix}e^{-y}$, which is exponentially small if y is large and positive and is exponentially large if y is large and negative. Thus it vanishes at the infinity of the upper half-plane and blows up at the infinity of the lower half-plane. Also, this function is oscillatory on the real axis that divides the region of the upper half-plane and region of the lower half-plane. We illustrate this in Figure 8.14, with the shaded area at the infinity of which the function is infinite. We note that each of the regions is of angular width π.

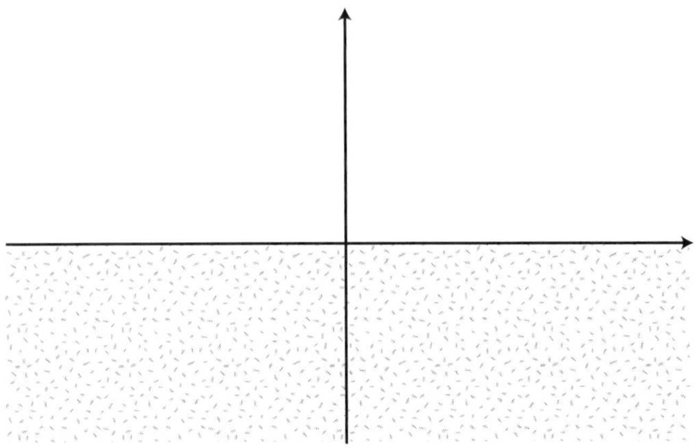

Figure 8.14

Consider next the function e^{iz^2}. Let $z = re^{i\theta}$; then $z^2 = r^2(\cos 2\theta + i\sin 2\theta)$. Hence we have

$$e^{iz^2} = e^{ir^2\cos 2\theta}e^{-r^2\sin 2\theta}.$$

From the expression above, we see that the function e^{iz^2} vanishes at the infinity with $\sin 2\theta > 0$, and blows up at the infinity with $\sin 2\theta < 0$. The angular width of each of the regions is $\pi/2$. We illustrate this below, with the shaded regions at the infinity of which e^{iz^2} is infinite.

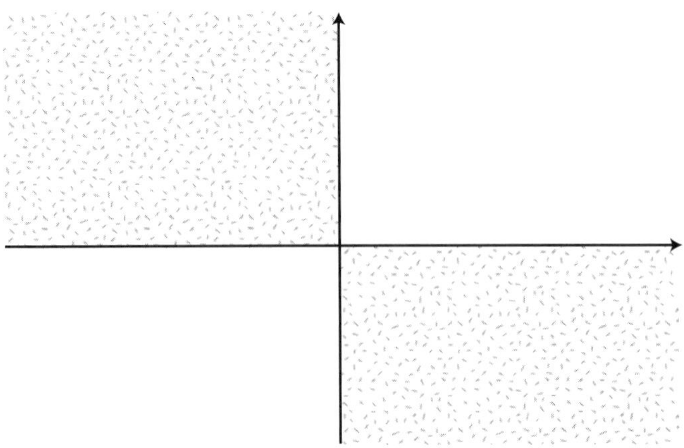

Figure 8.15

Similarly we illustrate the various regions of e^{iz^3} and e^{iz^4} in Figure 8.16. The shaded regions are the ones at the infinity of which the exponential function is infinite, and the unshaded regions are the ones at the infinity of which the exponential function vanishes. The angular width of each region for the former function is $\pi/3$, and that for the latter function is $\pi/4$.

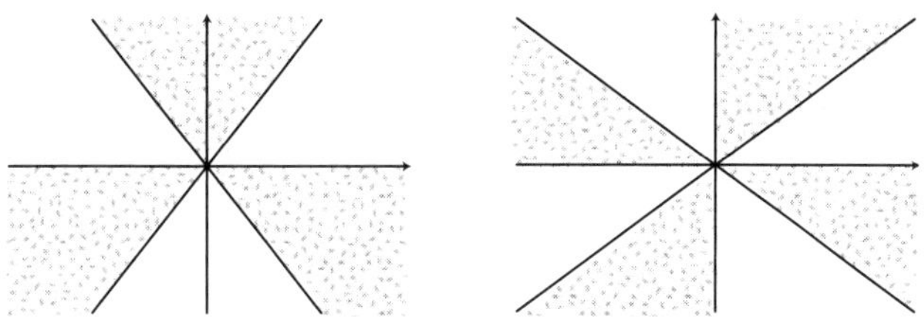

Figure 8.16

The integral over an arc that lies entirely at the infinity of an unshaded region is zero, as we shall show below. As an example, consider the integral

$$I = \int_{-\infty}^{\infty} e^{ix^3} dx.$$

Let us write it as

$$I = 2\text{Re } J,$$

where

$$J = \int_{0}^{\infty} e^{ix^3} dx.$$

The integrand of J is oscillatory. We would like to deform the contour into one on which the integrand is exponentially decreasing, for then we would be able to express it as a Gamma function. On such a contour,

$$z^3 = it^3,$$

where t is real and positive. This means that the contour should be

$$z = i^{1/3}t.$$

There are three roots for $i^{1/3}$: $e^{i\pi/6}$, $e^{5i\pi/6}$, and $e^{3i\pi/2}$. Which is the one we should use? We note from the figure above that we can change the contour from the positive real axis to the straight line making an angle $\pi/6$, but not to the other lines. Thus we have

$$J = e^{i\pi/6} \int_0^{\infty} e^{-t^3} dt.$$

Putting

$$t^3 = \tau,$$

we have

$$J = e^{i\pi/6} \int_0^{\infty} e^{-\tau} \tau^{-2/3} dt/3 = e^{i\pi/6} \frac{\Gamma(1/3)}{3}.$$

Thus

$$I = \frac{\Gamma(1/3)}{3}.$$

Next we consider the integral

$$I = \int_{-\infty}^{\infty} f(x) e^{ix} dx.$$

We shall prove that we may close the contour upstairs provided that $f(\infty) = 0$. The subtle point in this proof is that while e^{iz} vanishes exponentially in the upper half-plane with $y \to \infty$, it is finite when y is finite. Thus e^{iz} is finite if z is near the real axis. To show that this causes no difficulty, we close the contour by adding the three sides of a square at infinity, as shown in the figure below.

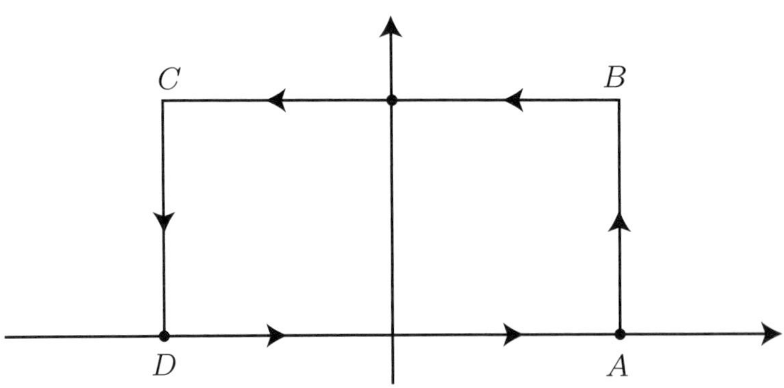

Figure 8.17

On the line BC, the integrand vanishes exponentially as L goes to infinity. Thus the contribution from this contour can be ignored. On the line CD, the magnitude of the integrand is less than $|f(z)|e^{-y}$. Since $f(z)$ vanishes at infinity by assumption, the factor $|f(z)|$ can be made arbitrarily small as L is chosen arbitrarily large. Also, the integral $\int_0^L e^{-y} dy$ is finite. Thus the contribution from CD is zero as L goes to infinity. Similarly, the integral over the line AB goes to zero as L goes to infinity. Thus the proof.

Homework Problems for This Chapter

Solutions to the Homework Problems can be found at www.lubanpress.com.

1. **a.** Find the integral representation of the solution of the differential equation

   $y'' + \alpha x y' + \beta y = 0,$

   where α and β are constants.

   Ans:

$$y(x) = \int_c e^{-xt+t^2/(2\alpha)} t^{\beta/\alpha - 1} dt.$$

   b. If α and β are negative, show that a possible choice of c is the positive real axis.

2. Find the leading term for the integral

$$\int_1^\infty \sqrt{x-1}\, e^{-\lambda \cosh x} dx,$$

   as well as that for the integral

$$\int_0^2 e^{\lambda x(1-x)} dx$$

   for $\lambda \gg 1$.

3. Find the leading term for each of the following integrals for $\lambda \gg 1$:

   a.

$$\int_{-1}^1 e^{-\lambda t^3} dt,$$

   b.

$$\int_1^\infty e^{-\lambda t^2} dt,$$

   c.

$$\int_{-2}^1 \sin x e^{-\lambda x^2} dx,$$

d.

$$\int_{-\pi}^{\pi} e^{-\lambda \sin x} dx,$$

e.

$$\int_{0}^{\infty} e^{-\lambda x} e^{-x^2} dx,$$

f.

$$\int_{0}^{\lambda} e^{t^3} dt,$$

g.

$$\int_{0}^{\infty} e^{-\lambda(t+t^5)} dt.$$

4. Consider the asymptotic series given by (8.24). Find the remainder R_n, the difference of $I(\lambda)$, and the sum of the first n terms of its asymptotic series, and prove that

$$|R_n| < M_n \Gamma(n+3/2) \lambda^{-n-3/2},$$

where M_n is independent of λ.

Hint: Prove that there is a M_n such that

$$\left| \frac{1}{\sqrt{1+y^2}} - \sum_{m=0}^{n} \frac{\Gamma(m+1/2)}{\Gamma(\frac{1}{2})m!} (-y^2)^m \right| < M_n y^{2n+2}.$$

5. Find the leading term for the integral

$$I(\lambda) = \int_{0}^{\infty} e^{-\lambda x - 1/x} dx, \lambda \gg 1,$$

which is the Laplace transform for $e^{-1/x}$. Can you explain qualitatively the asymptotic behavior of this integral?

Hint: If we identify $v(x)$ as x and $h(x)$ as $e^{-1/x}$, the minimum of $v(x)$ for x positive is $x = 0$, while the scale of the dominant region of contribution is $1/\lambda$. But when x is of the order of $1/\lambda$, $h(x)$ is exponentially small. Thus the region $x = O(1/\lambda)$ is not the dominant region of contribution. To determine the region of dominant contribution, we must take into account the factor $e^{-1/x}$.

6. Find the leading term for each of the integrals in

$$I(\lambda) = \int_1^\infty \sqrt{x-1}\ln(x-1)e^{-\lambda\cosh x}dx.$$

Give an estimate of the leading terms before you do the calculation.

7. Find the entire asymptotic series for each of the integrals in problem 3.

8. Find the leading term for each of the following integrals when $\lambda \gg 1$.

a. $\displaystyle\int_{-1}^{1} e^{i\lambda t^3}dt,$

b. $\displaystyle\int_{1}^{\infty} e^{i\lambda t^2}dt,$

c. $\displaystyle\int_{0}^{\pi} e^{i\lambda\cos x}dx,$

9. Find the third leading term of the Stirling formula.

10. Let (x_0, y_0) be a point at which the two curves $\xi(x,y) = 0$ and $\eta(x,y) = 0$ intersect. Show that if $\xi_x\eta_y - \xi_y\eta_x$ vanishes at (x_0, y_0), and if neither $\vec{\nabla}\xi(x_0, y_0)$ nor $\vec{\nabla}\eta(x_0, y_0)$ is a null vector, these two curves intersect tangentially.

11. Find the leading term for the integral

$$\int_0^\infty (1+x^2)^\lambda e^{-x^2}dx.$$

12. Let x_0 be a point of stationary phase for the integral (8.36). If the order of the lowest nonvanishing derivative of $u(x)$ at x_0 is n, give an estimate of the contribution of x_0 to the integral.

13. Expand $e^{ix\tau}$ in (8.65) into a Maclaurin series of x and show that $I(x)$ of (8.65) is a linear superposition of the two Maclaurin series solutions of the Airy equation given by (6.15) and (6.16).

14. Find the entire asymptotic series for the integral

$$I(\lambda) = \int_0^\infty e^{-\lambda x - \sqrt{x}} dx.$$

Is the series convergent?

15. Compute numerically the integrals for $\lambda = 1$, 3 and 10 in problem 3 and compare the values you obtain with the approximate values given by the leading terms.

16. Find the next leading term for the integral

$$I(\lambda) = \int_{-\infty}^\infty dx e^{i\lambda x}(e^{-1/x^2} - 1), \lambda \gg 1.$$

17. Find the leading term for

$$a_n = \int_0^{2\pi} \sqrt{2 + \cos\theta} e^{-in\theta} d\theta, n \gg 1,$$

which is the Fourier coefficient of $f(\theta) = \sqrt{2 + \cos\theta}$.

Hint: This integral has neither points of stationary phase nor contributions from the endpoints. Thus we need to transform it into a contour integral with the change of variable $z = e^{i\theta}$, getting

$$a_n = \frac{1}{2\pi i} \oint_C \frac{1}{z^{n+1}} \sqrt{\frac{z^2 + 4z + 1}{2z}} dz,$$

where c is the unit circle with its center at the origin. The integrand above is multivalued. You must make sure that you choose the correct branch for the integrand. In particular, since the point $\theta = 0$ and the point $\theta = 2\pi$ correspond to the same point $z = 1$ in the z-plane, the integrand considered as a function of z must be continuous across the real axis at $z = 1$. Now deform the contour of integration outward to make the absolute value of z as large as possible.

18. Find the leading term for the integral

$$I(\lambda) = \int_{-\infty}^\infty e^{i\lambda x} e^{ix^5/5} dx, \lambda \gg 1.$$

19. Find the asymptotic form of the integral representing the solution of the differential equation in problem 1. Do this for $x \to \infty$ as well as for $x \to -\infty$.

Hint: If x is positive, the magnitude of e^{-xt} is the largest if the real part of t is the smallest. Thus the magnitude of the integrand is largest at the leftmost point of c, which is the origin.

If $x < 0$, e^{-xt} is the largest when the real part of t is the largest. However, this does not mean that the magnitude of the integrand is the largest at the rightmost point of the contour c, which is the infinity on the positive real axis of t. This is because the integrand also has the factor $e^{t^2/(2\alpha)}$ which, since $\alpha < 0$, vanishes at $t = \infty$. Show that the integrand is maximum at $t = \alpha x$.

Reference

To find out more about the asymptotic forms as well as the integral representations of various special functions, see Erdelyi, Magnus, Oberhettinger, and Tricomi, *Higher Transcendental Functions* (Bateman Manuscript Project), Volumes 1, 2, and 3, McGraw-Hill (1953).

Chapter 9

Boundary Layers and Singular Perturbation

A. Regular Perturbation

In some physical problems, the solution depends on a parameter ϵ. When the parameter ϵ is very small, it is natural to expect that the solution will not be very different from the one with ϵ set to zero.

Consider for example the motion of a harmonic oscillator of unit mass with the spring constant equal to $(1 + \epsilon e^{-t})$. The equation of motion for this problem is

$$\ddot{x} + (1 + \epsilon e^{-t})x = 0, \tag{9.1}$$

with

$$x(0, \epsilon) = 1, \ \dot{x}(0, \epsilon) = 0, \tag{9.2}$$

where $x(t, \epsilon)$ is the displacement of the harmonic oscillator and $\dot{x}$ is the time derivative of x.

Equation (9.1) can be solved in a closed form and the exact solution is a Bessel function. But if ϵ is very small, the spring constant is approximately equal to unity at all times $t > 0$. Therefore, to the lowest order of approximation, we expect that we may ignore its small deviation from unity. Thus the differential equation for the harmonic oscillator is approximately

$$\ddot{x} + x \approx 0,$$

and the solution is approximately

$$x(t, \epsilon) \approx x(t, 0) = \cos t. \tag{9.3}$$

To find the small correction due to the small deviation of the spring constant from unity, one may anticipate that since the deviation is of the order of ϵ, this small correction is of the order of ϵ as well. Indeed, since all terms in the differential equation (9.1) are analytic functions of ϵ, one expects the solution of (9.1) to be in the form of a Taylor series of ϵ:

$$x(t,\epsilon) \approx x_0(t) + \epsilon x_1(t) + \cdots + \epsilon^n x_n(t) + \cdots, \tag{9.4}$$

where $x_n(t)$ is independent of ϵ. Substituting (9.4) into (9.1) and (9.2), we get

$$\left(\frac{d^2}{dt^2}+1\right)[x_0(t)+\epsilon x_1(t)+\cdots]+\epsilon e^{-t}[x_0(t)+\epsilon x_1(t)+\cdots]=0, \tag{9.5}$$

$$x_0(0)+\epsilon x_1(0)+\cdots = 1, \tag{9.6}$$

and

$$\dot{x}_0(0)+\epsilon \dot{x}_1(0)+\cdots = 0. \tag{9.7}$$

Setting ϵ in the equations above to zero, we get

$$\ddot{x}_0 + x_0 = 0,$$

with

$$x_0(0) = 1$$

and

$$\dot{x}_0(0) = 0.$$

The solution $x_0(t)$ is just $x(t,0)$ given by (9.3).

Setting to zero the coefficient of ϵ on the left side of (9.5), we get

$$\ddot{x}_1 + x_1 = -e^{-t}\cos t. \tag{9.8}$$

Similarly, we get from (9.6) and (9.7) that

$$x_1(0) = 0 \tag{9.9}$$

and

$$\dot{x}_1(0) = 0. \tag{9.10}$$

Using the operator method given in Chapter 1, we easily find that the solution of (9.8) satisfying (9.9) and (9.10) is

$$x_1(t) = \frac{1}{5}\cos t - \frac{3}{5}\sin t + e^{-t}\frac{2\sin t - \cos t}{5}.$$

Since both $x_1(t)$ and $x_0(t)$ are $O(1)$ at all times and since ϵ is very small, we have

$$|\epsilon x_1(t)| \ll |x_0(t)|$$

at all times, except when t is near the points where $x_0(t)$ happens to vanish. At such points, the solution $x(t, \epsilon)$ is small and of the order of ϵ.

One may further obtain the nth-order correction $x_n(t)$ from (9.5)–(9.7). We shall find that $x_n(t)$ is $O(1)$ at all times. Thus

$$|\epsilon^n x_n(t)| \ll |\epsilon^{n-1} x_{n-1}(t)|.$$

The inequalities above indicate that keeping more and more terms in the series of (9.4) makes the approximate solution better and better. The series (9.4) is a perturbation series, and the method given above in obtaining a perturbation series is called that of regular perturbation.

B. Boundary Layer Theory

It may be surprising, but not all problems with a small parameter can be solved by regular perturbation. Let us start the discussion with a simple example.

Consider the solution of the differential equation

$$\epsilon\frac{dy}{dx} + y = 0, \tag{9.11}$$

with the initial condition

$$y(0) = 1.$$

If we set ϵ in the differential equation above to zero, we get

$$y(x) = 0.$$

But this approximate solution does not satisfy the initial condition; in order to understand why this is so, we solve this differential equation in a closed form. We find that

$$y(x) = e^{-x/\epsilon}.$$

This solution of the differential equation satisfies the initial condition for all values of ϵ not equal to zero. But if we set ϵ to zero before we set x to zero, the solution above vanishes, and the initial condition is not satisfied.

Indeed, considered as a function of ϵ, the solution of the differential equation has an essential singularity at $\epsilon = 0$. It is therefore no wonder that it does not have a Maclaurin series of ϵ.

By way of comparison, consider the solution of the equation

$$\frac{dy}{dx} + \epsilon y = 0, \tag{9.12}$$

with the same initial condition. The exact solution is

$$y(x) = e^{-\epsilon x}.$$

The approximate solution obtained by setting ϵ in the differential equation to zero is

$$y(x) = 1,$$

which is a good approximation of the exact solution when ϵx is very small.

What makes the method of regular perturbation applicable for one but not for the other? We first observe that if we set ϵ to zero, eq. (9.12) remains a first-order differential equation, while eq. (9.11) turns into an algebraic equation. We also observe that when we drop $\epsilon y'$ from (9.11), we have implicitly assumed that $\epsilon y'$ is much smaller than y. But if $y(x)$ is a rapidly varying function of x, the magnitude of y' is much larger than that of y. In such a case, the magnitude of $\epsilon y'$ may not be small compared to that of y. As a matter of fact, we may easily verify that $\epsilon y'$ and y are of the same order of magnitude if $y(x) = e^{-x/\epsilon}$, the solution of (9.11).

We encounter similar difficulties with boundary-layer problems in fluid mechanics. Let us consider the second-order differential equation

$$\epsilon y''(x) + a(x)y'(x) + b(x)y(x) = 0, \tag{9.13}$$

which generally cannot be solved in a closed form. Without loss of generality, we shall choose the interval in which (9.13) holds to be $[0, 1]$. Typically, we require that the solution satisfy the boundary conditions

$$y(0) = A, y(1) = B. \tag{9.14}$$

The parameter ϵ is very small, while A and B are $O(1)$. Note that the small parameter ϵ is the coefficient of y''.

This problem is actually a simplified model of the boundary-layer problem in fluid mechanics. As we know, the highest-order term in the Navier-Stokes equation in fluid mechanics is equal to $\nu\nabla^2\vec{v}$, where $\vec{v}$ is the velocity of the fluid and ν is the kinematic viscosity. The kinematic viscosity is often a small parameter, and the order of the Navier-Stokes equation decreases if we ignore the viscosity.

Equation (9.13) can be solved with good accuracy with the use of the WKB method. (See homework problem 1.) However, we gain more insights into this problem by starting afresh.

Let us set $\epsilon = 0$. Then (9.13) becomes

$$a(x)y' + b(x)y = 0.$$

This equation is readily solved and we get

$$y_s(x) = c_s \exp\left[-\int \frac{b(x)}{a(x)}dx\right], \tag{9.15}$$

where c_s is a constant.

However, eq. (9.13) is of the second order and has two independent solutions. But the approximate equation we obtain is of the first order and has only one independent solution. Thus the approximate method we use yields only one of the independent solutions. This is to say that we lose one of the independent solutions as we set ϵ in the differential equation to zero. It is impossible to satisfy the two boundary conditions of (9.14) with only one solution.

How do we find the missing solution approximately? To get some insight into the answer, let us go to the much simpler case in which a and b are constants. Then the general solution of (9.13) is equal to

$$c_1 e^{r_1 x} + c_2 e^{r_2 x},$$

where c_1 and c_2 are arbitrary constants and r_1 and r_2 are the roots of the quadratic equation

$$\epsilon r^2 + ar + b = 0.$$

This quadratic equation can be solved easily. But if we set ϵ in this quadratic equation to zero, we get only one root

$$r \approx -b/a,$$

which gives the solution

$$e^{-bx/a}.$$

The solution is the one denoted by $y_s(x)$ in (9.15), with $a(x)$ and $b(x)$ replaced by the constants a and b, respectively. The other root of the quadratic equation is obtained by solving the equation exactly. When ϵ is very small, this root is approximately given by

$$-a/\epsilon.$$

This root tells us that the second solution of the differential equation is approximately

$$e^{-ax/\epsilon}.$$

We see that this solution is a rapidly varying function of x. Indeed, the magnitude of this solution is a rapidly increasing function of x if a is negative and a rapidly decreasing function of x if a is positive.

Note that we have been careful in stating that it is the magnitude of the solution, not the solution itself, that is rapidly increasing (decreasing) when a is negative (positive). This is because while $e^{x/\epsilon}$, for example, is a rapidly increasing function of x, $-e^{x/\epsilon}$ is a rapidly decreasing function of x. This shows that whether a solution is rapidly increasing or rapidly decreasing depends on the overall sign of the solution. Sometimes we do not make such a fine distinction, with the understanding that the overall sign of the solution is positive.

That the missing solution is a rapidly varying function of x also explains why it is not always legitimate to drop the y'' term in (9.13). Let us denote the rapidly varying solution by $y_r(x)$. Then $y_r'(x)$ is of the order of ϵ^{-1} times $y_r(x)$, and $y_r''(x)$ is of the order of ϵ^{-2} times $y_r(x)$. Therefore, the term $\epsilon y_r''(x)$ is bigger than $y_r(x)$ by a large multiple of ϵ^{-1}. Again, while it is true that ϵy, say, is always much smaller than y, it is not always true that $\epsilon y''$ is much smaller than y.

On the other hand, y_s'' and y_s' are both of the same order of magnitude as y_s. Thus it is justified to treat $\epsilon y_s''$ as a small perturbation term. Thus we write (9.13) as

$$a(x)y_s'(x) + b(x)y_s(x) = -\epsilon y_s''(x), \tag{9.16}$$

and seek the solution $y_s(x)$ in the perturbation series

$$y_s(x) = y_s^{(0)}(x) + \epsilon y_s^{(1)}(x) + \cdots .$$

Substituting this series into (9.16), we easily find that $y_s^{(0)}(x)$ is given by (9.15), and that

$$a(x)\frac{d}{dx}y_s^{(n)}(x) + b(x)y_s^{(n)}(x) = -\frac{d^2}{dx^2}y_s^{(n-1)}(x), n = 1, 2 \cdots . \tag{9.17}$$

Thus $y_s^{(n)}(x)$, $n = 1, 2, \cdots$, satisfies a first-order ODE with the inhomogeneous term equal to the negative of the second-order derivative of $y_s^{(n-1)}(x)$. Once $y_s^{(0)}$ is found, we may then solve (9.17) with $n = 1$ and obtain $y_s^{(1)}$. Similarly, all $y_s^{(n)}(x), n = 2, 3, \cdots$, can be obtained by successive iteration.

We remark that by treating $\epsilon y_s''$ as a small perturbation, we have implicitly assumed that the solution is slowly varying. Since all terms in the perturbation series so obtained are slowly varying, the assumption is justified and the method of regular perturbation produces successive approximations of the slowly varying solution. On the other hand, this method cannot produce the rapidly varying solution. Clearly, if the solution is rapidly varying, it is not justified to treat $\epsilon y''$ as a small perturbation term.

To find successive approximations of the rapidly varying solution, we note that when $a(x)$ and $b(x)$ are constants, the rapidly varying function is

$$\exp(-ax/\epsilon).$$

Thus we surmise that when $a(x)$ and $b(x)$ depend on x, the rapidly varying solution is approximately

$$\exp\left[-\epsilon^{-1}\int a(x)dx\right].$$

Therefore, we shall put

$$y_r(x) = c_r \exp\left[-\epsilon^{-1}\int a(x)dx\right]v(x). \tag{9.18}$$

With the rapidly varying behavior of $y_r(x)$ taken up by the exponential factor, we expect $v(x)$ to be slowly varying. Substituting (9.18) into (9.13), we get

$$av' + (a' - b)v = \epsilon v''. \tag{9.19}$$

Treating the term $\epsilon v''$ as a small perturbation we may solve (9.19) with regular perturbation. We put

$$v(x) = v_0(x) + \epsilon v_1(x) + \cdots, \tag{9.20}$$

and get

$$v_0(x) = \frac{1}{a(x)}\exp\left[\int \frac{b(x)}{a(x)}dx\right] \tag{9.21}$$

as well as

$$a\frac{d}{dx}v_n + (a' - b)v_n = \frac{d^2}{dx^2}v_{n-1}. \tag{9.22}$$

Note that, just like (9.17) for $y_s^{(n)}$, (9.22) is a first-order linear differential equation for v_n and can be solved in a closed form once v_{n-1} is given. Thus we may obtain v_n for all n by successive iterations.

We note that this prodecure yields us only the solution that is in the form of (9.18), where v is slowly varying. Hence it does not yield y_s.

In summary, the two independent solutions of (9.13) are a slowly varying solution $y_s(x)$ and a rapidly varying solution $y_r(x)$. The general solution of this equation is

$$y(x) = y_s(x) + y_r(x). \tag{9.23}$$

In the lowest order, $y_s(x)$ is given by (9.15) and

$$y_r(x) = \frac{c_r}{a(x)} \exp \int \left(-\frac{a}{\epsilon} + \frac{b}{a}\right) dx, \tag{9.24}$$

where c_r is a constant. We note that $|y_r(x)|$ is a rapidly increasing function of x if $a(x) < 0$, and is a rapidly decreasing function of x if $a(x) > 0$.

The solutions $y_s(x)$ and $y_r(x)$ are so far specified only up to multiplicative constants, which are fixed by the boundary conditions (9.14). In most problems of physical interest, the magnitudes of the boundary values A and B in (9.14) are of the order of unity.

If $|y_r(x)|$ is rapidly decreasing, we expect that it is appreciable only in a small neighborhood near the lower endpoint $x = 0$, and drops to very small values as x gets away from the origin. Thus, if c_r, the coefficient of the rapidly varying solution, is positive, the graph for the solution (9.23) goes up sharply as the origin is approached. And if c_r is negative, the curve goes down sharply as the origin is approached. (See Figure 9.1 below.) In either case, the solution is said to have a boundary layer near the lower endpoint.

Similarly, if $|y_r(x)|$ is rapidly increasing, we expect it to be appreciable only near the upper endpoint $x = 1$. Thus the solution of the boundary-value problem has a boundary layer near the upper endpoint $x = 1$. Therefore, if c_r is positive, the graph for the solution (9.23) goes up sharply as the endpoint $x = 1$ is approached. And if c_r is negative, the curve goes down sharply as the endpoint $x = 1$ is approached.

Let us verify these statements. We have from (9.14) and (9.23) that

$$y_s(0) + y_r(0) = A \tag{9.25}$$

and

$$y_s(1) + y_r(1) = B. \tag{9.26}$$

From these two equations we get

$$y_r(1)y_s(0) - y_r(0)y_s(1) = y_r(1)A - y_r(0)B. \tag{9.27}$$

Consider first the case $a(x) > 0$. In this case, $|y_r(x)|$ is rapidly decreasing and hence $|y_r(0)|$ is much larger than $|y_r(1)|$. Since $y_s(0)$ and $y_s(1)$ are of the same order of magnitude, we may drop the first term on the left side of (9.27) as well as the first term on the right side of (9.27) and get

$$y_s(1) \approx B. \tag{9.28}$$

Equation (9.28) determines $y_s(x)$ uniquely. We then have from (9.25) that

$$y_r(0) \approx A - y_s(0). \tag{9.29}$$

Equation (9.29) determines $y_r(x)$ uniquely. From (9.29), we see that $y_r(x)$ is of the order of unity at $x = 0$; hence it is appreciable only if x is close to the lower endpoint $x = 0$, dropping to very small values outside the small neighborhood known as a boundary layer.

Since $y_r(x)$ is negligible in the region outside the boundary layer, we have $y(x) \approx y_s(x)$ in this region. Let the solution in the region outside the boundary layer be denoted as y_{out}, and we have

$$y_{out}(x) \approx B \exp \int_x^1 \frac{b(x')}{a(x')} dx'. \tag{9.30}$$

This solution satisfies the boundary condition at $x = 1$.

Next we shall find the solution valid inside the boundary layer near $x = 0$. In this region, $y_r(x)$ is appreciable. We have from (9.24) that near $x = 0$, $y_r(x)$ is approximately equal to a constant times

$$\exp\left[-a(0)x/\epsilon\right].$$

Together with (9.29), we find that

$$y_r(x) \approx [A - y_s(0)] \exp\left[-\frac{a(0)x}{\epsilon}\right]. \tag{9.31}$$

Also, when x is small, we can approximate $y_s(x)$ by $y_s(0)$. Thus the solution (9.23) inside the boundary layer is given by

$$y_{in}(x) = y_s(0) + [A - y_s(0)] \exp\left[-\frac{a(0)x}{\epsilon}\right], \tag{9.32}$$

where

$$y_s(0) \approx B \exp \int_0^1 \frac{b(x')}{a(x')} dx'.$$

We combine (9.30) and (9.32) and obtain a solution valid for all values of x:

$$y_{uniform}(x) = B \exp \int_x^1 \frac{b(x')}{a(x')} dx' + [A - y_{out}(0)] \exp\left[-\frac{a(0)x}{\epsilon}\right]. \tag{9.33}$$

The second term on the right side of (9.33) is appreciable only if x is of the order of ϵ, dropping to zero when x is much larger than ϵ. Therefore, the solution has a boundary layer of width ϵ near the lower endpoint $x = 0$.

If $a(x) < 0$, $|y_r(x)|$ is rapidly increasing, and the magnitude of $y_r(1)$ is exponentially larger than that of $y_r(0)$. Since $y_s(0)$ and $y_s(1)$ are of the same order of magnitude, (9.27) gives

$$y_s(0) \approx A. \tag{9.34}$$

From (9.26) we get

$$y_r(1) = B - y_s(1), \tag{9.35}$$

which is $O(1)$. Since its magnitude is rapidly increasing, $y_r(x)$ is appreciable only when x is near the upper endpoint. The solution (9.23) is said to have a boundary layer of width ϵ near $x = 1$. Outside the boundary layer, $y_r(x)$ is negligible and (9.23) gives $y(x) \approx y_s(x)$. Thus the solution outside the boundary layer is approximately equal to

$$y_{out}(x) \approx A \exp\left[-\int_0^x \frac{b(x')}{a(x')} dx'\right]. \tag{9.36}$$

Inside the boundary layer near $x = 1$, the rapidly varying solution $y_r(x)$ is appreciable and is given by

$$y_r(x) \approx [B - y_s(1)] \exp\left[\frac{a(1)(1-x)}{\epsilon}\right]. \tag{9.37}$$

Also, inside the small region of the boundary layer, we may replace $y_s(x)$ by its value at $x = 1$. Thus (9.23) gives

$$y_{in}(x) = y_{out}(1) + [B - y_{out}(1)] \exp\left[\frac{a(1)(1-x)}{\epsilon}\right], \tag{9.38}$$

where

$$y_{out}(1) = A \exp\left[-\int_0^1 \frac{b(x)}{a(x)} dx\right].$$

Combining (9.36) and (9.38), we obtain an approximate solution that holds for all values of x:

$$\begin{aligned} y_{uniform}(x) &= A \exp\left[-\int_0^x \frac{b(x')}{a(x')} dx'\right] \\ &\quad + [B - y_{out}(1)] \exp\left[\frac{a(1)(1-x)}{\epsilon}\right]. \end{aligned} \tag{9.39}$$

In a specific calculation, we may, instead of plugging the functions a and b into the formulae for y_{out} and y_{in}, obtain both y_{out} and y_{in} directly from (9.13). First of all, the differential equation for y_{out} is obtained by neglecting the term $\epsilon y''$ in (9.13). The resulting equation is of the first order and can be solved in a closed form. The multiplicative constant in this solution can be fixed with the boundary condition at the endpoint where $y_r(x)$ is negligible.

As for $y_{in}(x)$, let us first consider the case with $a(x) < 0$. Then the boundary layer is at $x = 1$. Thus we seek the approximate solution of (9.13) valid in a small neighborhood near $x = 1$. In this neighborhood, we approximate $a(x)$ in (9.13) by $a(1)$. We shall also drop the term $b(x)y$ in (9.13). Then (9.13) becomes

$$\epsilon y_{in}'' + a(1) y_{in}' = 0.$$

The two independent solutions of the equation above are 1 and $e^{-a(1)x/\epsilon}$. The former solution is the slowly varying solution, which is approximately a constant in the small neighborhood. The latter solution agrees with the rapidly varying solution given by (9.37). Thus we have

$$y_{in}(x) = y_{out}(1) + c_2 e^{a(1)(1-x)/\epsilon}. \tag{9.40}$$

The coefficient c_2 is determined from the boundary condition at $x = 1$. We get

$$c_2 = B - y_{out}(1).$$

If $a(x) > 0$, the boundary layer is at $x = 0$ and we approximate (9.13) near $x = 0$ by

$$\epsilon y_{in}'' + a(0) y_{in}' = 0.$$

This equation yields

$$y_{in}(x) = y_{out}(0) + c_3 e^{-a(0)x/\epsilon}. \tag{9.41}$$

We determine c_3 from the boundary condition at $x = 0$:

$$c_3 = A - y_{out}(0).$$

In the literature, one of the constants in $y_{in}(x)$ is determined by matching $y_{in}(x)$ with $y_{out}(x)$. Such a matching is not needed, as both $y_s(x)$ and $y_r(x)$ are good approximations throughout the interval $[0, 1]$.

Take, for example, the case of $a(x) > 0$. In this case, the boundary layer is near $x = 0$, and $y_{in}(x)$ is the sum of a constant and a rapidly varying exponential function with a decreasing magnitude. The constant is equal to $y_{out}(0)$. This is true to all orders of perturbation. Thus the constant in $y_{in}(x)$ is easily found once $y_{out}(x)$ has been computed. The saving of labor is not much in the zeroth order, but is significant as we compute the next-order terms. In the literature, the next-order terms of $y_r(x)$ are obtained by treating as perturbation terms all the terms in (9.13) not kept in the approximate differential equation we just derived for $y_s(x)$. The perturbative solution in each order has two arbitrary constants. One of the constants is determined by matching $y_{in}(x)$ with $y_{out}(x)$ in the region $1 \gg x \gg \epsilon$, and the other by satisfying the boundary condition at $x = 0$. In contrast, we solve (9.22), which is easier to do as it is of the first order. The solution of this equation has only one arbitrary constant. This constant is determined by the boundary condition at $x = 0$ and no matching is required. (These remarks obviously apply equally well if $a(x)$ is negative and the boundary layer is at $x = 1$.)

Problem for the Reader

Solve approximately the equation

$$\epsilon y'' + (1 + x)y' + y = 0, \; 0 < x < 1,$$

with the boundary conditions

$$y(0) = y(1) = 1.$$

Solution

We have

$$a(x) = 1 + x > 0;$$

thus the magnitude of $y_r(x)$ is rapidly decreasing and there is a boundary layer near $x = 0$ with a width of order ϵ. For x outside the boundary layer, we have

$$(1 + x)y'_{out} + y_{out} = 0.$$

Thus we get

$$y_{out}(x) = \frac{c}{1 + x}.$$

The constant c is determined from the boundary condition at $x = 1$ and we have

$$y_{out}(x) = \frac{2}{1 + x}.$$

In particular,

$$y_{out}(0) = 2.$$

To obtain $y_{in}(x)$, we make the approximation

$$a(x) \approx a(0) = 1.$$

Thus

$$\epsilon y''_{in} + y'_{in} \approx 0.$$

Therefore we have

$$y_{in}(x) = 2 + c' e^{-x/\epsilon}.$$

The boundary condition

$$y(0) = 1$$

requires that

$$c' = -1.$$

Thus

$$y_{in}(x) = 2 - e^{-x/\epsilon}.$$

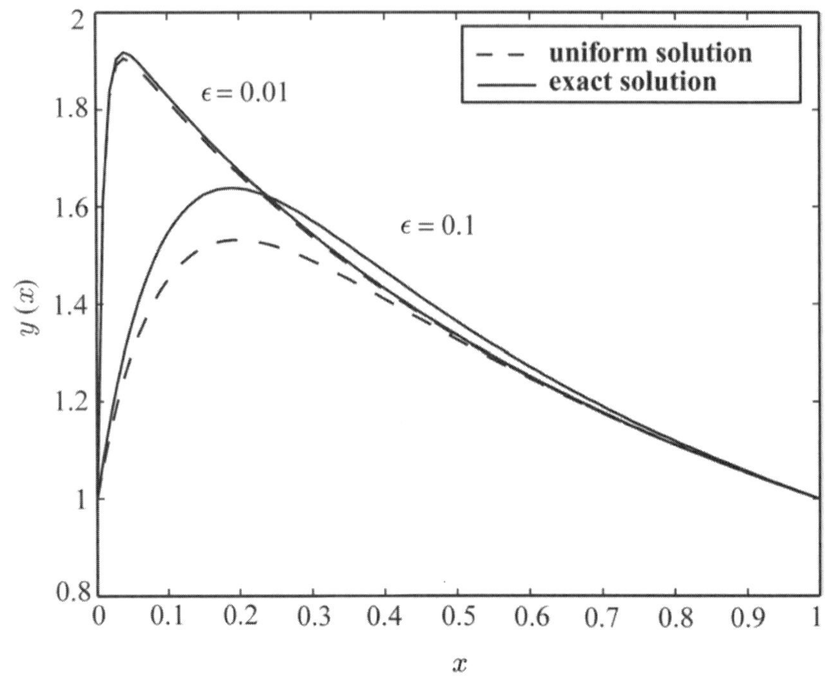

Figure 9.1

We also find that

$$y_{uniform} = \frac{2}{1+x} - e^{-x/\epsilon}.$$

The solution is plotted in the figure above. We see that it is given by $2/(1+x)$ almost everywhere. The function $2/(1+x)$ is equal to 2 at $x = 0$. But this function does not satisfy the boundary condition at $x = 0$. Thus the rapidly varying solution y_r kicks in near the point $x = 0$, and the solution y goes through a rapid transition inside the boundary layer, dipping to the value 1 at $x = 0$.

Problem for the Reader

Find the next-order terms for the differential equation in the preceding problem.

Solution

We first obtain y_{out} up to the next order by writing the ODE in the form

$$\frac{d}{dx}y_{out}(x) + \frac{1}{1+x}y_{out}(x) = -\frac{\epsilon}{1+x}\frac{d^2}{dx^2}y_{out}(x),$$

with

$$y_{out}(1) = 1,$$

treating the right side as a perturbation term. The solution y_{out} can be obtained from the equation above by iteration. Let

$$y_{out}(x) = y_0(x) + \epsilon y_1(x) + \cdots;$$

then

$$\frac{d}{dx}y_0(x) + \frac{1}{1+x}y_0(x) = 0,$$

with

$$y_0(1) = 1.$$

Thus

$$y_0(x) = \frac{2}{1+x},$$

as we have obtained previously. We also have

$$y_1'(x) + \frac{1}{1+x}y_1(x) = -\frac{4}{(1+x)^4},$$

with

$$y_1(1) = 0.$$

Thus we get

$$y_1(x) = \frac{4}{1+x}\int_x^1 \frac{dx'}{(1+x')^3} = -\frac{1}{2(1+x)} + \frac{2}{(1+x)^3}.$$

Therefore, up to first order in ϵ,

$$y_{out}(x) = \frac{2}{1+x} - \frac{\epsilon}{2(1+x)} + \frac{2\epsilon}{(1+x)^3}.$$

In particular,

$$y_{out}(0) = 2 + \frac{3\epsilon}{2}.$$

This is the approximate solution of the differential equation outside the boundary layer near $x = 0$.

The rapidly varying solution $y_r(x)$ is appreciable inside the boundary layer. Thus the solution of the differential equation inside the boundary layer is equal to

$$y_{in}(x) = y_r(x) + y_s(x).$$

While $y_s(x)$ has already been determined, we need to determine $y_r(x)$. First we note that the boundary condition $y(0) = 1$ gives

$$y_r(0) = -1 - \frac{3\epsilon}{2}.$$

Next we put

$$y_r(x) = v(x)\exp\left(-\int \frac{1+x}{\epsilon}dx\right) = v(x)\exp\left(-\frac{x + x^2/2}{\epsilon}\right).$$

By (9.19), the function $v(x)$ satisfies

$$(1+x)v' = \epsilon v''.$$

Since

$$y_r(0) = v(0),$$

we have

$$v(0) = -1 - \frac{3\epsilon}{2}.$$

We easily find

$$v_0(x) = -1$$

and

$$v_1(x) = -3/2.$$

Thus

$$y_r(x) = -\left(1 + \frac{3\epsilon}{2}\right)\exp\left(-\frac{x + x^2/2}{\epsilon}\right)$$

and

$$y_{in}(x) = 2 + \frac{3\epsilon}{2} - \left(1 + \frac{3\epsilon}{2}\right) \exp\left(-\frac{x + x^2/2}{\epsilon}\right).$$

We also have

$$y_{uniform}(x) = \frac{2}{1+x} - \frac{\epsilon}{2(1+x)} + \frac{2\epsilon}{(1+x)^3} - \left(1 + \frac{3\epsilon}{2}\right) \exp\left(-\frac{x + x^2/2}{\epsilon}\right).$$

The boundary-layer method can also be applied to some nonlinear problems. Below is an example.

Problem for the Reader

Solve approximately the nonlinear equation

$$\epsilon y'' + y' + e^{-y} = 0, \; 0 < x < 1,$$

with the boundary conditions

$$y(0) = y(1) = 0.$$

Solution

We have

$$a(x) = 1 > 0.$$

Thus the magnitude of the rapidly varying solution decreases with x and there is a boundary layer of width ϵ at $x = 0$.

Outside the boundary layer, we have

$$y'_{out} + e^{-y_{out}} = 0.$$

Together with the boundary condition at $x = 1$, we get

$$y_{out}(x) = \ln(2 - x).$$

Inside the boundary layer, we have

$$\epsilon y''_{in} + y'_{in} = 0.$$

Thus we get

$$y_{in}(x) = \ln 2 - \ln 2e^{-x/\epsilon},$$

where the boundary condition at $x = 0$ has been utilized. From the above we also get

$$y_{uniform}(x) = \ln(2 - x) - \ln 2e^{-x/\epsilon}.$$

C. Turning Points

In the preceding section, we assume that $a(x)$ is of the same sign throughout the interval in which the differential equation (9.13) holds. The solutions $y_r(x)$ and $y_s(x)$, given by (9.24) and (9.15), respectively, are good approximations of the two independent solutions of the differential equation.

But if x_0 is a simple zero of $a(x)$, the function $a(x)$ changes sign at x_0, and the approximations we have made in the preceding section fail in the neighborhood of x_0. Borrowing the terminology from quantum mechanics, we shall call x_0 a turning point of eq. (9.13).

To obtain approximate solutions near a turning point, we approximate $a(x)$ and $b(x)$ in (9.13) by algebraic functions of x, and solve the resulting equation exactly. This is the same approach as the one used to handle the turning point of the wave equation discussed in Chapter 7. The only difference is that the approximate solutions for (9.13) near a turning point are parabolic cylinder functions rather than Airy functions, as we shall show.

It will be convenient to get rid of the y' term in (9.13). Let us write (9.13) as

$$\left(D + \frac{a}{\epsilon}\right) Dy + \frac{b}{\epsilon} y = 0.$$

This tells us that to eliminate the y' term, we can put

$$y = \exp\left(-\int^x dx' \frac{a(x')}{2\epsilon}\right) Y(x). \tag{9.42}$$

This is because as we move the exponential factor to the left of D, we make the replacement

$$D \to D - \frac{a}{2\epsilon}.$$

Thus eq. (9.13) becomes

$$\left(D+\frac{a}{2\epsilon}\right)\left(D-\frac{a}{2\epsilon}\right)Y+\frac{b}{\epsilon}Y=0.$$

It is clear that the differential equation above does not have a Y' term. Indeed, it is equal to

$$D^2Y+\left[\frac{2b(x)-a'(x)}{2\epsilon}-\frac{a^2(x)}{4\epsilon^2}\right]Y=0. \tag{9.43}$$

Since $1/(4\epsilon^2)$ is very large, we see that an alternative way to solve eq. (9.43) is to use the WKB method, approximating $\eta(x)$ in (7.16) by

$$\sqrt{\frac{a^2+2\epsilon(a'-2b)}{4\epsilon^2}}\simeq\frac{|a|}{2\epsilon}+\frac{a'-2b}{2|a|}.$$

The WKB approximation is valid provided that $a(x)$ does not vanish. This approximation yields for Y two WKB solutions; the magnitude of one of them is a rapidly increasing function of x, while the magnitude of the other is a rapidly decreasing function of x. By (9.42), we multiply these two WKB solutions by the exponential function

$$\exp\left(-\frac{1}{2\epsilon}\int a dx\right)$$

and the WKB solutions become y_s and y_r. (See homework problem 1.)

The WKB approximation fails at a turning point. Let x_0 be a turning point. We shall assume that x_0 is a simple zero of $a(x)$. Then we have, near $x=x_0$,

$$a(x)\approx\alpha(x-x_0)$$

and

$$b(x)\approx\beta.$$

By (7.18), the WKB approximation holds only if

$$|x-x_0|\gg\sqrt{\epsilon}.$$

When x is close to x_0, or more precisely

$$|x-x_0|\ll 1,$$

eq. (9.43) is approximately

$$\frac{d^2Y}{dx^2} + \left[\frac{2\beta - \alpha}{2\epsilon} - \frac{\alpha^2(x - x_0)^2}{4\epsilon^2}\right] Y \approx 0, \tag{9.44}$$

which turns out to be exactly solvable. Before we show how to do this, we first note that when x is near x_0, we have

$$\exp\left(-\int_{x_0}^{x} dx' \frac{a(x')}{2\epsilon}\right) \simeq \exp\left(-\frac{\alpha(x - x_0)^2}{4\epsilon}\right).$$

Thus, for x close to x_0, the solution y is related to Y by

$$y \approx e^{-\alpha(x-x_0)^2/(4\epsilon)} Y. \tag{9.45}$$

We here take note that if α is positive, the factor multiplying Y in (9.45) drops exponentially as $|x - x_0|$ increases; on the other hand, if α is negative, this factor grows exponentially as $|x - x_0|$ increases.

We shall scale the variable $(x - x_0)$ in (9.44) by

$$x - x_0 = \sqrt{\frac{\epsilon}{|\alpha|}} X. \tag{9.46}$$

Then (9.44) becomes

$$\frac{d^2Y}{dX^2} + \left[\frac{\beta}{|\alpha|} - \frac{sign(\alpha)}{2} - \frac{X^2}{4}\right] Y = 0, \tag{9.47}$$

where

$$\text{sign}(\alpha) = \alpha/|\alpha|.$$

We note that with the independent variable scaled this way, the resulting equation (9.47) is independent of the small parameter ϵ. As a result, the solution of (9.47) varies by an amount $O(1)$ when X varies by an amount $O(1)$. Thus we say that the scale of X is unity. By (9.46), the scale of $(x - x_0)$ is $\sqrt{\epsilon}$. (We have made the implicit assumption that α is of the order of unity.) Therefore, there is a boundary layer of width $\sqrt{\epsilon}$ near the turning point x_0. Before we solve (9.47) in a closed form, let us digress with a few examples to help the readers gain a little more familiarity with what we have done.

First, there is a more direct way to see from (9.13) what the width of the boundary layer near $x = x_0$ is. Without loss of generality, we choose $x_0 = 0$ for convenience and write eq. (9.13) near the turning point as

$$\epsilon y'' + \alpha x y' + \beta y \approx 0.$$

To determine the scale of x, we make the change of variable

$$x = \delta X.$$

Then we have

$$\frac{\epsilon}{\delta^2}\frac{d^2}{dX^2}y + \alpha X\frac{d}{dX}y + \beta y = 0.$$

We see that if we choose

$$\delta = \sqrt{\epsilon},$$

the differential equation above is independent of the small parameter ϵ. Thus both solutions expressed as functions of X are independent of ϵ. This means that the scale of X for these solutions is unity, or the scale of x for these solutions is $\sqrt{\epsilon}$, as concluded earlier.

This argument can also be applied to the case $a(x) \neq 0$, which we have studied earlier. It gives us another way to deduce that for the case of $a(x) > 0$, for example, the width of the boundary layer of (9.13) near $x = 0$ is of the order of ϵ, as we had concluded earlier. It can also be used to justify that to calculate $y_{in}(x)$ near $x = 0$, we may drop the term $b(x)y(x)$ from (9.13).

Problem for the Reader

Justify the above with a scaling argument.

Solution

Near $x = 0$, (9.13) is approximately

$$\epsilon y'' + a(0)y' + b(0)y \simeq 0.$$

To study the scale of the solution of this equation near $x = 0$, we put

$$x = \delta X, \ \delta \ll 1.$$

The differential equation above, expressed by the independent variable X, is

$$\frac{\epsilon}{\delta^2}\frac{d^2y}{dX^2} + \frac{a(0)}{\delta}\frac{dy}{dX} + b(0)y = 0.$$

We shall choose δ so that all small parameters are eliminated from the differential equation. After this is done, y, dy/dX, and d^2y/dX^2 are of the same order of magnitude.

Since δ is small, the last term, $b(0)y$, in the differential equation above is much smaller than the second term,

$$\frac{a(0)}{\delta}\frac{dy}{dX}.$$

Therefore, it is justified to drop the last term in the differential equation. The remaining two terms in the differential equation are of the same order of magnitude if

$$\delta = \epsilon.$$

With such a choice for δ, all small parameters disappear from the differential equation. This means that the scale of the solution as a function of X is unity, which implies that the scale of the solution as a function of x is ϵ.

To help the readers further digest the method we have introduced in this chapter, let us consider the fourth-order linear differential equation

$$\epsilon y'''' + a(x)y'' + b(x)y = 0, 0 < x < 1,$$

with $a(x) < 0$. The values of y and y' are specified at the endpoints and are of the order of unity.

Problem for the Reader

a. How many slowly varying solutions does this equation have?

b. How many rapidly varying solutions does this equation have?

c. Where are the boundary layers? What is the width of each of the boundary layers?

Solution

a. There are two independent slowly varying solutions that satisfy the second-order differential equation

$$a(x)y_s'' + b(x)y_s \simeq 0.$$

The solution y_{out} is a linear superposition of these two solutions.

b. There are two rapidly varying solutions that satisfy the following near a point near a point x_0:

$$\epsilon y'''' + a(x_0)y'' \simeq 0.$$

c. The solution with rapidly decreasing magnitude near $x = 0$ is

$$y_-(x) = c \exp\left(-\frac{\sqrt{|a(0)|}}{\epsilon^{1/2}} x\right).$$

The solution $y_{in}(x)$ near the lower endpoint $x = 0$ is a linear superposition of $y_-(x)$ and the two slowly varying solutions, the latter being approximately constant inside the boundary layer near $x = 0$.

Similarly, the solution of rapidly increasing magnitude is, near the upper endpoint $x = 1$,

$$y_+(x) = d \exp\left[-\frac{\sqrt{|a(1)|}}{\epsilon^{1/2}}(1 - x)\right].$$

The solution $y_{in}(x)$ near the endpoint $x = 1$ is a linear superposition of $y_+(x)$ and the two slowly varying solutions.

Since a rapidly varying solution is smaller than its derivative by a factor of $\epsilon^{1/2}$, it is as small as $\epsilon^{1/2}$ if its derivative is of the order of unity. Therefore, as a crude approximation, we may ignore the rapidly varying solutions as we make $y_{in}(x)$ satisfy the boundary values $y(0)$ and $y(1)$. The rapidly varying solutions come into play as we make $y_{in}(x)$ satisfy the boundary values $y'(0)$ and $y'(1)$.

There is a boundary layer of width $\epsilon^{1/2}$ near each of the endpoints. This can be deduced either from the forms of $y_\pm(x)$ given above or from a scaling argument. Since the rapidly varying solutions are as small as $\epsilon^{1/2}$ at the endpoints, $y(x)$ itself does not have a conspicuous bump or dip near the boundary layers. But $y'(x)$ has a bump or dip near each of the endpoints.

Let us return to (9.47). Equation (9.47) is a parabolic cylinder equation, which is usually written in the generic form

$$\frac{d^2\phi}{dz^2} + (\nu + 1/2 - z^2/4)\phi = 0. \tag{9.48}$$

One of the solutions of eq. (9.48) is the parabolic cylinder function denoted by $D_\nu(z)$. Since equation (9.48) is unchanged with the replacement of $z \to -z$, $D_\nu(-z)$ also satisfies equation (9.48). In addition, (9.48) is unchanged if we make the replacements $z \to iz$ as well as $\nu \to -\nu - 1$. Thus $D_{-\nu-1}(iz)$ and $D_{-\nu-1}(-iz)$ are also solutions of eq. (9.48).

We shall choose the two independent solutions of y near the turning point to be

$$y_1 = e^{-\alpha x^2/(4\epsilon)} D_\nu\left(\sqrt{\frac{|\alpha|}{\epsilon}}x\right) \tag{9.49}$$

and

$$y_2 = e^{-\alpha x^2/(4\epsilon)} D_\nu\left(-\sqrt{\frac{|\alpha|}{\epsilon}}x\right), \tag{9.50}$$

where

$$\nu = \frac{\beta}{|\alpha|} - \frac{sign(\alpha)+1}{2}.$$

The asymptotic forms of the parabolic cylinder function have been assigned as homework problem 19 of Chapter 8. They are

$$\begin{aligned} D_\nu(X) &\approx X^\nu e^{-X^2/4},\ X \to \infty, \\ &\approx \frac{\sqrt{2\pi}}{\Gamma(-\nu)}|X|^{-\nu-1}e^{X^2/4},\ X \to -\infty. \end{aligned}$$

The first asymptotic form given above is valid even when X is complex. Indeed, it holds when $|X| \gg 1$ and

$$-\frac{3}{4}\pi < \arg X < \frac{3}{4}\pi.$$

When $\arg X$ is outside this range, we add to the first asymptotic form the term

$$-\frac{\sqrt{2\pi}}{\Gamma(-\nu)}e^{i\pi\nu}X^{-\nu-1}e^{X^2/4}.$$

Such a sum is the asymptotic form of $D_\nu(X)$ in the range

$$\pi/4 < \arg X < 5\pi/4.$$

When $\arg X$ is equal to π, this added term is the same as the second asymptotic form given above for $D_\nu(X)$ as $X \to -\infty$. We shall replace the parabolic cylinder functions in (9.49) and (9.50) by these asymptotic forms when $|X|$ is large.

The qualitative behaviors of y_1 and y_2 depend on the sign of α. This is because the exponential factors in (9.49) and (9.50) are either growing exponentially or vanishing exponentially as $|X|$ becomes large, depending on the sign of α. Thus there are two separate cases to discuss.

Case of $\alpha > 0$

For $\alpha > 0$, we have

$$\nu = \frac{\beta}{\alpha} - 1.$$

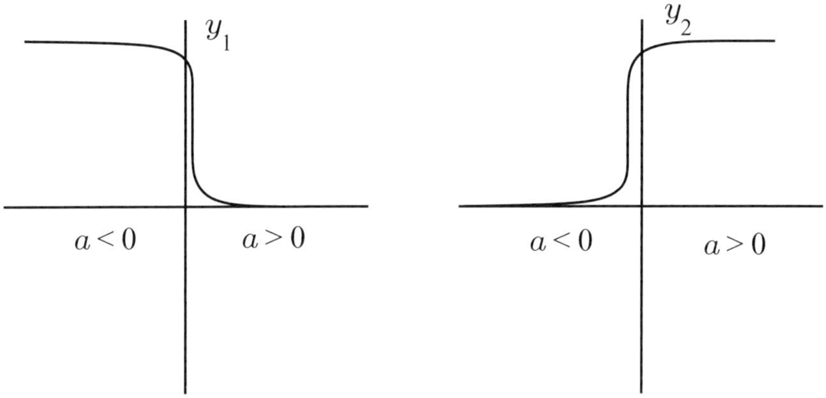

Figure 9.2

Also, the exponential function vanishes as $|X|$ increases. Thus we have

$$\begin{aligned} y_1 &\to (X)^{-1+\beta/\alpha} e^{-X^2/2}, \ X \to \infty, \\ &\to \frac{\sqrt{2\pi}}{\Gamma(1-\beta/\alpha)} (|X|)^{-\beta/\alpha}, \ X \to -\infty \end{aligned} \tag{9.51}$$

and

$$\begin{aligned} y_2(x) &\to \frac{\sqrt{2\pi}}{\Gamma(1-\beta/\alpha)} X^{-\beta/\alpha}, \ X \to \infty \\ &\to |X|^{-1+\beta/\alpha} e^{-X^2/2}, \ X \to -\infty. \end{aligned} \tag{9.52}$$

The behaviors of these two functions are schematically illustrated above; In the region $x > 0$, y_1 is the rapidly varying function, and y_2 is the slowly varying solution, while in the region $x < 0$, y_1 is the slowly varying solution and y_2 is the rapidly varying solution. The role of y_1 interchanges with that of y_2 as one crosses the turning point.

Case of $\alpha < 0$

We have

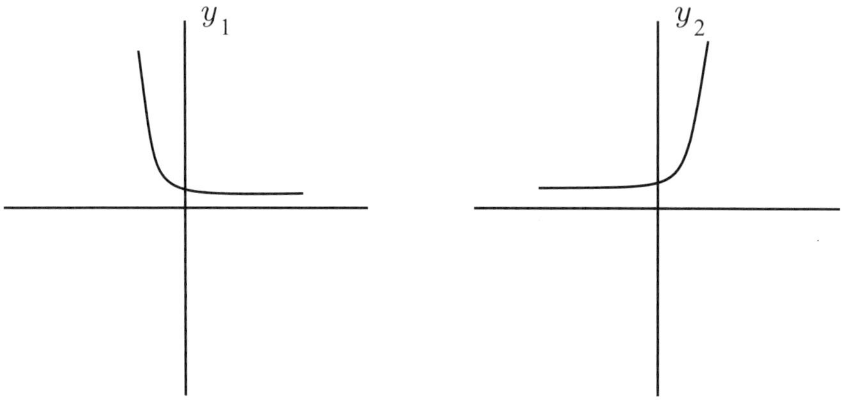

Figure 9.3

$$\nu = -\frac{\beta}{\alpha}.$$

Also, the exponential function blows up as $|X|$ becomes large. Thus

$$\begin{aligned} y_1 &\to X^{-\beta/\alpha},\ X \to \infty, \\ &\to \frac{\sqrt{2\pi}}{\Gamma(\beta/\alpha)}|X|^{-1+\beta/\alpha}e^{X^2/2}, \quad X \to -\infty, \end{aligned} \tag{9.53}$$

and

$$\begin{aligned} y_2 &\to \frac{\sqrt{2\pi}}{\Gamma(\beta/\alpha)}X^{-1+\beta/\alpha}e^{X^2/2},\ X \to \infty, \\ &\to |X|^{-\beta/\alpha},\ X \to -\infty. \end{aligned} \tag{9.54}$$

We illustrate in the figure above the behaviors of y_1 and y_2 for $\alpha < 0$. In the region $x > 0$, y_2 is the rapidly varying solution and y_1 the slowly varying solution. And in the region $x < 0$, y_1 is the rapidly varying solution and y_2 is the slowly varying solution.

In the next three sections, we shall show how to match these approximate solutions valid inside the neighborhood of the turning point with the approximate solutions valid outside the neighborhood of the turning point.

D. Turning Point at an Endpoint

We begin with examples of boundary-layer problems with a turning point at one of the endpoints. As before, the first order of business is to determine where the boundary layers are, and whether the rapidly varying function is increasing or decreasing. A good strategy is to start in a region where y_r is negligible. It is also preferable if this region contains one of the endpoints. This is because in this region, the solution has only one arbitrary constant and the solution can be uniquely determined with the help of one of the boundary conditions. Then we extend the solution to other regions.

We emphasize that in the region near the turning point x_0, the scale of $(x - x_0)$ is $\sqrt{\epsilon}$ and hence both independent solutions of the equation change appreciably within the distance of $\sqrt{\epsilon}$. It is then not correct to approximate $y_s(x)$ by a constant in this region. Instead, inside the boundary layer of a turning point, Y is approximately equal to a linear superposition of parabolic cylinder functions. Thus a more elaborate procedure is needed.

Problem for the Reader

Solve approximately

$$\epsilon y'' + 2xy' + (1 + 3x^3)y = 0, \ 0 < x < 1, \epsilon \ll 1,$$

with the boundary conditions

$$y(0) = 0, \ y(1) = 1.$$

Solution

Since $a(x) = 2x \geq 0$, the magnitude of the rapidly varying solution is a decreasing function of x. There is a boundary layer at $x = 0$. Since $x = 0$ is also a turning point, the width of the boundary layer near $x = 0$ is $\sqrt{\epsilon}$.

The solution y_r is negligible outside the boundary layer. Thus, when $x \gg \sqrt{\epsilon}$, the solution is equal to y_{out}, which satisfies

$$2xy'_{out} + (1 + 3x^3)y_{out} = 0.$$

This equation yields

$$y_{out}(x) = \frac{1}{\sqrt{x}} e^{-x^3/2},$$

where we have made use of the boundary condition at $x = 1$.

Inside the boundary layer near $x = 0$, we have

$$y_{in}(x) = e^{-x^2/(2\epsilon)} \left[c_1 D_{-1/2}\left(\sqrt{\frac{2}{\epsilon}}x\right) + c_2 D_{-1/2}\left(-\sqrt{\frac{2}{\epsilon}}x\right)\right].$$

These solutions are good approximations if $x \ll 1$.

There are two arbitrary constants, c_1 and c_2. We determine one of these constants by matching $y_{in}(x)$ with $y_{out}(x)$ in the overlapping region where both approximations are valid. The parabolic cylinder function solution is valid when $x \ll 1$, while by (7.18), the WKB solutions are valid when

$$\left|\frac{d}{dx}\frac{\epsilon}{x}\right| \ll 1,$$

or

$$x \gg \sqrt{\epsilon}.$$

Thus the overlapping region is

$$\sqrt{\epsilon} \ll x \ll 1.$$

When $x \gg \sqrt{\epsilon}$ we have

$$y_{in}(x) \approx c_2 (2\epsilon)^{1/4} x^{-1/2}.$$

Also, when $x \ll 1$, we have

$$y_{out}(x) \approx x^{-1/2}.$$

We note that both $y_{in}(x)$ and $y_{out}(x)$ are equal to a constant times $x^{-1/2}$ in the region $1 \gg x \gg \sqrt{\epsilon}$. This is an indication that this region is indeed the overlapping region in which both approximations hold. Joining $y_{in}(x)$ with y_{out} in this region, we get

$$c_2 = (2\epsilon)^{-1/4}.$$

From the boundary condition at $x = 0$, we get

$$c_1 = -c_2.$$

Thus we have

$$y_{in}(x) = e^{-x^2/(2\epsilon)} (2\epsilon)^{-1/4} \left[-D_{-1/2}\left(\sqrt{\frac{2}{\epsilon}}x\right) + D_{-1/2}\left(-\sqrt{\frac{2}{\epsilon}}x\right)\right].$$

As a final observation, we note that $y_{out}(0)$ is infinite. But as we continue $y_{out}(x)$ into the region of the boundary layer, it turns into

$$e^{-x^2/(2\epsilon)}(2\epsilon)^{-1/4}D_{-1/2}\left(-\sqrt{\frac{2}{\epsilon}}x\right).$$

At $x = 0$, the expression above is equal to $(2\epsilon)^{-1/4}D_{-1/2}(0)$, which is a large number but is not infinity.

Problem for the Reader

Solve approximately

$$\epsilon y'' - 2xy' + (1 + 3x^3)y = 0, \ 0 < x < 1, \epsilon \ll 1,$$

with the boundary conditions

$$y(0) = \gamma, \ y(1) = 1.$$

Solution

Since $a(x) = -2x \leq 0$, the magnitude of the rapidly varying solution is increasing with x. Thus there is a boundary layer of width ϵ near the endpoint $x = 1$. Also, since $a(x)$ has a simple zero at $x = 0$, there is a boundary layer of width $\sqrt{\epsilon}$ near the endpoint $x = 0$.

We start by seeking the solution $y_{in}^{(a)}(x)$ valid inside the boundary layer near $x = 0$. This is because $y_r(x)$ is negligible in this region. Thus the solution in this region has only one arbitrary constant that can be determined from the boundary condition at $x = 0$.

The solution inside the boundary layer near $x = 0$ is a linear combination of the solutions given by (9.49) and (9.50) respectively, with $\alpha = -2$ and $\nu = 1/2$. The solution y_2 is the rapidly increasing solution that is negligible inside the boundary-layer region near $x = 0$. Thus we are left with only the solution y_1. We get

$$y_{in}^{(a)}(x) = \gamma e^{x^2/(2\epsilon)}D_{1/2}\left(\sqrt{\frac{2}{\epsilon}}x\right)/D_{1/2}(0),$$

where the boundary condition at $x = 0$ has been utilized.

Incidentally, the value $D_\nu(0)$ can be expressed by Gamma functions. Indeed, one may derive from the integral representation of $D_\nu(x)$ that

$$D_\nu(0) = 2^{\nu/2} \frac{\sqrt{\pi}}{\Gamma(1/2 - \nu/2)}.$$

Thus we have

$$y_{in}^{(a)}(x) = 2^{-1/4}\Gamma(1/4)\pi^{-1/2}\gamma e^{x^2/(2\epsilon)} D_{1/2}\left(\sqrt{\frac{2}{\epsilon}x}\right).$$

From the equation above, we get, for $x \gg \sqrt{\epsilon}$,

$$y_{in}^{(a)}(x) \approx \Gamma(1/4)\pi^{-1/2}\gamma \left(\sqrt{\frac{1}{\epsilon}x}\right)^{1/2}.$$

Outside the boundary layers, we have

$$-2xy'_{out} + (1 + 3x^3)y_{out} = 0.$$

Thus we get

$$y_{out}(x) = c\sqrt{x}e^{x^3/2}.$$

Matching $y_{out}(x)$ with $y_{in}^{(a)}(x)$ in the region $1 \gg x \gg \sqrt{\epsilon}$, we get

$$c = \Gamma(1/4)\pi^{-1/2}\gamma\epsilon^{-1/4}.$$

Thus

$$y_{out}(x) = \Gamma(1/4)\pi^{-1/2}\gamma\epsilon^{-1/4}\sqrt{x}e^{x^3/2}.$$

In particular,

$$y_{out}(1) = \Gamma(1/4)\pi^{-1/2}\gamma\epsilon^{-1/4}e^{1/2}.$$

Finally, we seek the solution inside the boundary layer near $x = 1$. Since $a(1) = -2$, we have

$$y_r(x) = [1 - y_{out}(1)]\, e^{-2(1-x)/\epsilon}.$$

Thus we have

$$y_{in}^{(b)}(x) = y_{out}(1) + [1 - y_{out}(1)]\, e^{-2(1-x)/\epsilon}.$$

Problem for the Reader

Solve

$$\epsilon y'' - x(1+x)y' + \frac{1}{2}y = 0, 0 < x < 1, \tag{9.55}$$

with the boundary conditions

$$y(0) = 1$$

and

$$y(1) = 2.$$

Solution

Since

$$a(x) = -x(1+x) \leq 0,$$

the magnitude of the rapidly varying solution grows with x. Thus there is a boundary layer at $x = 1$ of width ϵ. Also, since $x = 0$ is a turning point, there is a boundary layer with width $\sqrt{\epsilon}$ near the origin. Thus there is a boundary layer at each of the endpoints.

We start by seeking the solution inside the boundary layer near $x = 0$. Since in this region $y_r(x)$ is negligible, the solution is equal to a constant times $y_1(x)$, which is given by (9.49). We easily find that $y_{in}^{(a)}$ is equal to

$$y_{in}^{(a)}(x) = e^{X^2/4} D_{1/2}(X)/D_{1/2}(0), \tag{9.56}$$

where $X = x/\sqrt{\epsilon}$. When X is large, we may use the asymptotic form for the parabolic cylinder function and we get

$$y_{in}^{(a)}(x) \approx (2\pi^2\epsilon)^{-1/4}\Gamma(1/4)x^{1/2}, \; x \gg \sqrt{\epsilon},$$

where we have replaced $D_{1/2}(0)$ by $2^{1/4}\sqrt{\pi}/\Gamma(1/4)$.

Next we find y_{out} that holds in the region outside the boundary layers. We have

$$-x(1+x)y'_{out} + y_{out}/2 = 0.$$

Thus we get

$$y_{out}(x) = c\sqrt{\frac{x}{1+x}}. \tag{9.57}$$

Now y_{out} remains a good approximation of the solution as long as $x \gg \sqrt{\epsilon}$. In the region $\sqrt{\epsilon} \ll x \ll 1$, both $y_{out}(x)$ and $y_{in}(x)$ are equal to a constant times $\sqrt{x}$. This indicates that this region is the overlapping region for these two solutions. Matching y_{out} with $y_{in}^{(a)}$ in the region $\sqrt{\epsilon} \ll x \ll 1$, we get

$$c = (2\pi^2\epsilon)^{-1/4}\Gamma(1/4).$$

Thus we determine y_{out} to be

$$y_{out}(x) = (2\pi^2\epsilon)^{-1/4}\Gamma(1/4)\sqrt{\frac{x}{1+x}},$$

which is of the order of $\epsilon^{-1/4}$, a large number. In particular,

$$y_{out}(1) = (8\pi^2\epsilon)^{-1/4}\Gamma(1/4).$$

Finally, we seek the solution inside the boundary layer near $x = 1$. In this region, we have $a(x) \approx -2$. Thus the rapidly varying solution is equal to $ce^{-2(1-x)/\epsilon}$. The constant c is easily determined by the use of the boundary condition at $x = 1$. We find the solution inside the boundary layer near $x = 1$ as

$$y_{in}^{(c)}(x) = (8\pi^2\epsilon)^{-1/4}\Gamma(1/4) + \left[2 - (8\pi^2\epsilon)^{-1/4}\Gamma(1/4)\right] e^{-2(1-x)/\epsilon}.$$

The solution valid for all x is

$$\begin{aligned} y_{uniform}(x) = {} & e^{x^2/(4\epsilon)}\sqrt{\frac{1}{1+x}} D_{1/2}(x/\sqrt{\epsilon})/D_{1/2}(0) \\ & + \left[2 - \left(8\pi^2\epsilon\right)^{-1/4}\Gamma(1/4)\right] e^{-(1-x)/\epsilon}. \end{aligned}$$

To close this section, we consider an example in which the order of the parabolic cylinder function is an integer.

Problem for the Reader

Solve

$$\epsilon y'' - xy' + x^3 y = 0,\ 0 < x < 1,\ \epsilon \ll 1,$$

with the boundary conditions

$$y(0) = 1$$

and

$$y(1) = 2.$$

Solution

Since $a(x) \leq 0$, the magnitude of the rapidly varying solution grows as x increases. Thus there is a boundary layer of width ϵ near $x = 1$. There is also a boundary layer of width $\sqrt{\epsilon}$ near the turning point at $x = 0$.

Since the rapidly varying solution is negligible inside the boundary layer near $x = 0$, we begin with the solution there. We have

$$\alpha = -1$$

and

$$\beta = 0.$$

Thus the order of the parabolic cylinder equation is of the order

$$\nu = 0.$$

Since

$$D_0(X) = D_0(-X) = e^{-X^2/4},$$

by (9.49) and (9.50), both $y_1(x)$ and $y_2(x)$ are the same and equal to unity and slowly varying. Thus neither of them is the rapidly varying solution, which is $D_{-1}(iX)$. But since the rapidly varying solution is negligible, there is no need to use it. We get from (9.49) that

$$y_{in}^{(a)}(x) = 1,$$

where the boundary condition at $x = 0$ has been utilized.

In the region outside the boundary layers, we have

$$y'_{out}(x) = x^2 y_{out}(x).$$

Solving this differential equation and matching the solution obtained with $y_{in}^{(a)}(x)$ in the overlapping region $1 \gg x \gg \sqrt{\epsilon}$, we get

$$y_{out}(x) = e^{x^3/3}.$$

From this solution we get

$$y_{out}(1) = e^{1/3}.$$

Finally, inside the boundary layer near $x = 1$, $a(x)$ is approximately equal to -1. Thus we have

$$\epsilon \frac{d^2}{dx^2} y_{in}^{(b)} - \frac{d}{dx} y_{in}^{(b)} = 0.$$

Hence

$$y_{in}^{(b)}(x) = e^{1/3} + (2 - e^{1/3}) e^{-(1-x)/\epsilon}.$$

E. Interior Turning Points

Next we'll examine some examples in which there is a turning point inside the interval where (9.13) holds. This is to say that the turning point is an interior point.

This is a good time to make a subtle elaboration on how to continue a solution through a turning point. We shall put, without loss of generality, the turning point at the origin.

Let us first consider the case in which $\alpha > 0$. The rapidly varying solution increases as x increases for $x < 0$ and decreases as x increases for $x > 0$. Thus the rapidly varying function decreases as we move away from the origin in either direction.

Let us take the rapidly varying solution (9.24) of $x < 0$. We start at the lower endpoint $x = -1$ and continue this solution to the right. Before we reach the origin, indeed at a distance $\sqrt{\epsilon}$ away from the origin, (9.24) fails to be a good approximation. To continue this solution through the origin, we must join it with $y_2(x)$ which, by the second asymptotic form of (9.52), is rapidly varying when x is negative with $|x| \gg \sqrt{\epsilon}$. As we continue this solution to positive values of x, it turns into the slowly varying solution when $x \gg \sqrt{\epsilon}$, as can be seen from the first asymptotic formula of (9.52).

The second independent solution cannot be obtained by starting with the slowly varying solution at $x = -1$ and continuing it to the right. To see this, consider the solution near $x = -1$, which is a sum of a slowly varying solution and a rapidly varying solution. Let the magnitude of the latter be very small compared to that of the former at $x = -1$. As we continue this solution to the right, the magnitude of the rapidly varying solution grows exponentially and may become comparable to or even larger than that of the slowly varying solution. This says that if we start at $x = -1$, it is difficult to distinguish a slowly varying solution from the sum of a slowly varying solution and a rapidly varying solution.

To obtain the second independent solution of the differential equation, we still start with the rapidly varying solution, but this time at the upper endpoint $x = 1$. We continue this solution to the left. Inside the boundary layer, it is equal to $y_1(x)$, as we can see from the first asymptotic formula of (9.51). As we continue this solution to the region $x < 0$, it turns into the slowly varying solution, as we can see from the second asymptotic formula of (9.51).

As we have mentioned, for $\alpha > 0$, the magnitude of a rapidly varying solution always decreases as we move away from the origin. Therefore, if a rapidly varying solution and a slowly varying solution are of the same order of magnitude at $x = 0$, then the former becomes much smaller than the latter as one leaves the origin. We shall call the former the small solution and the latter the large solution. It is meaningful to take a small solution at one side of the turning point and continue it across the turning point. The small solution turns into the large solution at the other side of the turning point. But it is not meaningful to start with the large solution and continue it across the turning point.

If $\alpha < 0$, the magnitude of the rapidly varying solution decreases as x increases in the region $x < 0$ and increases as x increases in the region $x > 0$. Thus the magnitude of a rapidly varying solution always increases as we leave the origin in either direction. If a rapidly varying solution and a slowly varying solution are of the same order of magnitude at the origin, then the former becomes much larger than the latter as we move away from the origin. Therefore, the rapidly varying solution is the large solution and the slowly varying solution is the small solution. As before, we must start with the small solution and continue it across the turning point. Thus we start with the slowly varying solution at an endpoint and continue it across the origin. This solution turns into the rapidly varying solution as one crosses the origin.

We shall give a couple of examples to illustrate how to use this concept to solve a boundary-layer problem with an interior turning point. It is also possible to use this concept to solve boundary-layer problems with two or more turning points. (See homework problem 12.)

Problem for the Reader

Solve

$$\epsilon y'' + 2(\tan x)y' - y = 0, \quad -1 < x < 1,$$

with

$$y(-1) = 1$$

and

$$y(1) = 3.$$

Solution

We have

$$a(x) > 0, \; x > 0;$$
$$a(x) < 0, \; x < 0.$$

There is a boundary layer at the turning point $x = 0$, the width of which is of the order of $\sqrt{\epsilon}$. On either side of the origin, the rapidly varying solution is always the small solution, and the slowly varying solution is always the large solution.

Let us start at $x = -1$ with the small solution, which is the rapidly varying solution. The magnitude of this solution rapidly increases as we move toward the origin. It turns into the large and slowly varying solution in the region to the right of the origin.

Similarly, if we start with the rapidly varying solution at $x = 1$, its magnitude increases rapidly as we move toward the origin. In the region at the left of the origin, it turns into the large and slowly varying solution.

Since the rapidly varying solution is always the small solution, the solution of the boundary-value problem is approximately given by the large and slowly varying solution for all values of x outside the boundary layer. The slowly varying solution satisfies

$$2(\tan x)y_s' - y_s \simeq 0,$$

or

$$y_s(x) = c\sqrt{\sin x}.$$

Thus we have

$$y_{out}^{(a)} = 3\sqrt{\frac{\sin x}{\sin 1}}, \quad 0 < x < 1,$$

where the boundary condition $y(1) = 3$ has been taken into account. Also

$$y_{out}^{(b)} = \sqrt{-\sin x / \sin 1}, \quad -1 < x < 0.$$

From the forms of the solutions given above, we note that when $|x| \ll 1$, $y_{out}(x)$ is of the order of $\sqrt{x}$. It is therefore of the order of $\epsilon^{1/4}$ when x is of the order of $\sqrt{\epsilon}$. While $y_{out}(x)$ fails when we get inside the boundary layer, the parabolic cylinder function solutions are joined with $y_{out}(x)$ in the region $|x| \gg \epsilon^{1/2}$. Thus we expect the value of the solution inside the boundary layer to be of the order of $\epsilon^{1/4}$, which is a relatively small number.

To complete the discussion, we determine the solution $y_{in}(x)$ inside the boundary layer near $x = 0$. By (9.49) and (9.50), we have

$$X = \sqrt{2/\epsilon}x$$

and

$$y_{in}(x) = e^{-x^2/(2\epsilon)}\left[c_1 D_{-3/2}\left(\sqrt{\frac{2}{\epsilon}}x\right) + c_2 D_{-3/2}\left(-\sqrt{\frac{2}{\epsilon}}x\right)\right].$$

As $X \to \infty$, we have

$$y_{in}(x) \to c_2 2^{7/4}\epsilon^{-1/4}\sqrt{x}.$$

Matching with $y_{out}^{(a)}$ in the region $x > 0$, we get

$$c_2 \approx 3\sqrt{\frac{1}{2^{7/2}\sin 1}}\epsilon^{1/4}.$$

As $X \to -\infty$, we have

$$y_{in}(x) \to c_1 2^{7/4}\epsilon^{-1/4}\sqrt{-x}.$$

Matching with $y_{out}^{(b)}(x)$, we get

$$c_1 \approx \sqrt{\frac{1}{2^{7/2}\sin 1}}\epsilon^{1/4}.$$

Thus

$$y_{in}(x) = e^{-x^2/(2\epsilon)}\sqrt{\frac{1}{2^{7/2}\sin 1}}\epsilon^{1/4}\left[D_{-3/2}\left(\sqrt{\frac{2}{\epsilon}}x\right) + 3D_{-3/2}\left(-\sqrt{\frac{2}{\epsilon}}x\right)\right],$$

which is of the order of $\epsilon^{1/4}$ when X is of the order of unity, as expected.

Problem for the Reader

Solve approximately

$$\epsilon y'' - 3xy' + (1 + x^4)y = 0, \ -1 < x < 1, \ \epsilon \ll 1.$$

The boundary conditions are $y(-1) = y(1) = 1$.

Solution

We have

$$\alpha = -3 < 0.$$

It is then not difficult to show that near the origin, the small solution is the slowly varying solution, and the rapidly varying solution is the large solution.

There is a boundary layer of width ϵ near each of the endpoints. There is an additional boundary layer of width $\sqrt{\epsilon}$ near the origin, which is a turning point. Solving this problem of three boundary layers is facilitated by the use of the concept of small solution and large solution.

We start with the small and slowly varying solution at the left of the origin. This solution becomes the large and rapidly varying solution in the region $x > 0$. Thus this solution is sharply peaked at the upper endpoint $x = 1$, and the value of this solution is exponentially small in other regions of x.

To obtain the second independent solution of the differential equation, we start with the slowly varying solution at the right of the origin, and continue it to negative values of x, where it becomes the large and rapidly varying solution in the region $x < 0$. Thus this solution is sharply peaked at the lower endpoint $x = -1$, and the value of this function is exponentially small outside a small region near $x = -1$.

The solution of the problem is a linear superposition of the two solutions described above. The first solution is appreciable only near the endpoint $x = 1$, and the second solution is appreciable only near the endpoint $x = -1$. Thus the linear superposition is appreciable only if x is near either one of the endpoints.

Near $x = 1$, we have $a(x) \approx -3$. Therefore we have, for $x > 0$,

$$y_{in}^{(a)}(x) \approx e^{-3(1-x)/\epsilon}.$$

Similarly, for $x < 0$, we have

$$y_{in}^{(b)}(x) \approx e^{-3(1+x)/\epsilon}.$$

And

$$y_{uniform}(x) = e^{-3(1-x)/\epsilon} + e^{-3(1+x)/\epsilon}.$$

F. Other Problems

In this section we shall solve several problems that either have not been solved correctly in the book by Bender and Orszag or have been considered unsolvable. The boundary-layer problems that have not been correctly solved are the ones in which $a(x)$ has a multiple zero at the turning point. Actually, problems with this kind of turning point are easier to solve than the one in which the zero of $a(x)$ is a simple zero. This is because the WKB solutions are good approximate solutions for the former problems. The new feature is that the behavior of the solution near such a turning point can be expressed by the integral $I_{2n}(X)$ given below instead of the parabolic cylinder functions. This integral has many scales near the turning point.

For example, let $a(x)$ have a double zero at $x = 0$. More precisely, let

$$a(x) \approx \alpha x^2$$

and

$$b(x) \approx \beta$$

near $x = 0$. The solution inside the boundary layer is expressed not by the parabolic cylinder function but by

$$I_4(X) \equiv \int_0^X \sqrt{1 + X'^4} dX', \tag{9.58}$$

where

$$X \equiv \left(\frac{\alpha^2}{4\beta\epsilon}\right)^{1/4} x.$$

We'll demonstrate this method with two examples.

Problem for the Reader

Solve

$$\epsilon y'' - x^2 y' - y = 0, \quad y(0) = y(1) = 1. \tag{9.59}$$

(Bender and Orszag, *Advanced Mathematical Methods for Scientists and Engineers,* McGraw-Hill, Section 9.4, Example 1.)

Solution

Since $a(x) = -x^2 < 0$, the rapidly varying solution is an increasing function of x. Thus there is a boundary layer of width ϵ near $x = 1$. Also, since $a(x)$ vanishes at $x = 0$, there is a turning point at $x = 0$. The zero of $a(x)$ at $x = 0$ is a double zero.

We put

$$y(x) = e^{x^3/(6\epsilon)} Y(x).$$

By (9.43) we have

$$\frac{d^2}{dx^2} Y - \left(\frac{1 - x}{\epsilon} + \frac{x^4}{4\epsilon^2}\right) Y = 0.$$

We shall make a change of variable

$$x \equiv (4\epsilon)^{1/4} X.$$

The differential equation becomes

$$\frac{d^2}{dX^2} Y - \lambda^2 \eta^2(x) Y = 0,$$

where

$$\lambda = \left(\frac{4}{\epsilon}\right)^{1/4},$$

and

$$\eta = \sqrt{1 + X^4 - \frac{2}{\lambda} X}.$$

This equation can be solved by the WKB method. This is because the magnitude of

$$\frac{d}{dX}\left(\frac{1}{\lambda\eta}\right)$$

is always small compared to unity, as we may readily verify.

The term

$$-\frac{2}{\lambda} X$$

is small compared to $1 + X^4$ for all X. This is because it is much smaller than unity if X is much less than $\lambda/2$, and is much less than X^4 if X is much larger than $(2/\lambda)^{1/3}$. Thus we may treat

$$-\frac{2}{\lambda} X$$

as a small perturbation and get

$$\eta \approx \sqrt{1 + X^4} - \lambda^{-1} X (1 + X^4)^{-1/2}.$$

Therefore,

$$\int_0^X \eta(X') dX' \approx I_4(X) - \frac{1}{2\lambda} \ln\left(X^2 + \sqrt{1 + X^4}\right).$$

The rapidly varying solution of the equation is appreciable only inside the boundary layer near $x = 1$. Outside this boundary layer, we may ignore the rapidly varying solution. Therefore, outside the boundary layer, Y is given by

$$Y(x) = \frac{1}{\sqrt{\eta(x)}} \exp\left(-\lambda \int_0^X \eta(X')dX'\right).$$

Thus we have

$$y_{out}(x) = \left(1 + \frac{x^4}{4\epsilon}\right)^{-1/4} \left(\frac{x^2}{\sqrt{4\epsilon}} + \sqrt{1 + \frac{x^4}{4\epsilon}}\right)^{1/2}$$
$$\exp\left[\frac{x^3}{6\epsilon} - \left(\frac{4}{\epsilon}\right)^{1/4} I_4(X)\right], \tag{9.60}$$

where $y(0) = 1$ has been utilized. The solution (9.60) is a good approximation everywhere outside the boundary layer near $x = 1$. In particular, it is a good approximation even inside the boundary layer near $x = 0$.

The behavior of the solution inside the boundary layer near $x = 0$ is in fact quite complicated. Indeed, the solution has many scales. To begin, when $x \gg \epsilon^{1/5}$, it satisfies the equation

$$x^2 y_{out}' + y_{out} = 0.$$

To see this, let us perform for $I_4(X)$ an integration by parts. We then get

$$I(X) = \frac{1}{3}\sqrt{X^6 + X^2} + \frac{2}{3}\int_0^X \frac{dX'}{\sqrt{1 + X'^4}}$$

Since

$$\int_0^\infty \frac{dX}{\sqrt{1 + X^4}} = \frac{\Gamma^2(1/4)}{4\sqrt{\pi}},$$

we have

$$I_4(X) = \frac{1}{3}\sqrt{X^6 + X^2} + \frac{\Gamma^2(1/4)}{6\sqrt{\pi}} - \frac{2}{3}\int_X^\infty \frac{dX'}{\sqrt{1 + X'^4}}.$$

When $x \gg \epsilon^{1/4}$, X is much larger than unity. Thus we have, for $x \gg \epsilon^{1/4}$,

$$\left(\frac{4}{\epsilon}\right)^{1/4} I(X) \approx \frac{x^3}{6\epsilon} - \frac{1}{x} + \left(\frac{4}{\epsilon}\right)^{1/4} \frac{\Gamma^2(1/4)}{6\sqrt{\pi}} - \frac{\epsilon}{15x^5}. \tag{9.61}$$

The last term in the expression above is small if $x \gg \epsilon^{1/5}$. Thus we get

$$y_{out}(x) = \sqrt{2} \exp\left[\frac{1}{x} - \left(\frac{4}{\epsilon}\right)^{1/4} \frac{\Gamma^2(1/4)}{6\sqrt{\pi}}\right], \quad x \gg \epsilon^{1/5}, \tag{9.62}$$

which is the solution of the differential equation for y_{out}.

When $x \ll \epsilon^{1/4}$, we have

$$I_4(X) = X + \frac{X^5}{10} + \cdots .$$

Thus the solution (9.60) reduces to

$$y_{in}(x) = e^{-x/\sqrt{\epsilon} + x^3/(6\epsilon) - x^5/(40\epsilon^{3/2})}. \tag{9.63}$$

In the region $x \ll \epsilon^{1/3}$, we need to keep only the first term in the exponent of (9.63).

In the region where x is $O(\epsilon^{1/3})$, the second term in the exponent becomes appreciable. And when x is $O(\epsilon^{3/10})$, the third term in the exponent becomes appreciable.

We mention that we have dropped a term of the order of $x^9/\epsilon^{5/2}$ from the exponent of (9.63). This term is appreciable when $x = O(\epsilon^{5/18})$.

Near the boundary layer $x = 1$, the rapidly varying WKB solution comes into play. But since this solution is appreciable only when x is close to unity, we may approximate the differential equation near $x = 1$ by

$$\epsilon y'' - y = 0.$$

Thus the rapidly varying solution is equal to

$$ce^{-(1-x)/\epsilon},$$

where

$$c = 1 - y_{out}(1),$$

and $y_{out}(1)$ can be obtained from (9.60). The solution inside the boundary layer near $x = 1$ is

$$y_{in}(x) = y_{out}(1) + ce^{-(1-x)/\epsilon}.$$

We also have

$$y_{uniform}(x) = y_{out}(x) + ce^{-(1-x)/\epsilon}. \tag{9.64}$$

We give another example below.

Problem for the Reader

Solve

$$\epsilon y'' + x^2 y' - y = 0, \; y(0) = y(1) = 1. \tag{9.65}$$

(Bender and Orszag, *Advanced Mathematical Methods for Scientists and Engineers,* McGraw-Hill, Section 9.4, Example 2.)

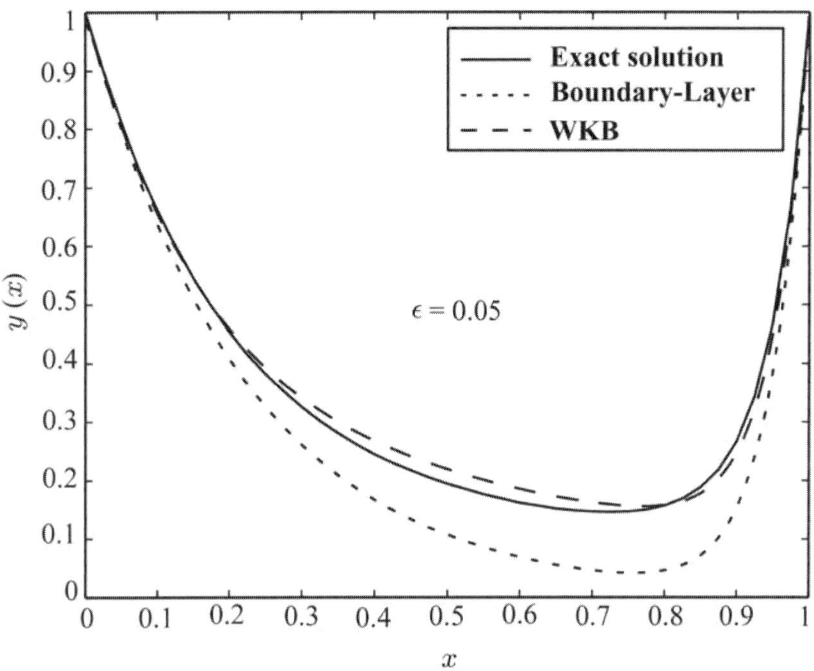

Figure 9.4

Solution

Since $a(x) = x^2 > 0$, the rapidly varying solution is a decreasing function of x. Thus there is a boundary layer near $x = 0$. The point $x = 0$ is also a turning point where $a(x)$ has a double zero.

We first find the solution outside the boundary layer. The solution y_{out} satisfies

$$y'_{out} = \frac{1}{x^2} y_{out}.$$

Thus we have

$$y_{out}(x) = \exp\left(1 - \frac{1}{x}\right), \tag{9.66}$$

which satisfies the boundary condition at $x = 1$. This solution is a good approximation as long as x is not inside the boundary layer near $x = 0$.

The WKB solutions for this problem are good approximations throughout the interval $[0, 1]$, either outside or inside the boundary layer. To see this, we put

$$y(x) = e^{-x^3/(6\epsilon)} Y(x).$$

By (9.43) we have

$$\frac{d^2}{dX^2}Y - \lambda^2\eta^2(x)Y = 0,$$

where

$$\eta = \sqrt{1 + X^4 + \frac{2}{\lambda}X},$$

with

$$x = (4\epsilon)^{1/4}X.$$

That the WKB solutions are good approximations can be verified, as we have done in the preceding example. And as before, the term $2X/\lambda$ can be treated as a perturbative term. Thus we have

$$\eta \approx \sqrt{1+X^4} + \lambda^{-1}X(1+X^4)^{-1/2}.$$

We get

$$\int_0^X \eta(X')dX' = I_4(X) + \frac{1}{2\lambda}\ln\left(X^2 + \sqrt{1+X^4}\right). \qquad (9.67)$$

The solution y of (9.65) is equal to

$$e^{-x^3/(6\epsilon)}$$

times a linear superposition of the two WKB solutions of Y. The rapidly increasing WKB solution of Y, after being multiplied by the rapidly decreasing factor

$$e^{-x^3/(6\epsilon)},$$

is the solution y_s. Thus we have

$$y_s(x) = c\left(1 + \frac{x^4}{4\epsilon}\right)^{-1/4}\left(\frac{x^2}{\sqrt{4\epsilon}} + \sqrt{1 + \frac{x^4}{4\epsilon}}\right)^{1/2}$$
$$\exp\left[-\frac{x^3}{6\epsilon} + \left(\frac{4}{\epsilon}\right)^{1/4} I_4(X)\right]. \qquad (9.68)$$

When $x \gg \epsilon^{1/5}$, we have, by (9.61),

$$y_s(x) \approx c\sqrt{2}\exp\left[-\frac{1}{x} + \left(\frac{4}{\epsilon}\right)^{1/4} \frac{\Gamma^2(1/4)}{6\sqrt{\pi}}\right]. \tag{9.69}$$

Comparing (9.69) with (9.66), we find that

$$c = 1\sqrt{2}\exp\left[1 - \left(\frac{4}{\epsilon}\right)^{1/4} \frac{\Gamma^2(1/4)}{6\sqrt{\pi}}\right].$$

When $x \ll \epsilon^{1/4}$ we have

$$y_s(x) \approx c\exp\left(-\frac{x^3}{6\epsilon} + \frac{x}{\epsilon^{1/2}} + \frac{x^5}{40\epsilon^{3/2}}\right). \tag{9.70}$$

The rapidly decreasing solution near $x = 0$ is equal to

$$y_r(x) = [1 - y_s(0)]\left(1 + \frac{x^4}{4\epsilon}\right)^{-1/4}\left(\frac{x^2}{\sqrt{4\epsilon}} + \sqrt{1 + \frac{x^4}{4\epsilon}}\right)^{-1/2}$$
$$\exp\left[-\frac{x^3}{6\epsilon} - \left(\frac{4}{\epsilon}\right)^{1/4} I_4(X)\right]. \tag{9.71}$$

When $x \ll \epsilon^{1/4}$, it reduces to

$$y_r(x) = [1 - y_s(0)]\exp\left(-\frac{x^3}{6\epsilon} - \frac{x}{\epsilon^{1/2}} - \frac{x^5}{40\epsilon^{3/2}}\right). \tag{9.72}$$

We thus have

$$y_{in}(x) = y_s(0)\exp\left(-\frac{x^3}{6\epsilon} + \frac{x}{\epsilon^{1/2}} + \frac{x^5}{40\epsilon^{3/2}}\right) + [1 - y_s(0)]$$
$$\exp\left(-\frac{x^3}{6\epsilon} - \frac{x}{\epsilon^{1/2}} - \frac{x^5}{40\epsilon^{3/2}}\right), \tag{9.73}$$

which is valid for $x \ll \epsilon^{5/18}$. We also have

$$y_{uniform}(x) = y_s(x) + y_r(x).$$

Before we close the discussion of turning points with multiple zeroes, let's touch on the general case when $a(x)$ has an nth-order zero near $x = 0$, i.e.,

$$a(x) \approx \alpha x^n,\ n > 1,$$

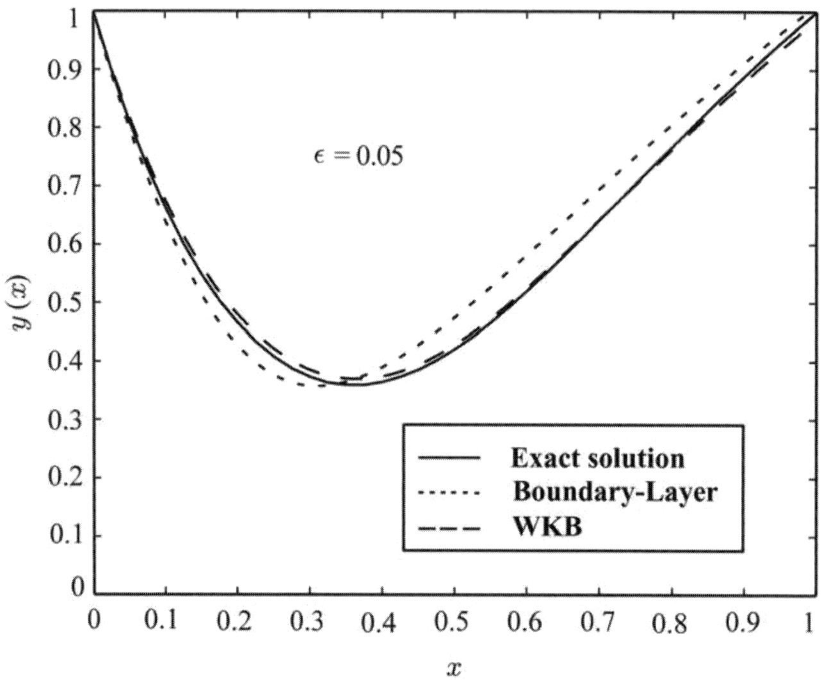

Figure 9.5

and

$$b(x) \approx \beta \neq 0.$$

The WKB solutions are good approximations throughout the interval and they are described by

$$I_{2n}(X) = \int_0^X \sqrt{1 + X'^{2n}} dX',$$

where

$$X \equiv \left(\frac{\alpha^2}{4\beta\epsilon}\right)^{1/2n} x.$$

On the other hand, if $b(x)$ turns out to have a simple zero at the turning point, and if $a(x)$ has a double zero at the turning point, then the WKB method does not apply.

This exceptional case can be solved by approximating the equation near the turning points with the confluent hypergeometric function. (See homework problem 6.)

The second kind of problems we will treat are the ones with an interior turning point and with the order of the parabolic cylinder equation equal to an integer. Let $a(x)$ have a simple zero at $x = 0$. Near $x = 0$, let

$$a(x) \approx \alpha x,$$

and

$$b(x) \approx \beta.$$

Let

$$\frac{\beta}{\alpha} = n + 1,$$

where n is an integer. There are two separate cases to consider.

Case of $\alpha > 0$

In this case, ν is equal to the integer n. In the region $x > 0$, the rapidly varying solution decreases with x. In the region $x < 0$, the rapidly varying solution increases with x.

Near the origin, one of the solutions is approximately equal to

$$e^{-\alpha x^2/(4\epsilon)} D_n(\sqrt{\alpha/\epsilon x}).$$

Since

$$D_n(\sqrt{\alpha/\epsilon x}) = 2^{-1/(2n)} e^{-\alpha x^2/(4\epsilon)} H_n(\sqrt{\alpha/2\epsilon x}),$$

where $H_n(z)$ is a polynomial named the Hermite polynomial of order n, this solution is approximately equal to

$$2^{-1/(2n)} e^{-\alpha x^2/(2\epsilon)} H_n(\sqrt{\alpha/2\epsilon x}).$$

This solution decreases exponentially as we leave the origin in either direction. This is the small, rapidly decreasing solution for all values of x.

Incidentally, the other independent solution of the equation near the turning point is

$$e^{-\alpha x^2/(4\epsilon)} D_{-n-1}(i\sqrt{\alpha/\epsilon}x).$$

Since

$$D_{-n-1}(i\sqrt{\alpha/\epsilon}x) \approx (i\sqrt{\alpha/\epsilon}x)^{-n-1} e^{\alpha x^2/(4\epsilon)},$$

this solution is the large, slowly varying solution for all values of x.

Since the small solution continues to be the small solution as we move past the turning point, there is a difficulty. We may start with the small solution from the left of the turning point. Or we may start with the small solution from the right of the turning point. But both solutions are the same solution. Therefore, one is inclined to conclude that it is impossible to obtain an approximate solution that satisfies the two boundary conditions.

Case of $\alpha < 0$

In this case, the order ν is equal to $(n+1)$. One of the solutions is

$$e^{-\alpha x^2/(4\epsilon)} D_{n+1}(\sqrt{|\alpha|/\epsilon}x).$$

This solution is slowly varying as x leaves the origin in either direction. Thus it is the only small solution regardless of whether x is positive or negative. Therefore, one is inclined to conclude that it is not possible to obtain an approximate solution that satisfies the boundary conditions.

This conclusion is incorrect. It is only in the zeroth order of ϵ that the small solution remains the small solution as it is continued across the turning point. In the higher orders of ϵ the solution picks up a component of the large solution as it is continued to the other side of the turning point. This means that even if ν is an integer, the small solution still becomes the large solution as it is continued to the other side of the turning point. Thus we may obtain two independent solutions of the problem by starting with the small solution from either side of the turning point. Indeed, the solutions of $\alpha < 0$ are always peaked around the endpoints, and are exponentially small outside the boundary layers near the endpoints. This is true whether ν is an integer or not. As for the case of $\alpha > 0$, the solution outside the boundary layer near the turning point is always given by the slowly varying solution (9.15), whether ν is an integer or not. The difference is with the solution inside the boundary layer near $x = 0$, as we shall show with an example below.

Problem for the Reader

Solve

$$\epsilon y'' + x(1+x^2)y' + y = 0, \tag{9.74}$$

with

$$y(-1) = A,\ y(1) = B. \tag{9.75}$$

Solution

Since the equation is linear, we may solve separately the problem with $B = 0$ and the problem with $A = 0$. The solution with neither A nor B equal to zero is the sum of these two solutions.

The origin is a turning point with the order ν equal to zero. The rapidly varying solution increases with x for $x < 0$, and decreases with x for $x > 0$. This means that the rapidly varying solution decreases as x leaves the origin in either direction. In the zeroth-order approximation, equation (9.74) can be reduced to the parabolic cylinder equation of order zero. Thus in the zeroth order, the small and rapidly varying solution is given by

$$e^{-x^2/(2\epsilon)}. \tag{9.76}$$

This solution decreases like a Gaussian as x moves away from the origin in either direction.

We start with the solution that is the small solution in the region $x > 0$ with $x \ll 1$. This solution is equal to a constant times the solution of (9.76).

We shall show that as we continue this solution into negative values of x, it is not just equal to the small and rapidly varying solution of (9.76). Instead, it picks up an extra term, which is a constant times the large and slowly varying solution. This extra term owes its origin to the terms in the differential equation that we have neglected when we approximate $a(x)$ of (9.74) by x. As we continue the small and rapidly varying solution through the origin, the neglected term x^3y' generates a term in the solution that is equal to a constant times the large and slowing varying solution. While this constant is proportional to ϵ, this component of the solution becomes the larger one as x becomes negative. This is because the rapidly varying solution decreases exponentially as x becomes negative, and the slowly varying solution becomes much larger than the rapidly varying solution.

Let us first find the solution satisfying the boundary conditions $y(-1) = A$ and $y(1) = 0$. We easily find y_{out} in the region $x < 0$ as

$$y_{out}(x) = -\frac{A\sqrt{(1+x^2)/2}}{x}. \tag{9.77}$$

Next we subtantiate the statement that the small solution for positive x becomes the large solution for negative x. We write the differential equation near $x = 0$ as

$$y'' + \frac{1}{\epsilon}xy' + \frac{1}{\epsilon}y = -\frac{1}{\epsilon}x^3 y'. \tag{9.78}$$

When x is small, the term on the right side of (9.78) is small compared to the terms on the left side of the equation. Hence it can be treated as a perturbative term. If we set the right side of (9.78) to zero, the two independent unperturbed solutions obtained are

$$y_1(x) = e^{-x^2/(2\epsilon)} \tag{9.79}$$

and

$$y_2(x) = e^{-x^2/(2\epsilon)} \int_0^x e^{x'^2/(2\epsilon)} dx'. \tag{9.80}$$

The solution $y_1(x)$ is the one given by (9.76) and is rapidly varying. The solution $y_2(x)$ is slowly varying. Indeed, in the limit $|x| \gg \sqrt{\epsilon}$, we have

$$y_2(x) \approx \frac{\epsilon}{|x|}. \tag{9.81}$$

To take care of the term on the right side of (9.78), we use the method of variation of parameters, treating the term on the right side of eq. (9.78) as the source term $f(x)$ in (1.30) and making use of (1.31). The solution y of (1.30) is equal to the right side of (1.31) plus a complementary solution. This complementary solution is undetermined unless we specify the conditions imposed on the solution. The boundary condition for our solution here is that it is the small solution when x is positive. This means that it has only a rapidly varying term when x is positive and much larger than $\sqrt{\epsilon}$. The most general solution satisfying this boundary condition is equal to a constant times y, where y satisfies the integral equation

$$\begin{aligned} y(x) = e^{-x^2/(2\epsilon)} &+ \frac{1}{\epsilon} e^{-x^2/(2\epsilon)} \int_0^x y_2(x') e^{x'^2/(2\epsilon)} x'^3 \frac{d}{dx'} y(x') dx' \\ &+ \frac{1}{\epsilon} y_2(x) \int_x^\infty x'^3 \frac{d}{dx'} y(x') dx'. \end{aligned} \tag{9.82}$$

Note that the complementary solution is

$$e^{-x^2/(2\epsilon)}.$$

Note also that the lower limit of integration for the second integral above is x. (This accounts for a difference of the sign with the corresponding term in (1.31).) The solution of the integral equation satisfies the boundary condition imposed on y in the region $x \gg \sqrt{\epsilon}$.

Replacing $y(x')$ in the integrals above by

$$e^{-x'^2/(2\epsilon)}$$

as the first approximation, we get

$$y(x) \approx e^{-x^2/(2\epsilon)} - \frac{1}{\epsilon^2} e^{-x^2/(2\epsilon)} \int_0^x y_2(x') x'^4 dx' \\ - \frac{1}{\epsilon^2} y_2(x) \int_x^\infty x'^4 e^{-x'^2/(2\epsilon)} dx'. \tag{9.83}$$

When x is positive and is much larger than $\sqrt{\epsilon}$, the last term of (9.83) is negligible and we get

$$y(x) \approx e^{-x^2/(2\epsilon)} - \frac{x^4}{4\epsilon} e^{-x^2/(2\epsilon)} \approx e^{-x^2/(2\epsilon) - x^4/(4\epsilon)}, \tag{9.84}$$

which is indeed rapidly varying. Note that the factor x^4/ϵ is much smaller than unity when x is much smaller than $\epsilon^{1/4}$. Indeed, the last expression in (9.84) is just the factor

$$e^{-\frac{1}{\epsilon} \int a(x) dx}$$

in (9.24).

But when x is negative and $|x|$ is much larger than $\sqrt{\epsilon}$, the last term in (9.83) is the largest term and we get

$$y(x) \approx -3\sqrt{2\pi}\epsilon^{1/2} y_2(x) \approx \frac{3\sqrt{2\pi}\epsilon^{3/2}}{x}, \tag{9.85}$$

which matches with (8.77) if we multiply the solution of (9.83) by a factor

$$-\frac{A\epsilon^{-3/2}}{3\sqrt{\pi}}.$$

Thus we have

$$y_{in}(x) = -\frac{A}{6\sqrt{\pi}\epsilon^{3/2}} e^{-x^2/(2\epsilon)} + \frac{A}{6\sqrt{\pi}\epsilon^{7/2}} e^{-x^2/(2\epsilon)} \int_0^x y_2(x') x'^4 dx' \\ + \frac{A}{6\sqrt{\pi}\epsilon^{7/2}} y_2(x) \int_x^\infty x'^4 e^{-x'^2/(2\epsilon)} dx'. \tag{9.86}$$

When x is negative, $y_{in}(x)$ approaches $-A/(\sqrt{2}x)$ when $|x|/\sqrt{\epsilon}$ is large. This term comes from the last term in (9.86). When x is of the order of $\sqrt{\epsilon}$, $y_{in}(x)$ is approximately

$$-\frac{A}{6\sqrt{\pi}\epsilon^{3/2}}e^{-x^2/(2\epsilon)}, \tag{9.87}$$

which is the first term on the right side of (9.86). It is as large as $\epsilon^{-3/2}$ at the origin. When x is positive and much larger than $\sqrt{\epsilon}$, the solution vanishes exponentially.

In summary, the solution is given by (9.77) when x is negative and is outside the boundary layer near $x = 0$, and is given by (9.86) inside the boundary layer. There is a bump of the order of $\epsilon^{-3/2}$ inside the boundary layer. When ν is not an integer, the value of y at $x = 0$ is of the order of $\epsilon^{-(\nu+1)/2}$. The result for $\nu = 0$ is larger than what is expected by a factor ϵ^{-1}. Needless to say, the rapidly varying solution vanishes exponentially when x is positive and outside the boundary layer.

Next we solve the problem with the boundary condition $y(-1) = 0$ and $y(1) = B$. We'll give the results here without much exposition. We have, in the region $x > 0$,

$$y_{out}(x) = \frac{B\sqrt{(1+x^2)/2}}{x}. \tag{9.88}$$

The counterpart of (9.83) is

$$\begin{aligned} y(x) \approx\ & e^{-x^2/(2\epsilon)} - \frac{1}{\epsilon^2}e^{-x^2/(2\epsilon)}\int_0^x y_2(x')x'^4dx' \\ & + \frac{1}{\epsilon^2}y_2(x)\int_{-\infty}^x x'^4 e^{-x'^2/(2\epsilon)}dx'. \end{aligned} \tag{9.89}$$

In the region $1 \gg x \gg \sqrt{\epsilon}$, we have

$$y \approx \frac{3\sqrt{2\pi}\epsilon^{3/2}}{x}.$$

Thus the solution matching with (9.88) is

$$\begin{aligned} y_{in}(x) = \ & \frac{B\epsilon^{-3/2}}{6\sqrt{\pi}}e^{-x^2/(2\epsilon)} - \frac{B\epsilon^{-7/2}}{6\sqrt{\pi}}e^{-x^2/(2\epsilon)}\int_0^x y_2(x')x'^4dx' \\ & - \frac{B\epsilon^{-7/2}}{6\sqrt{\pi}}y_2(x)\int_x^{\infty} x'^4e^{-x'^2/(2\epsilon)}dx'. \end{aligned}$$

From (9.89), we see that $y_{in}(x)$ is approximately

$$\frac{B}{6\sqrt{\pi}\epsilon^{3/2}}e^{-x^2/(2\epsilon)}, \tag{9.90}$$

when the magnitude of x is of the order of $\sqrt{\epsilon}$. The solution is exponentially small when x is negative and $|x|$ is much larger than $\sqrt{\epsilon}$. We will give another example below.

Problem for the Reader

Solve

$$\epsilon y'' - x(1 + x/2)y' + y = 0, \tag{9.91}$$

with

$$y(-1) = 0$$

and

$$y(1) = 1. \tag{9.92}$$

Solution

In this problem,

$$a(x) = -x(1 + x/2).$$

There is a turning point at $x = 0$ with $\nu = 1$, which is an integer. The rapidly varying solution for $x > 0$ increases as x increases. The rapidly varying solution for $x < 0$ decreases as x increases.

Although ν is an integer, the small and slowly varying solution at the left of the turning point still turns into the large and rapidly increasing solution after we pass the turning point. Thus there is a solution that is sharply peaked around $x = 1$. Let the value of this solution be the order of unity at $x = 1$. Then this solution is exponentially small outside the boundary layer at $x = 1$. The solution near $x = 1$ satisfies approximately

$$\epsilon y'' - \frac{3}{2}y' = 0.$$

Thus

$$y_{in}(x) = e^{-3(1-x)/(2\epsilon)} \tag{9.93}$$

is a good approximation of the solution of (9.91) that satisfies the boundary condition at $x = 1$. Since the solution (9.93) becomes exponentially small once x is outside the boundary layer near $x = 1$, it also approximately satisfies the boundary condition at $x = -1$. Thus it is a good approximation of the solution of the problem for all x.

Problem for the Reader

Solve (9.91) with the boundary conditions $y(-1) = 1$ and $y(0) = 0$.

Solution

We start with the slowing varying solution y_{out} in the region $x > 0$, and continue it to the left. After we pass the origin and get to the region of negative x, it becomes a linear superposition of a rapidly varying solution and a slowly varying solution. At the origin, the rapidly varying solution is by a factor ϵ smaller than the slowly varying solution. But it grows exponentially as we continue the solution to negative values of x, eventually overcoming the factor ϵ and becoming larger than the slowly varying solution.

With the same considerations as those given previously, we conclude that the approximate solution satisfying the boundary conditions is

$$y = e^{-(1+x)/(2\epsilon)}.$$

This solution is sharply peaked near the lower endpoint $x = -1$, and is exponentially small outside the boundary layer of width ϵ near the lower endpoint. It satisfies exactly the boundary condition at $x = -1$ and satisfies approximately the boundary condition at $x = 1$.

Consider more generally the solution of (9.91) with the boundary conditions $y(-1) = A$ and $y(1) = B$. The solution of this problem has two sharp peaks, one at each of the endpoints. It is exponentially small outside the two boundary layers near the endpoints, the width of each of which is of the order of ϵ. Specifically, the solution of this boundary-value problem is

$$Be^{-3(1-x)/(2\epsilon)} + Ae^{-(1+x)/(2\epsilon)}.$$

Homework Problems for This Chapter

Solutions to the Homework Problems can be found at www.lubanpress.com.

1. Eliminate the y' term in (9.13) by a change of the dependent variable, and obtain the WKB solutions of the resulting equation. Prove that the two solutions obtained are the same as y_s and y_r given by (9.15) and (9.24), respectively.

2. Solve approximately the following two equations:

a. $\epsilon y'' + (1+x)^2 y' + y = 0,\ 0 < x < 1,$ with $y(0) = 0$ and $y(1) = 2$.

b. $\epsilon y'' - (1+x^2)y' + y = 0, 0 < x < 1,$ with $y(0) = 0$ and $y(1) = 2$.

3. Solve approximately

$$\epsilon y'' - 2(\sin x)y' + (\cos x)y = 0,\ 0 < x < 1, \epsilon \ll 1,$$

with the boundary conditions

$$y(0) = y(1) = 1.$$

4. Solve approximately

$$\epsilon y'' + x(1+x)y' + \frac{1}{2}y = 0,\ 0 < x < 1,$$

with the boundary conditions

$$y(0) = 1$$

and

$$y(1) = 2.$$

5. Solve approximately

$$\epsilon y'' - 2xy' + (1+3x^3)y = 0,\ -1 < x < 1,$$

with the boundary conditions

$$y(-1) = 2 \quad \text{and} \quad y(1) = 3.$$

6. Solve approximately $\epsilon y'' + x^3 y' + (1+x)y = 0$ with $y(0) = y(1) = 1$.

7. Solve approximately

$$\epsilon y'' + xy' + (\sin x)y = 0,\ 0 < x < 1, \epsilon \ll 1.$$

The boundary conditions are

$$y(0) = 1, y(1) = 2.$$

8. Solve approximately

$$\epsilon y'' - (\sin x)y' + (\cos x)y = 0,\ 0 < x < 1, \epsilon \ll 1,$$

with the boundary conditions

$$y(0) = y(1) = 1.$$

9. Solve approximately

$$\epsilon y'' + x^2 y' + y = 0,\ y(0) = y(1) = 1.$$

Hint: The WKB method fails at the point $X = 1$, where $X \equiv (4\epsilon)^{-1/4}x$. Show that the equation near $X = 1$ can be approximated by the Airy equation. Use the Airy functions to join the WKB solutions for $X < 1$ to the WKB solutions for $X > 1$.

10. Show that if $a(x) = \alpha x^2$ and $b(x) = \beta x$ near the point $x = 0$, equation (9.13) can be transformed into the confluent hypergeometric equation. Discuss how you would solve (9.13) with the boundary conditions $y(-1) = A$ and $y(1) = B$ if $a(x)$ has a double zero and $b(x)$ has a simple zero at $x = 0$.

Hint: The dimensions of the terms in the differential equation

$$\epsilon \frac{d^2y}{dx^2} + \alpha x^2 \frac{dy}{dx} + \beta x y = 0$$

are -2 and 1. This suggests that one makes the change of variable

$$x^3 = t.$$

11. Describe how you would solve the fourth-order linear differential equation

$$\epsilon y'''' + a(x)y' + b(x)y = 0,\ 0 < x < 1,$$

where $a(x) > 0$. The values of y and y' at the endpoints are specified to be of the order of unity. Use the scaling argument to prove that the width of a boundary layer is of the order of $\epsilon^{1/3}$. Show also that outside the boundary layer,

$$y_{out}(x) = y(0) \exp\left(-\int_0^x \frac{b(x')}{a(x')}dx'\right) + O(\epsilon^{1/3}).$$

Hint: There are one slowly varying solution, two rapidly increasing solutions, and one rapidly decreasing solution. One uses the sum of the slowly varying solution and the rapidly decreasing solution to satisfy the two boundary conditions at $x = 0$. (How do you satisfy the boundary conditions at $x = 1$?)

12. Describe how you would solve the boundary-layer problem with two or more interior turning points.

Hint: The large solution with respect to a turning point is the small solution with respect to the next turning point.

Chapter 10

Small Nonlinear Oscillations

In several of the preceding chapters, we discuss how to obtain approximate solutions of linear equations. Here, we present methods to treat nonlinear problems of small oscillations.

A. Summing Leading Terms

We'll start with an example of nonlinear problems. Consider the pendulum consisting of a bob of mass m at the end of a rod of negligible mass. We will first derive the equation of motion for this problem.

One of the best ways to do that is using the Lagrangian approach. The velocity of the mass is equal to $l\dot{\Theta}$, where l is the length of the rod and Θ is the angle the rod makes with a vertical line. Thus the kinetic energy of the mass is

$$T = \frac{1}{2}m(l\dot{\Theta})^2,$$

and the gravitational potential of the mass is

$$V = -mgl\cos\Theta.$$

The Lagrangian $L = T - V$ is

$$L = \frac{1}{2}m(l\dot{\Theta})^2 + mgl\cos\Theta.$$

The equation of motion is given by the Euler-Lagrange equation

$$\frac{d}{dt}\left(\frac{\delta L}{\delta \dot{\Theta}}\right) = \frac{\delta L}{\delta \Theta}.$$

Since

$$\frac{\delta L}{\delta \dot{\Theta}} = ml^2\dot{\Theta}, \; \frac{\delta L}{\delta \Theta} = -mgl\sin\Theta,$$

the equation of motion for the pendulum is

$$\ddot{\Theta} = -\frac{g}{l}\sin\Theta. \tag{10.1}$$

We note that the independent variable t does not appear explicitly in (10.1). Such an equation is called autonomous. An autonomous differential equation is invariant under a translation of the time variable. This is to say that this equation remains the same if we replace t by $t+c$. Thus if $\Theta(t)$ satisfies this equation, so does $\Theta(t+c)$.

An autonomous second-order differential equation can be converted into a first-order differential equation. We will do this in two steps.

In the first step, we denote $\dot{\Theta}$ by Π and (10.1) becomes

$$\begin{aligned} \frac{d\Theta}{dt} &= \Pi \\ \frac{d\Pi}{dt} &= -\frac{g}{l}\sin\Theta. \end{aligned} \tag{10.2}$$

The equations in (10.2) form a system of two autonomous first-order differential equations. They are, taken together, equivalent to the single second-order differential equation (10.1) and are called Hamilton's equations for the pendulum motion.

In the second step, we eliminate dt from (10.2) and get

$$\frac{d\Pi}{d\Theta} = -\frac{g\sin\Theta}{l\Pi}.$$

This equation is a first-order differential equation with Π the dependent variable and Θ the independent variable. The variable t does not appear in this equation.

The differential equation is separable, and the solution is easily found by an integration to be

$$\frac{1}{2}\Pi^2 + \frac{g}{l}(1-\cos\Theta) = E,$$

where E is a constant. Indeed, E is equal to the energy of the pendulum divided by ml^2. This solution gives a relationship between Θ and Π, and can be graphed as a closed curve in the Θ–Π plane, which is also known as the phase plane.

As a side remark, the system of equations (10.2) is a special case of the system of autonomous equations

$$\frac{dx}{dt} = F(x, y)$$

and

$$\frac{dy}{dt} = G(x, y).$$

If both $G(x, y)$ and $F(x, y)$ vanish at the point (x_0, y_0), then the constant solution

$$x(t) = x_0,\ y(t) = y_0$$

solves this system of equations. Such a point (x_0, y_0) is called a critical point of the equation.

Let us return to the pendulum problem. We see that the points $\Pi_0 = 0$ and $\Theta_0 = n\pi$, $n = 0, \pm\pi, \pm 2\pi, \cdots$, are the critical points of the equations of (10.2). Physically, a critical point of (10.2) is a point of equilibrium for the pendulum. From the first equation of (10.2), the pendulum is stationary if $\Pi = 0$. And from the second equation of (10.2), there is no force on the pendulum motion if $\sin\Theta = 0$. Thus the constant solution $\Theta(t) = n\pi$, $\Pi(t) = 0$ satisfies (10.2).

Of these points of equilibrium, the ones with n an even integer are points of stable equilibrium. Consider the critical point of $n = 0$. At the point $\Theta = \Pi = 0$, the pendulum assumes the lowest angular position possible and is at rest. Let us lift the pendulum to a small angle δ and release it. When Θ is small, we approximate $\sin\Theta$ by Θ. Then (10.1) is reduced to the equation of motion of a harmonic oscillator, and the pendulum motion is the harmonic motion $\delta\cos\omega_0 t$ with $\omega_0 = \sqrt{g/l}$. The angular position of the pendulum is less than δ at all times, and the motion is stable.

While this approximation is good initially, after a long time period it invariably fails. The nonlinear terms in the Maclaurin series expansion of $\sin\Theta$ are small compared to Θ when Θ is small, but over a long period of time the contributions from these small terms add up and alter the harmonic motion significantly. We shall show how to determine the motion of the pendulum as a function of time with the nonlinear terms taken into account.

The solution in the phase plane gives us $\dot{\Theta}$ as a function of Θ or vice versa, but it does not give us Θ as a function of t. To find the latter, we need to do one more integration. First, we find $\dot{\Theta}$ as a function of Θ:

$$\frac{d\Theta}{dt} = \pm\sqrt{2E - 2g(1 - \cos\Theta)/l},$$

which is rewritten as

$$\frac{d\Theta}{\sqrt{2E - 2g(1 - \cos\Theta)/l}} = \pm dt.$$

Calling the integral of the left side of the equation above $F(\Theta)$, we get

$$F(\Theta) = \pm(t + c),$$

where c is a constant. The function $F(\Theta)$ is equal to a constant times the elliptic integral of the first kind. This solution gives Θ as an implicit function of t.

When Θ and $\dot{\Theta}$ are initially small, we may obtain the pendulum motion by perturbative methods instead. Such a perturbative solution describes Θ as an explicit function of t, and gives us a better physical picture of the motion.

Let us first specify the initial conditions of this motion. As we have seen, if the initial values of $\Theta(t)$ and $\dot{\Theta}(t)$ are sufficiently small, the pendulum motion is periodic with the value of Θ small at all times. In such a periodic motion, $\dot{\Theta}(t)$ must vanish at some repeated values of t. Physically, this happens when the pendulum is at its highest angular position allowed. Without losing generality, we shall assume that this happens at $t = 0$. Then the initial conditions are

$$\Theta(0) = \delta$$

and

$$\dot{\Theta}(0) = 0,$$

where δ is a small number.

Next we expand $\sin\Theta$ into the Maclaurin series

$$\sin\Theta = \Theta - \Theta^3/6 + \Theta^5/120 + \cdots,$$

and change the dependent variable from Θ to θ,where

$$\theta \equiv \Theta/\delta.$$

Then eq. (10.1) becomes

$$\ddot{\theta} + \omega_0^2\theta = \epsilon\theta^3/6 - \epsilon^2\theta^5/(120\omega_0^2) + \cdots, \tag{10.3}$$

where

$$\epsilon \equiv \omega_0^2 \delta^2.$$

The initial conditions for $\theta(t)$ are

$$\theta(0) = 1 \tag{10.4a}$$

and

$$\dot{\theta}(0) = 0, \tag{10.4b}$$

both being independent of δ. Note that after being scaled, the dependent variable θ is no longer small and is of the order of unity, while the coefficient of the cubic force in (10.3) is the small constant ϵ, which is of the order of δ^2.

If we set ϵ to zero, (10.3) is reduced to the linear equation for the motion of the harmonic oscillator. In this approximation, the solution is $\cos \omega_0 t$. We shall calculate the corrections to the free harmonic motion generated by the nonlinear terms.

We first apply the method of regular perturbation to this problem. In this method, we express $\theta(t)$ by the perturbation series

$$\theta_0(t) + \epsilon\theta_1(t) + \cdots.$$

Since ϵ is a small number, one may expect $\epsilon\theta_1(t)$ to be small compared to θ_0. Thus one may expect θ_0 to be a good zeroth-order approximation of the pendulum motion. Similarly, one may expect $\epsilon^2\theta_2$ to be small compared to $\epsilon\theta_1$. Thus one may expect $\theta_0 + \epsilon\theta_1$ to be an even better approximation than θ_0. And one may expect that the approximation gets better and better as one includes more and more terms in the perturbation series.

Let us see if this expectation is borne out. With θ replaced by its perturbation series, (10.3) becomes

$$\begin{aligned}\left(\frac{d^2}{dt^2} + \omega_0^2\right)(\theta_0 + \epsilon\theta_1 + \cdots) &= \frac{\epsilon}{6}\left[\theta_0 + \epsilon\theta_1 + \cdots\right]^3 \\ &\quad - \frac{\epsilon^2}{120\omega_0^2}\left[\theta_0 + \epsilon\theta_1 + \cdots\right]^5 + \cdots.\end{aligned} \tag{10.5}$$

The initial condition (10.4a) says that

$$\theta_0(0) + \epsilon\theta_1(0) + \cdots = 1.$$

By equating the coefficients of ϵ^n, $n = 0, 1, \cdots$, on the two sides of the equation above, we get

$$\theta_0(0) = 1,$$

as well as

$$\theta_1(0) = \theta_2(0) = \cdots = 0.$$

Similarly, the initial condition (10.4b) gives

$$\dot{\theta}_0(0) = \dot{\theta}_1(0) = \dot{\theta}_2(0) = \cdots = 0.$$

We shall determine $\theta_n(t)$ successively.

The differential equation for $\theta_0(t)$ is obtained by equating the ϵ^0 terms on the two sides of eq. (10.5). We get

$$\ddot{\theta}_0 + \omega_0^2 \theta_0 = 0.$$

The solution of this equation satisfying the initial conditions is

$$\theta_0(t) = \frac{1}{2}(e^{i\omega_0 t} + e^{-i\omega_0 t}). \tag{10.6}$$

The differential equation for $\theta_1(t)$ is obtained by equating the terms linear in ϵ on the two sides of (10.5). We get

$$\ddot{\theta}_1 + \omega_0^2 \theta_1 = \frac{\theta_0^3}{6}. \tag{10.7}$$

Eq. (10.7) has a source term on the right. This equation governs the motion of a harmonic oscillator driven by the force $\theta_0^3/6$. Making use of (10.6), this force can be expressed as

$$\frac{\theta_0^3}{6} = \frac{1}{48} e^{3i\omega_0 t} + \frac{1}{16} e^{i\omega_0 t} + c.c., \tag{10.8}$$

where $c.c.$ is the abbreviation of the complex conjugate.

Of the terms above, we pay particular attention to the terms

$$e^{i\omega_0 t}/16 + e^{-i\omega_0 t}/16. \tag{10.9}$$

This is because these terms are also complementary solutions of (10.7). Therefore, the phase of each of the terms in (10.9) is coherent with that of the free harmonic motion. The magnitude of the motion of a harmonic oscillator subject to such coherent forces grows linearly with time. To verify this, let θ_P be a particular solution of

$$\left(\frac{d^2}{dt^2}+\omega_0^2\right)\theta_P = e^{i\omega_0 t}.$$

Using the operator method given in Chapter 1, we have

$$\theta_p(t) = \frac{1}{(D-i\omega_0)(D+i\omega_0)}e^{i\omega_0 t},$$

where $D \equiv d/dt$. By (1.5), we move the factor $e^{i\omega_0 t}$ to the leftmost position and compensate for this by replacing D with $D+i\omega_0$. We get

$$\theta_P = e^{i\omega_0 t}\frac{1}{D(D+2i\omega_0)}1 = e^{i\omega_0 t}\frac{1}{D(2i\omega_0)}1 = \frac{e^{i\omega_0 t}}{2i\omega_0}t. \tag{10.10}$$

In the above, the factor t comes from

$$\frac{1}{D}1 = t,$$

as the inverse of differentiation is integration. The expression in (10.10) grows linearly with time. Thus the force $\epsilon e^{i\omega_0 t}/16$ generates for $\epsilon\theta_1(t)$ a term

$$-\frac{i\epsilon e^{i\omega_0 t}}{32\omega_0}t, \tag{10.11a}$$

which grows linearly with time. Similarly, the force $\epsilon e^{-i\omega_0 t}/16$ generates for $\epsilon\theta_1(t)$ a term

$$\frac{i\epsilon e^{-i\omega_0 t}}{32\omega_0}t, \tag{10.11b}$$

which also grows linearly with time.

In contrast, the force $e^{3i\omega_0 t}\epsilon/48$ generates for $\epsilon\theta_1$ a term

$$\frac{1}{D^2+\omega_0^2}\frac{\epsilon e^{3i\omega_0 t}}{48} = -\frac{\epsilon e^{3i\omega_0 t}}{384\omega_0^2},$$

the magnitude of which is a constant. We also note that the frequency associated with the term above is three times the frequency of the harmonic oscillator, and is hence called a term of the third harmonic. Generally, higher harmonic terms are produced when nonlinear interactions are present.

When t is very large, the term in (10.11a) and its complex conjugate (10.11b) are large compared to the third harmonic terms. Thus the former are the leading terms of $\epsilon\theta_1(t)$. While $\epsilon\theta_1(t)$ is initially small, its leading terms grow with time, and it becomes of the order of $\theta_0(t)$ when $\epsilon t/\omega_0$ is $O(1)$. As a consequence, while the lowest-order term of regular perturbation is a good approximation when $\epsilon t/\omega_0$ is small, it is not a good approximation when t is of the order of $\omega_0\epsilon^{-1}$, and regular perturbation fails.

The terms of (10.9) are called the secular terms of (10.7). These secular terms are the source terms in the differential equation (10.7) that produce the growing leading terms of θ_1, rendering regular perturbation invalid when t is sufficiently large.

We shall carry this argument to higher orders. From (10.5), we easily show that the equation for $\theta_2(t)$ is

$$\ddot{\theta}_2 + \omega_0^2\theta_2 = \frac{\theta_0^2\theta_1}{2} - \frac{\theta_0^5}{120\omega_0^2}.$$

This equation is that of the motion of a harmonic oscillator driven by the force

$$\frac{\theta_0^2\theta_1}{2} - \frac{\theta_0^5}{120\omega_0^2}. \tag{10.12}$$

The leading terms of (10.12) come from the first term, and are of the order of $te^{\pm i\omega_0 t}$. (Note the factor t, which comes from θ_1.) They are the secular terms of the differential equation satisfied by θ_2. The phase of each of these secular terms is coherent with the free harmonic motion, and the magnitude of each of these secular terms grows linearly with t. Such secular terms generate a motion with a magnitude that grows like t^2, as

$$\frac{1}{D}t = t^2/2.$$

Indeed, it is straightforward to find that these terms of θ_2 are

$$\frac{1}{2^9\omega_0^2}\frac{(it)^2}{2!}e^{i\omega_0 t} + c.c. \tag{10.13}$$

Because of the existence of such terms, $\epsilon^2\theta_2(t)$ is comparable to $\theta_0(t)$ when t is $O(\epsilon^{-1})$. (From now on we regard ω_0 as $O(1)$.)

It is also easy to demonstrate that since

$$\frac{1}{D}\frac{t^{n-1}}{(n-1)!}=\frac{t^n}{n!},$$

the leading terms of $\theta_n(t)$ are of the order of $t^n e^{i\omega_0 t}$ or $t^n e^{-i\omega_0 t}$. Thus $\epsilon^n\theta_n(t)$ is also comparable to $\theta_0(t)$ when t is $O(\epsilon^{-1})$.

We will spare the reader the details of calculating such leading terms. This is because we'll present in the next sections methods that enable us to obtain these terms a little more easily.

With (10.11a) and (10.13), we find that

$$\theta_0(t)+\epsilon\theta_1(t)+\epsilon^2\theta_2(t)=\frac{1}{2}e^{i\omega_0 t}\left(1+i\omega_1\epsilon t+\omega_1^2\epsilon^2\frac{i^2t^2}{2!}\right)+c.c.,$$

where

$$\omega_1=-\frac{1}{16\omega_0},$$

and where we have included only the leading terms. The sum above suggests that summing the leading terms of $\theta_n(t)$, $n=1,2,\cdots$, gives

$$\theta(t)\approx\frac{1}{2}e^{i(\omega_0+\omega_1\epsilon)t}+\frac{1}{2}e^{-i(\omega_0+\omega_1\epsilon)t}. \tag{10.14}$$

This will be verified in the next section. Equation (10.14) is equal to the lowest-order term $\theta_0(t)$ of (10.6) with ω_0 replaced by $\omega_0+\omega_1\epsilon$. It demonstrates clearly why (10.6) is a good approximation of the pendulum motion only for t not very large. In plain words, a pendulum that is off by a few seconds a day, say, is fairly accurate over a period of a few days. But over a period of years the errors accumulate and the time given by the errant pendulum is not accurate enough to be useful.

We point out that the right side of (10.14) contains all the leading terms that are of the order of $\epsilon^n t^n e^{\pm i\omega_0 t}$. All these terms are $O(1)$ when t is $O(\epsilon^{-1})$. The next-leading terms, not included in (10.14), are of the order of $\epsilon^n t^{n-1}e^{\pm i\omega_0 t}$. When $t=O(\epsilon^{-1})$, these next-leading terms are of the order of ϵ, small compared to the leading terms. Therefore, the sum of the leading terms of regular perturbation is a good approximation even when t is as large as $O(\epsilon^{-1})$. But when t is much larger than ϵ^{-1}, the nonleading terms may become large and the sum of leading terms is not a good approximation of the pendulum motion.

Before we close this section, we remark that the example above makes a lie of the assertion that dropping a "small" term in a differential equation is always justified. In order to clarify this point further, we illustrate with another example. Consider the differential equation

$$\frac{d^2x}{dt^2} + (1+\epsilon)x = 0.$$

The exact solutions of this equation are

$$e^{\pm i\sqrt{1+\epsilon}\,t}.$$

Now the term $(1 + \epsilon)t$ in the differential equation above is approximately equal to t, as ϵt is always small compared to t. But if we drop the small term ϵt from the equation, the solutions of the resulting equation are $e^{\pm it}$, which are not good approximations of the exact solutions given above when t is of the order of ϵ^{-1}. This shows small errors may add up and become considerable over a long time.

More generally, consider a differential equation with a small parameter ϵ. Let the series solution of regular perturbation be

$$x = x_0 + \epsilon x_1 + \epsilon^2 x_2 + \cdots .$$

This perturbation series is useful when $\epsilon^n x_n$ is much smaller than $\epsilon^{n-1} x_{n-1}$ for all n. But near a singular point of the differential equation, x_n may blow up and become much larger than x_{n-1}. Thus truncating the perturbation series may not be justified when t is sufficiently close to such a singular point.

B. Renormalized Perturbation—The Improved Poincare Method

In the example of the pendulum, we've seen that regular perturbation fails when ϵt is of the order of unity. Fortunately, there exist other perturbation methods that enable us to handle this contingency. These perturbative methods are generally known as methods of singular perturbation.

The method we'll demonstrate next is called that of renormalized perturbation. We shall first verify, with the use of this method, that the sum of leading terms is equal to (10.14). This sum is a good approximation of the solution for all values of t up to the values of the order of ϵ^{-1}. Then we shall show how to improve this approximation. The improved approximation will be a good approximation for all t.

The pendulum motion, while not that of a pure harmonic oscillator, repeats itself every so often and is hence periodic. A periodic function can be expressed as a Fourier series. Let the period of this motion be P, and let

$$\omega \equiv 2\pi/P.$$

Then the Fourier series of $\Theta(t)$ is of the form

$$\sum A_n e^{in\omega t},$$

where A_n is a constant called the Fourier coefficient. As ϵ approaches zero, the pendulum motion approaches that of the harmonic oscillator, and ω approaches ω_0. Thus ω differs only slightly from ω_0 when ϵ is small. Indeed, (10.14) shows that to the first order, we have

$$\omega \simeq \omega_0 + \omega_1 \epsilon.$$

We shall find ω as an infinite series of ϵ.

It will be convenient to put the relation between ω_0 and ω in the form

$$\omega_0^2 = \omega^2(1 + a_1\epsilon + a_2\epsilon^2 + \cdots), \tag{10.15}$$

where the coefficients a_n, $n = 1, 2, 3, \cdots$, are real constants to be determined. We rewrite eq. (10.3) as

$$\ddot{\theta} + \omega^2\theta = (\omega^2 - \omega_0^2)\theta + \frac{\epsilon\theta^3}{6} - \frac{\epsilon^2\theta^5}{120\omega_0^2} + \cdots. \tag{10.16}$$

The left side of (10.16) is equal to that of (10.3) with ω_0 replaced by ω. As a result, the right side of (10.16) has the additional term $(\omega^2 - \omega_0^2)\theta$. This term is called the counter term, which we will use to eliminate secular terms of the equation. We shall regard ω as of the order of unity and put

$$\theta(t) = \theta_0(t) + \epsilon\theta_1(t) + \cdots;$$

then (10.16) is, explicitly,

$$\begin{aligned}\left(\frac{d^2}{dt^2} + \omega^2\right)&(\theta_0 + \epsilon\theta_1 + \cdots) = \\ &- \epsilon\omega^2(a_1 + a_2\epsilon + \cdots)(\theta_0 + \epsilon\theta_1 + \cdots) + \frac{\epsilon}{6}(\theta_0 + \epsilon\theta_1 + \cdots)^3 \\ &- \frac{\epsilon^2}{120\omega_0^2}(\theta_0 + \epsilon\theta_1 + \cdots)^5 + \cdots.\end{aligned} \tag{10.17}$$

We shall consider the solution of (10.17) satisfying the initial conditions

$$\theta(0) = 1, \quad \dot{\theta}(0) = 0.$$

Then the initial conditions for $\theta_n(t), n = 0, 1, 2, \cdots$, are

$$\theta_0(0) = 1, \ \theta_1(0) = \theta_2(0) = \theta_3(0) = \cdots = 0,$$

and

$$\dot{\theta}_0(0) = \dot{\theta}_1(0) = \dot{\theta}_2(0) = \cdots = 0.$$

We require the constants a_n to be so chosen that there are no secular terms in the equation for θ_n.

The equation for $\theta_0(t)$ is

$$\ddot{\theta}_0 + \omega^2\theta_0 = 0. \tag{10.18}$$

The solution of (10.18) is

$$\theta_0(t) = A_0 e^{i\omega t} + A_0^* e^{-i\omega t},$$

where A_0 is a constant. That $\dot{\theta}_0(0) = 0$ gives

$$A_0 - A_0^* = 0,$$

which says that A_0 is real. From the initial condition $\theta_0(0) = 1$ we get $A_0 = 1/2$. Thus

$$\theta_0(t) = \frac{1}{2}(e^{i\omega t} + e^{-i\omega t}). \tag{10.19}$$

We note that $\theta_0(t)$ given by (10.19) is the same as (10.6), with ω_0 replaced by ω.

By equating the ϵ terms on the two sides of (10.17), we obtain the differential equation for $\theta_1(t)$ as

$$\ddot{\theta}_1(t) + \omega^2\theta_1(t) = \frac{\theta_0^3}{6} - \omega^2 a_1\theta_0. \tag{10.20}$$

Note that the last term on the right side of (10.20) is a term absent in (10.7). This term comes from the counter term $(\omega^2 - \omega_0^2)\theta$. Substituting the expression (10.19) for θ_0 into the right side of (10.20), we find that the secular terms in (10.20) are

$$e^{i\omega t}(1 - 8\omega^2 a_1)/16 \tag{10.21}$$

and

$$e^{-i\omega t}(1 - 8\omega^2 a_1)/16. \tag{10.22}$$

Note that in the secular terms, the coefficient of $e^{i\omega t}$ is always the complex conjugate of the coefficient of $e^{-i\omega t}$. This is regardless of the initial conditions imposed on θ_0, as the sum of secular terms is real. That in the present problem A_0 is real makes these two coefficients real and therefore equal to each other, as we can see from (10.21) and (10.22).

The secular terms generate growing terms for the solution of (10.20) unless their sum is equal to zero, which is the case if we choose

$$a_1 = 1/(8\omega^2). \tag{10.23}$$

We see that this single choice of a_1 makes both the expression of (10.21) and that of (10.22) vanish. With the right side of (10.20) having no secular terms, the solution of (10.20) has no terms growing with time. Therefore, $\epsilon\theta_1$ is smaller than $\theta_0(t)$ for all times. Furthermore, that $\theta_1(t)$ has no $te^{\pm i\omega t}$ terms implies that the equation for $\theta_2(t)$ has no $te^{\pm i\omega t}$ secular terms. This in turn implies that $\theta_2(t)$ has no $t^2e^{\pm i\omega t}$ terms. Indeed, this choice of a_1 eliminates in one stroke the $t^n e^{\pm i\omega t}$ terms in $\theta_n(t)$ for all n. This means that all such leading terms have been included in $\theta_0(t)$, which is explicitly given by (10.19).

From (10.15) and (10.23) we get, to the order of ϵ,

$$\omega = \omega_0 - \frac{1}{16\omega_0}\epsilon; \tag{10.24}$$

hence we have, as promised, proven that (10.14) is valid. We note that since $\epsilon = \omega_0^2\delta^2$, the angular frequency of the pendulum decreases as δ increases. This is a result we expect, as it takes a little longer for the pendulum to make one complete cycle if the initial angular position $\Theta(0)$ is a little larger.

Borrowing the terminology from quantum field theory, we shall call ω_0 the bare angular frequency and ω the renormalized angular frequency. The former is a parameter in the equation of motion for $x(t)$, while the latter is a parameter of the motion $x(t)$ itself. The regular perturbation series of the pendulum motion, expressed with the bare angular frequency ω_0, is called the unrenormalized perturbation series of the pendulum motion. The perturbation series of the pendulum motion expressed with the renormalized angular frequency ω is called the renormalized perturbation series of the pendulum motion.

Problem for the Reader

Find $\theta_1(t)$ that satisfies (10.20) with the initial conditions $\theta_1(0) = \dot{\theta}_1(0) = 0$.

Solution

The particular solution of (10.20) generated by the source terms is the Fourier sum

$$-\frac{e^{3i\omega t} + e^{-3i\omega t}}{384\omega^2}.$$

For a real-valued function, the Fourier coefficient of $e^{3i\omega t}$ is always the complex conjugate of the coefficient of $e^{-3i\omega t}$. Because the Fourier coefficients of the source terms of (10.20) are real, the Fourier coefficients of the particular solution of (10.20) are also real. As a result, the two coefficients given above are equal and the derivative of the particular solution of (10.20) vanishes at $t = 0$.

The complementary solution of (10.20) is

$$A_1 e^{i\omega t} + A_1^* e^{-i\omega t}.$$

The initial condition $\dot{\theta}_1(0) = 0$ gives

$$A_1 - A_1^* = 0.$$

Thus A_1 is real. The initial condition $\theta_1(0) = 0$ then gives

$$\theta_1(t) = \frac{e^{i\omega t} - e^{3i\omega t} + e^{-i\omega t} - e^{-3i\omega t}}{384\omega^2}.$$

In a problem in which there are multiple bare parameters in the equation of motion, one expresses the perturbation series for the solution of the equation in terms of the corresponding renormalized parameters. The renormalized parameters are the ones directly related to the motion itself, rather than the ones that appear in the differential equation governing the motion. The lowest-order term in the renormalized perturbation series often includes all of the leading terms in the unrenormalized perturbation series.

As an example, consider two harmonic oscillators coupling with each other. We will denote their coordinates by x and y, and their bare angular frequencies by ω_0 and W_0, respectively. The kinetic energy for this system is

$$T = \frac{\dot{x}^2 + \dot{y}^2}{2},$$

where, for simplicity, we have taken the masses of both particles to be unity. The potential energy of this system is

$$V = \frac{\omega_0^2 x^2 + W_0^2 y^2 + \epsilon x^2 y^2}{2},$$

where $\epsilon x^2 y^2/2$ is the energy of interaction between the two oscillators, with ϵ a small positive constant. The Lagrangian L of the system is equal to $T - V$. We thus have

$$\frac{\delta L}{\delta \dot{x}} = \dot{x}, \ \frac{\delta L}{\delta x} = -\omega_0^2 x - \epsilon x y^2.$$

The Euler-Lagrange equation for the motion of the first oscillator is

$$\frac{d}{dt}\frac{\delta L}{\delta \dot{x}} = \frac{\delta L}{\delta x}.$$

Therefore we have

$$\ddot{x} = -\omega_0^2 x - \epsilon x y^2.$$

Similarly, the equation of motion for the second oscillator is

$$\ddot{y} = -W_0^2 y - \epsilon x^2 y.$$

There are secular terms in both differential equations. These secular terms can be eliminated by renormalizing both angular frequencies in the problem. Physically, it means that the periods of oscillation for both particles are modified by the coupling between them.

Problem for the Reader

Find the renormalized angular frequencies of these two oscillators.

Solution

We write the renormalized angular frequencies for these two oscillators as

$$\omega_0^2 = \omega^2(1 + \epsilon a_1 + \cdots + \epsilon^n a_n + \cdots)$$

and

$$W_0^2 = W^2(1 + \epsilon b_1 + \cdots + \epsilon^n b_n + \cdots),$$

where a_n and b_n are real constants. We put

$$x(t) = x_0(t) + \epsilon x_1(t) + \epsilon^2 x_2(t) + \cdots$$

and

$$y(t) = y_0(t) + \epsilon y_1(t) + \epsilon^2 y_2(t) + \cdots .$$

The constants a_n and b_n are used to eliminate the secular terms in the differential equations for $x_n(t)$ and $y_n(t)$.

We rewrite the differential equations as

$$\ddot{x} + \omega^2 x = -\epsilon x y^2 - \omega^2(\epsilon a_1 + \cdots + \epsilon^n a_n + \cdots)x$$

and

$$\ddot{y} + W^2 y = -\epsilon x^2 y - W^2(\epsilon b_1 + \cdots + \epsilon^n b_n + \cdots)y,$$

where the last term in each of the equations is a counter term of frequency renormalization. The zeroth-order solutions are

$$x_0(t) = Ae^{i\omega t} + A^* e^{-i\omega t}$$

and

$$y_0(t) = Be^{iWt} + B^* e^{-iWt}.$$

The differential equation satisfied by $x_1(t)$ is

$$\ddot{x}_1 + \omega^2 x_1 = -x_0 y_0^2 - \omega^2 a_1 x_0.$$

The secular terms in this equation are

$$-A(2BB^* + a_1\omega^2)e^{i\omega t} + c.c.$$

To eliminate the secular terms above, we choose

$$a_1 = -2BB^*/\omega^2.$$

Similarly,

$$b_1 = -2AA^*/W^2.$$

Since a_1 and b_1 are both negative, the renormalized frequencies are larger than the respective bare frequencies. Therefore, as a result of the interaction between the harmonic oscillators, it takes a shorter time for each of the oscillators to complete a full cycle.

A slightly different method used to treat this kind of problem is that of Poincare. In the Poincare method, one introduces for the equation of motion a new time variable T rather than a new parameter ω. The variable T is related to t by

$$T = t(1 + c_1\epsilon + \cdots).$$

Thus

$$\frac{d^2}{dt^2} = (1 + c_1\epsilon + \cdots)^2\frac{d^2}{dT^2}.$$

The real constants $c_1, \cdots$ are used to eliminate the secular terms in the differential equation.

Problem for the Reader

Apply the Poincare method to the problem of pendulum motion.

Solution

Expressing (10.3) with the independent variable T, we get

$$(1 + c_1\epsilon + \cdots)^2\frac{d^2}{dT^2}\theta + \omega_0^2\theta = \epsilon\theta^3/6 - \epsilon^2\theta^5/(120\omega_0^2) + \cdots .$$

The lowest-order solution is

$$\theta_0(t) = \frac{1}{2}(e^{i\omega_0 T} + e^{-i\omega_0 T}).$$

The differential equation for $\theta_1(t)$ is

$$\left(\frac{d^2}{dT^2} + \omega_0^2\right)\theta_1 = \theta_0^3/6 - 2c_1\frac{d^2}{dT^2}\theta_0.$$

The secular terms of the equation above are

$$\left(\frac{1}{16} + \omega_0^2 c_1\right) e^{i\omega_0 T} + c.c.$$

We choose

$$c_1 = -\frac{1}{16\omega_0^2},$$

and the secular terms of the differential equation for θ_1 are eliminated. We then have

$$T = \left(1 - \frac{\epsilon}{16\omega_0^2}\right) t.$$

Therefore,

$$\omega_0 T = \omega t.$$

Thus the zeroth-order solution yielded by the Poincare method is the same as that yielded by the method of renormalized perturbation.

The Poincare method is slightly less convenient to use than the method of renormalized perturbation. This is particularly true if there is more than one renormalized constant in the problem.

Let us return to the pendulum problem. The zeroth-order solution $\theta_0(t)$ of (10.19) with ω given by (10.24) includes all leading terms of the regular perturbation series, and is a good approximation of the solution for all values of

t up to $O(\epsilon^{-1})$. However, such an approximate solution does not account for the nonleading terms, and fails when t is of the order of ϵ^{-2}. We'll now show that it is possible to choose a_m and eliminate the secular terms in the differential equation for θ_m. The solution $\theta_m(t)$ is then in the form of a Fourier sum with Fourier coefficients that are real. As the solution $\theta_m(t)$ has no term the magnitude of which grows with time, $\epsilon^n\theta_n(t)$ is much smaller than $\epsilon^{n-1}\theta_{n-1}(t)$ for all times.

This we will show by induction. We have already found that this is true for $m = 0$ and 1. We will now assume that this is true for $m < n-1$, and prove that this is true for $m = n$.

The source terms in the differential equation for θ_n are polynomials of θ_m where m is less than n. The coefficients of these polynomials are real. Since the Fourier coefficients of θ_m with $m < n$ are by assumption real, the Fourier coefficients of the source terms are real. The secular terms are part of the source terms. Thus they are in the form

$$s_n(e^{i\omega t} + e^{-i\omega t}),$$

where s_n is a real constant. Choosing a_n to make s_n vanish, we eliminate all secular terms.

With secular terms absent from the differential equation it satisfies, $\theta_n(t)$ is equal to a Fourier sum. The initial condition $\dot{\theta}_n(0) = 0$ then ensures that all coefficients in the Fourier sum of $\theta_n(t)$ are real, as we have argued above. Thus $\epsilon^n\theta_n(t)$, having one more factor of the small parameter ϵ, is much smaller than $\epsilon^{n-1}\theta_{n-1}$ at all times, and renormalized perturbation works for all times.

Therefore, in the theory of renormalized perturbation, θ_0 of (10.19) is the zeroth-order approximation of the solution good for all t, provided that ω is chosen to be the exact renormalized angular frequency of the motion, which we obtain in the form of a perturbation series of ϵ. If we add to $\theta_0(t)$ the next-order approximations $\epsilon\theta_1(t)$, $\epsilon^2\theta_2(t)\cdots$, the coefficient $1/2$ in (10.19) is turned into a perturbation series of ϵ. Indeed, we may show that

$$\theta(t) = \frac{1}{2}\left(1 + \frac{2^{-6}\epsilon}{3\omega^2} + \cdots\right)e^{i\omega t} - \frac{2^{-7}\epsilon}{3\omega^2}e^{3i\omega t} + c.c.$$

As we have seen, ω is a perturbation series of ϵ. If we truncate this infinite series and keep only terms up to the order of ϵ^n, then (10.19) is a good zeroth-order approximation only if the first neglected term in the exponents of (10.19) is much less than unity. Since this term is of the order of $\epsilon^{n+1}t$, (10.19) is a good zeroth-order approximation only if t is much less than ϵ^{-n-1}, if the perturbation series for ω is so truncated.

Thus we may say that the pendulum motion has infinitely many time scales. But this is so only because we measure time in the unit ω_0^{-1}. If we measure time in the unit ω^{-1}, then there are no other time scales but the period of the pendulum motion.

The pendulum motion is a special case of the motion of a particle in a potential, which we will briefly discuss. Let the motion of the particle be of one spatial dimension. We denote the coordinate of the particle by X, and the potential by $V(X)$. The force on the particle is equal to $-dV(X)/dX$. Let the force vanish at a point of X, which we will take without losing generality to be the origin $X = 0$. Then $dV(X)/dX$ is equal to zero at the origin. Let the potential $V(X)$ near the origin have the Maclaurin series expansion

$$V(0) + \frac{a}{2}X^2 + \frac{b}{3}X^3 + \cdots .$$

If the particle is initially located at the origin and is at rest, then it will stay at the origin forever, there being no force to make it move. Thus the origin is a point of equilibrium of the particle.

Let us displace the particle slightly from the origin and release. To be specific, let X be equal to ϵ and dX/dt be equal to zero at the initial time. The small motion that follows obeys the differential equation

$$\frac{d^2X}{dt^2} = -V'(X) = -aX - bX^2 + \cdots ,$$

where we have taken the mass of the particle to be unity. Note that there are no small parameters in the differential equation, the only small parameter being the initial displacement of X. If a is positive, the motion satisfies approximately

$$\frac{d^2X}{dt^2} \simeq -\omega_0^2 X$$

where $\omega_0^2 = a$. Thus the motion is approximately the harmonic motion $\epsilon \cos \omega_0 t$, which is periodic and remains small at all times. Such a motion is called stable. We shall soon prove that if ϵ is sufficiently small, the presence of the nonlinear force does not alter the stability of the motion, which will remain small and periodic for all times. It will be convenient to scale the dependent variable by

$$x = \frac{X}{\epsilon}.$$

Problem for the Reader

Find the differential equation as well as the initial conditions satisfied by x. Discuss how you would solve this problem.

Solution

The differential equation satisfied by $x(t)$ is

$$\frac{d^2x}{dt^2} + \omega_0^2 x = -\epsilon b x^2 + \cdots . \tag{10.26a}$$

The initial conditions for $x(t)$ are

$$x(0) = 1, \dot{x}(0) = 0. \tag{10.26b}$$

The initial conditions for $x(t)$ are independent of the small constant ϵ, and x is $O(1)$. The small parameter ϵ appears in the coefficient of x^2 on the right side of (10.26a). We see that the small motion X around a critical point with $a > 0$ is, by a scaling, the motion x of order unity satisfying a differential equation with a small parameter ϵ. If we set ϵ to zero, the unperturbed differential equation is that for a harmonic oscillator. The perturbed differential equation is solved by the method of renormalized perturbation, in the same way as we have done for the pendulum problem. It is, in the lowest order, a harmonic motion with a renormalized angular frequency. In the higher orders the motion has higher harmonic terms, and the solution is a Fourier series.

We shall now prove that the solution of (10.26a) satisfying the conditions in (10.26b) is stable and periodic for all times provided that ϵ is sufficiently small. To prove this, we note that if we replace t by $-t$, d^2x/dt^2 remains the same. Thus eq. (10.26a) remains the same under the replacement $t \to -t$. As a consequence, $y(t) \equiv x(-t)$ is a solution of (10.26a) if $x(t)$ is a solution of (10.26a). Note that $dy(t)/dt$ at the time $-t$ is equal to the negative of $dx(t)/dt$ at the time t. Now since $x(0)$ is equal to 1, $y(0)$ is equal to 1. And since $dx(t)/dt$ vanishes at $t = 0$, $dy(t)/dt$ also vanishes at $t = 0$. Thus $y(t)$ also satisfies both of the conditions of (10.26b). Since the solution satisfying both (10.26a) and (10.26b) is unique, we have $x(t) = y(t)$, or

$$x(t) = x(-t).$$

This says that $x(t)$ is an even function of t, which implies that $\dot{x}(t)$ is an odd function of t.

Let ϵ be sufficiently small so that the motion $x(t)$ is approximately the harmonic motion $x(t) = \cos \omega_0 t$ for a period of time. To be definite, let this period of time be $-2\pi/\omega_0 < t < 2\pi/\omega_0$. Now the time derivative of $\cos \omega_0 t$ vanishes at $t = \pi/\omega_0$. Since $x(t)$ is approximately equal to $\cos \omega_0 t$, $\dot{x}(t)$ vanishes at a time approximately equal to π/ω_0. Let the time at which $\dot{x}(t)$ vanishes be $P/2$, i.e., $\dot{x}(P/2) = 0$. Since $\dot{x}(t)$ is an odd function of t, it follows that $\dot{x}(-P/2) = 0$. And since $x(t)$ is an even function of t, $x(-P/2) = x(P/2)$. Therefore, $x(t)$ and $\dot{x}(t)$ at $t = -P/2$ are respectively equal to those at $t = P/2$. Since (10.26a) is autonomous, the motion is periodic, with period P.

These considerations are readily extended to motions under forces that depend on $\dot{x}$, as the argument we just presented goes through as long as the force is invariant under $t \to -t$. There are forces that depend on $\dot{x}$ and have this property, an example of which is given below.

Consider the equation

$$\ddot{x} + (\omega_0^2 + \epsilon \dot{x}^4)x = 0, \tag{10.27}$$

in which the nonlinear force $-\epsilon \dot{x}^4 x$ depends on $\dot{x}$. While $\dot{x}$ changes sign as we replace t by $-t$, $\dot{x}^4$ remains the same. Therefore, the differential equation (10.27) is invariant under $t \to -t$. By the same argument we just presented above, the motion satisfying (10.27) with the initial conditions (10.26b) is periodic provided that ϵ is sufficiently small.

Problem for the Reader

Obtain an approximate solution of (10.27) satisfying the initial conditions in (10.26b).

Solution

While we know that the motion is periodic, the period of the motion is not necessarily $2\pi/\omega_0$. Therefore, to obtain an approximate solution that holds for all t, one first performs a frequency renormalization.

Let the renormalized angular frequency of the motion be given by (10.15), and let

$$x(t) = x_0(t) + \epsilon x_1(t) + \epsilon^2 x_2(t) + \cdots .$$

Then the differential equation for $x(t)$ becomes

$$\left(\frac{d^2}{dt^2} + \omega^2\right)(x_0 + \epsilon x_1 + \cdots) = -\epsilon(\dot{x}_0 + \epsilon \dot{x}_1 + \cdots)^4 (x_0 + \epsilon x_1 + \cdots) \\ - \omega^2 (a_1 \epsilon + a_2 \epsilon^2 + \cdots)(x_0 + \epsilon x_1 + \cdots).$$

The last term in the equation above is the frequency counter term.

The lowest-order solution $x_0(t)$ is

$$x_0(t) = A_0 e^{i\omega t} + A_0^* e^{-i\omega t},$$

where A_0 is a constant. From the initial condition $x_0(0) = 0$, we get

$$A_0 + A_0^* = 0.$$

This says that the Fourier coefficient A_0 is purely imaginary, and the Fourier coefficient A_0^* is the negative of A_0. (In the pendulum problem treated previously, the Fourier coefficients of $\theta(t)$ are real. This is because in the pendulum problem treated, we set to zero the initial value of $\dot{\theta}(t)$, not that of $\theta(t)$.)

We determine the value of A_0 from the initial condition $\dot{x}_0(0) = 1$ and get

$$x_0(t) = -\frac{i}{2\omega}(e^{i\omega t} - e^{-i\omega t}).$$

Before we move on, we point out that since the derivative of $e^{\pm i\omega t}$ is equal to $\pm i\omega$ times $e^{\pm i\omega t}$, the Fourier coefficients of $\dot{x}_0(t)$ are real. The differential equation for $x_1(t)$ is

$$\left(\frac{d^2}{dt^2} + \omega^2\right) x_1 = -\dot{x}_0{}^4 x_0 - \omega^2 a_1 x_0.$$

Since the Fourier coefficients of $\dot{x}_0(t)$ are real, so are those of $\dot{x}_0^4$. Together with the fact that the Fourier coefficients of x_0 are imaginary, the Fourier coefficients of $-\dot{x}_0{}^4 x_0$ are imaginary. In particular, the secular terms from the nonlinear force $-\dot{x}_0{}^4 x_0$ are

$$\frac{i}{16\omega}(e^{i\omega t} - e^{-i\omega t}).$$

The secular terms from $-\omega^2 a_1 x_0$ are

$$\frac{i\omega}{2}a_1(e^{i\omega t} - e^{-i\omega t}).$$

Note that all Fourier coefficients above are imaginary, and the coefficient of $e^{i\omega t}$ is the negative of that of $e^{-i\omega t}$. The sum of these terms vanishes if we choose a_1 to be the real number

$$a_1 = -\frac{1}{8\omega^2}.$$

With an argument similar to the one we have presented for the pendulum problem, we may prove that the secular terms in the differential equation satisfied by $x_m(t)$ with m any positive integer can be eliminated by a choice of a_m. The only difference here is that the Fourier coefficients of $x_m(t)$ are imaginary rather than real, but the arguments go through in the same way. With ω determined, the approximate solution $x_0(t)$ is good for all t.

The method of renormalized perturbation generally fails when the nonlinear force has a term proportional to an odd power of $\dot{x}$. This is because such an equation is not invariant under $t \to -t$. As an example, consider the motion governed by the equation

$$\ddot{x} + \omega_0^2 x = -\epsilon\dot{x}^3, \tag{10.28}$$

with the initial conditions

$$x(0) = 1$$

and

$$\dot{x}(0) = 0, \tag{10.29}$$

where ϵ is positive and is very small. We shall verify that we will be unable to eliminate the secular terms in this equation by choosing ω in the form (10.15).

Let us first make an observation about this equation. If we set ϵ to zero, the equation is that of a harmonic oscillator, the solution of which is periodic. However, the nonlinear force $-\epsilon\dot{x}^3$ is dissipative and the solution with $\epsilon \neq 0$ vanishes at $t = \infty$. Such a solution is stable but not periodic; hence a renormalized angular frequency is not meaningful.

To prove that the solution with $\epsilon \neq 0$ vanishes at $t = \infty$, we multiply (10.28) with $\dot{x}$ and get

$$\frac{dE}{dt} = -\epsilon \dot{x}^4,$$

where

$$E = \frac{\dot{x}^2 + \omega_0^2 x^2}{2}.$$

From the above, we see that $E(t)$ is positive and decreases monotonically with time. Thus $E(t)$ must approach a nonnegative limit as $t \to \infty$. And as E approaches a limit, dE/dt must approach zero. But dE/dt is proportional to $\dot{x}^4$, so $\dot{x}$ must approach zero. This, in turn, implies that $\ddot{x}$ approaches zero. We then obtain from (10.28) that x approaches zero.

We shall show directly that the secular terms in this problem cannot be eliminated by frequency renormalization; let us apply renormalized perturbation and express (10.28) as

$$\ddot{x} + \omega^2 x = -\epsilon \dot{x}^3 - \omega^2 \epsilon (a_1 + a_2 \epsilon + \cdots) x. \tag{10.30}$$

We substitute

$$x = x_0 + \epsilon x_1 + \cdots$$

into (10.30). The zeroth-order solution is

$$x_0(t) = \frac{e^{i\omega t} + e^{-i\omega t}}{2}.$$

Note that the Fourier coefficients of $x_0(t)$ are real. Since the dissipative force depends on $\dot{x}(t)$, we calculate the time derivative of $x_0(t)$ and get

$$\dot{x}_0(t) = i\omega \frac{e^{i\omega t} - e^{-i\omega t}}{2}.$$

Note that the Fourier coefficients of $\dot{x}_0(t)$ are imaginary.

The differential equation for $x_1(t)$ is

$$\ddot{x}_1 + \omega^2 x_1 = -\dot{x}_0^3 - \omega^2 a_1 x_0.$$

The right side of the equation above is equal to

$$i\omega^3 \left(\frac{e^{i\omega t} - e^{-i\omega t}}{2} \right)^3 - \omega^2 a_1 \frac{e^{i\omega t} + e^{-i\omega t}}{2}. \tag{10.31}$$

Note that the Fourier coefficients of the first term in (10.31) are imaginary, while those of the second term in (10.31) are real. The secular terms are

$$\left(-\frac{3i\omega^3}{8} - \frac{\omega^2 a_1}{2} \right) e^{i\omega t}$$

and

$$\left(\frac{3i\omega^3}{8} - \frac{\omega^2 a_1}{2} \right) e^{-i\omega t}. \tag{10.32}$$

It is not possible to eliminate the secular terms with any choice of a_1. We shall solve (10.28) with the two-scale method.

C. The Two-Scale Method

Before we show how to obtain a solution of (10.28) valid over a time period as long as ϵ^{-1}, let us first get some sense of orientation by looking at the unrenormalized perturbation series of the solution for this problem. Just as in the pendulum problem, the leading terms of $x_n(t)$ for the present problem in the unrenormalized perturbation theory are of the order of

$$(\epsilon t)^n e^{i\omega_0 t} + c.c. = \tau^n e^{i\omega_0 t} + c.c., \tag{10.33}$$

where

$$\tau = \epsilon t.$$

The sum of these leading terms over n is therefore of the form

$$A_0(\tau)e^{i\omega_0 t} + c.c. \tag{10.34}$$

The difference from the pendulum problem is that the nonlinear forces being dissipative, $A_0(\epsilon t)$ decreases with time and is not in the form $e^{i\omega_1 \epsilon t}$. We shall show that the form of $A_0(\tau)$ can be determined by the two-scale method given next.

As a side remark, we have shown in Chapter 7 that each of the WKB solutions (7.4) is a product of two factors. One of these factors is a phase factor, which is rapidly varying. The other factor is $p^{-1/2}$, which is not rapidly varying. Thus these two factors vary with different scales. The solution $A_0(\tau)e^{i\omega_0 t}$ in (10.34) is also a product of two factors, which are functions of τ and t, respectively. These two time variables are of different scales. In particular, the function $A_0(\tau)$ is approximately a constant over the time scale of unity, and changes only over the time scale of ϵ^{-1}. We shall call t the short-time variable and τ the long-time variable.

While the solution (10.34) is a function of two variables, these two variables are related. This is to say that τ varies as t varies and we have

$$\frac{d}{dt} = \frac{\partial}{\partial t} + \frac{d\tau}{\partial t}\frac{\partial}{\partial \tau} = \frac{\partial}{\partial t} + \epsilon\frac{\partial}{\partial \tau}. \tag{10.35}$$

Thus

$$\frac{d^2}{dt^2} = \frac{\partial^2}{\partial t^2} + 2\epsilon\frac{\partial^2}{\partial \tau \partial t} + \epsilon^2\frac{\partial^2}{\partial \tau^2}. \tag{10.36}$$

We put

$$x(t) = x_0(t,\tau) + \epsilon x_1(t,\tau) + \epsilon^2 x_2(t,\tau) + \cdots,$$

where $x_n(t,\tau)$ is a function of two variables. Then (10.28) becomes

$$\begin{aligned}&\left(\frac{\partial^2}{\partial t^2} + 2\epsilon\frac{\partial^2}{\partial \tau \partial t} + \epsilon^2\frac{\partial^2}{\partial \tau^2} + \omega_0^2\right)\left(x_0 + \epsilon x_1 + \epsilon^2 x_2 + \cdots\right)\\ &= -\epsilon\left(\frac{\partial x_0}{\partial t} + \epsilon\frac{\partial x_0}{\partial \tau} + \epsilon\frac{\partial x_1}{\partial t} + \cdots\right)^3.\end{aligned} \tag{10.37}$$

From the initial condition $x(0) = 1$, we get as before

$$x_0(0,0) = 1 \tag{10.38}$$

and

$$x_1(0,0) = x_2(0,0) = \cdots = 0. \tag{10.39}$$

But the initial condition $\dot{x}(0) = 0$ must be handled somewhat differently. This is because the second initial condition is

$$\left(\frac{\partial}{\partial t} + \epsilon \frac{\partial}{\partial \tau}\right)(x_0 + \epsilon x_1 + \cdots)|_{t=\tau=0} = 0.$$

Thus the initial conditions for the time derivative of $x_n(t,\tau)$ are

$$\frac{\partial}{\partial t} x_0|_{t=\tau=0} = 0 \tag{10.40}$$

and

$$\left(\frac{\partial x_n}{\partial \tau} + \frac{\partial x_{n+1}}{\partial t}\right)|_{t=\tau=0} = 0, n = 0, 1, \cdots. \tag{10.41}$$

Equating the terms independent of ϵ on the two sides of (10.37), we get

$$\left(\frac{\partial^2}{\partial t^2} + \omega_0^2\right) x_0 = 0.$$

We see that the solution of the equation above is indeed in the form (10.34). The coefficient $A_0(\tau)$ is a function of τ, not a constant. From the initial conditions, we have

$$A_0(0) = 1/2. \tag{10.42}$$

The complex-valued coefficient $A_0(\tau)$ is a function of τ, which is determined by the requirement that the sum of secular terms in the differential equation satisfied by $x_1(t,\tau)$ be zero. This equation is obtained by equating the terms linear in ϵ on the two sides of (10.37). We get

$$\left(\frac{\partial^2}{\partial t^2} + \omega_0^2\right) x_1 = -2\frac{\partial^2 x_0}{\partial \tau \partial t} - \left(\frac{\partial x_0}{\partial t}\right)^3. \tag{10.43}$$

The secular terms in (10.43) are

$$(-2i\omega_0 A_0' - 3i\omega_0^3 A_0^2 A_0^*)e^{i\omega_0 t} + c.c., \tag{10.44}$$

where A_0' is $dA_0(\tau)/d\tau$. If we set the first term of (10.44) to zero, then its complex conjugate is automatically equal to zero, and eq. (10.43) has no secular terms. Thus we require

$$-3\omega_0^2 A_0^2 A_0^* - 2A_0' = 0, \tag{10.45}$$

which is a differential equation for $A_0(\tau)$.

Let us put

$$A_0 \equiv Re^{i\theta}.$$

Since

$$D(Re^{i\theta}) = e^{i\theta}(D + i\theta')R,$$

where $D = d/d\tau$, we get, after eliminating the common factor $e^{i\theta}$ from each term of (10.45),

$$-3\omega_0^2 R^3 - 2(R' + iR\theta') = 0.$$

Setting the real part and the imaginary part of the equation above separately to zero, we obtain

$$R' = -3\omega_0^2 R^3/2$$

and

$$\theta' = 0.$$

Thus we get

$$\theta = 0,$$

as well as

$$\frac{1}{R^2} = 3\omega_0^2\tau + 4.$$

We have

$$A_0(\tau) = R(\tau) = \frac{1}{\sqrt{4 + 3\omega_0^2\tau}}. \tag{10.46}$$

Hence

$$x_0(t,\tau) = \frac{\cos \omega_0 t}{\sqrt{1 + 3\omega_0^2 \tau/4}}. \tag{10.47}$$

We note that $A_0(\tau)$ is a decreasing function of τ, approaching zero like $\tau^{-1/2}$ as $\tau \to \infty$. Therefore, the magnitude of $x_0(t,\tau)$ approaches zero like $\tau^{-1/2}$ as $\tau \to \infty$.

We shall show that $x_1(t,\tau)$ also vanishes like $\tau^{-1/2}$ as $\tau \to \infty$. With its secular terms eliminated, (10.43) is

$$\left(\frac{\partial^2}{\partial t^2} + \omega_0^2\right) x_1 = i\omega_0^3 R^3(\tau) e^{3i\omega_0 t} + c.c.$$

The solution of the equation above is

$$x_1(t,\tau) = A_1(\tau) e^{i\omega_0 t} - \frac{i\omega_0}{8} R^3(\tau) e^{3i\omega_0 t} + c.c., \tag{10.48}$$

where the first term on the right side of (10.48) is a complementary solution of the differential equation. The coefficient $A_1(\tau)$ is a function of τ, which is determined by the requirement that the differential equation satisfied by $x_2(t,\tau)$ have no secular terms. From (10.37), the differential equation satisfied by $x_2(t,\tau)$ is

$$\left(\frac{\partial^2}{\partial t^2} + \omega_0^2\right) x_2 = -2\frac{\partial^2 x_1}{\partial \tau \partial t} - \frac{\partial^2 x_0}{\partial \tau^2} - 3\left(\frac{\partial x_0}{\partial t}\right)^2 \left(\frac{\partial x_0}{\partial \tau} + \frac{\partial x_1}{\partial t}\right).$$

Setting the sum of secular terms in the equation above to zero, we get the differential equation satisfied by $A_1(\tau)$. It is

$$2A_1' + 6\omega_0^2 R^2 A_1 + 3\omega_0^2 R^2 A_1^* = iR''/\omega_0 + 3i\omega_0 R^2 R'.$$

As $\tau \to \infty$, each of the source terms in the equation above vanishes like $\tau^{-5/2}$. Let

$$A_1(\tau) = B_1(\tau) + iC_1(\tau),$$

where $B_1(\tau)$ and $C_1(\tau)$ are the real part and the imaginary part of $A_1(\tau)$, respectively. Then we have

$$2B_1' + 9\omega_0^2 R^2 B_1 = 0 \tag{10.49a}$$

and

$$2C_1' + 3\omega_0^2 R^2 C_1 = R''/\omega_0 + 3\omega_0 R^2 R'. \tag{10.49b}$$

Equation (10.49a) for $B_1(\tau)$ is a linear first-order differential equation the solution of which is

$$B_1(\tau) = \frac{B_1(0)R^3(\tau)}{R^3(0)},$$

which shows that $B_1(\tau)$ vanishes like $\tau^{-3/2}$ as $\tau \to \infty$. Equation (10.49b) for $C_1(\tau)$ is also a linear first-order differential equation. The solution of this equation is

$$C_1(\tau) = \frac{C_1(0)R(\tau)}{R(0)} + R(\tau) \int_0^\tau \frac{R''(\tau') + 3\omega_0^2 R^2(\tau')R'(\tau')}{R(\tau')} d\tau'.$$

As $\tau' \to \infty$, the integrand above vanishes like τ'^{-2}. Thus, as $\tau \to \infty$, the integral above is convergent and we have

$$C_1(\tau) \approx R(\tau) \left[\frac{C_1(0)}{R(0)} + \int_0^\infty \frac{R''(\tau') + 3\omega_0^2 R^2(\tau')R'(\tau')}{R(\tau')} d\tau' \right],$$

which shows that $C_1(\tau)$ vanishes like $\tau^{-1/2}$ as $\tau \to \infty$.

From these calculations we conclude that $A_1(\tau)$ vanishes like $\tau^{-1/2}$ as $\tau \to \infty$, and is hence of the same order of magnitude as $A_0(\tau)$ as τ is large.We see from (10.48) that the higher-harmonic terms in $x_1(t, \tau)$ vanish even faster, like $\tau^{-3/2}$, as $\tau \to \infty$. Thus $x_1(t, \tau)$ vanishes like $\tau^{-1/2}$ as $\tau \to \infty$, which is of the order of $x_0(t, \tau)$. In addition, both $x_0(t, \tau)$ and $x_1(t, \tau)$ are of the order of unity when τ is of the order of unity. Thus $x_0(t, \tau)$ and $x_1(t, \tau)$ are of the same order of magnitude for all times. Therefore, $\epsilon x_1(t, \tau)$ is much smaller than $x_0(t, \tau)$ at all times.

We will use induction to show that $x_n(t)$ for any n is of the order of $x_0(t)$ at all times, and hence $\epsilon^n x_n(t, \tau)$ is much smaller than $\epsilon^{n-1} x_{n-1}(t, \tau)$ at all times. We will sketch briefly how this is done. Let us write

$$x_n(t, \tau) = A_n(\tau)e^{i\omega_0 t} + A_n^*(\tau)e^{-i\omega_0 t} + \text{ higher harmonic terms},$$

and

$$A_n(\tau) = B_n(\tau) + iC_n(\tau).$$

We assume that both $B_m(\tau)$ and $C_m(\tau)$ for $m < n$ vanish like $\tau^{-1/2}$ as $\tau \to \infty$. We prove that both $B_n(\tau)$ and $C_n(\tau)$ vanish like $\tau^{-1/2}$ as $\tau \to \infty$.

The differential equations satisfied by $B_n(\tau)$ and $C_n(\tau)$ are obtained by requiring that the sum of secular terms in the differential equation for $x_n(t,\tau)$ be zero. We find that the differential equation satisfied by $B_n(\tau)$ is the one given by (10.49a), with a source term added to the right side of the equation. This source term vanishes like $\tau^{-3/2}$ as $\tau \to \infty$. The differential equation for $C_n(\tau)$ is in the form of (10.49b), with the source terms vanishing like $\tau^{-5/2}$ as $\tau \to \infty$. We may then argue as we have done above that both $B_n(\tau)$ and $C_n(\tau)$ vanish like $\tau^{-1/2}$ as $\tau \to \infty$. (See homework problem 4.) Thus $x_0(x,\tau)$, the sum of all of the leading terms of regular perturbation, is the zeroth-order approximation of $x(t)$ good for all times. Also, the sum $x_0(t,\tau)+ \epsilon x_1(t,\tau)$ is the first-order approximation of $x(t)$ good for all times. The two-scale method gives us the perturbation series for the solution of (10.28) that is good for all times.

This is not always the case. For other problems, the approximate solution yielded by the two-scale method is often good only for values of t up to the order of ϵ^{-1}. We will give an example below.

Problem for the Reader

Find, with the two-scale method, the lowest-order approximate solution of

$$\ddot{x} + (1 + \epsilon\dot{x}^2)x = 0, \tag{10.50}$$

satisfying

$$x(0) = 0, \dot{x}(0) = 1.$$

For what values of t is this approximate solution good?

Solution

We put

$$x(t) = x_0(t,\tau) + \epsilon x_1(t,\tau) + \cdots ,$$

where

$$\tau = \epsilon t.$$

With the use of the two-scale method, (10.50) becomes

$$\left(\frac{\partial^2}{\partial t^2}+1+2\epsilon\frac{\partial^2}{\partial t\partial\tau}+\epsilon^2\frac{\partial^2}{\partial\tau^2}\right)(x_0+\epsilon x_1+\cdots)$$
$$+\epsilon\left(\frac{\partial}{\partial t}x_0+\epsilon\frac{\partial}{\partial\tau}x_0+\epsilon\frac{\partial}{\partial t}x_1+\cdots\right)^2(x_0+\epsilon x_1+\cdots)=0.$$

The initial conditions are

$$x_0(0,0)=\ x_1(0,0)=\cdots=0,\ \frac{\partial}{\partial t}x_0(t,\tau)\mid_{t=\tau=0}=1,$$
$$\frac{\partial}{\partial t}x_1(t,\tau)\mid_{t=\tau=0}=-\frac{\partial}{\partial\tau}x_0(t,\tau)\mid_{t=\tau=0}.$$

It is straightforward to find that

$$x_0(t,\tau)=A_0(\tau)e^{it}+A_0^*(\tau)e^{-it}.$$

The initial condition $y_0(0,0)=0$ gives

$$A_0(0)=-A_0^*(0).$$

And the initial condition $\frac{\partial}{\partial t}x_0(t,\tau)\mid_{t=\tau=0}=1$ gives

$$iA_0(0)-iA_0^*(0)=1.$$

Thus we get

$$A_0(0)=-i/2.$$

The equation satisfied by x_1 is

$$\left(\frac{\partial^2}{\partial t^2}+1\right)x_1+2\frac{\partial^2}{\partial t\partial\tau}x_0+\left(\frac{\partial}{\partial t}x_0\right)^2x_0=0.$$

Setting the sum of secular terms in this equation to zero, we get

$$2iA_0'+A_0^2A_0^*=0.$$

Let

$$A_0=e^{i\theta}R.$$

We have

$$2i(R' + i\theta' R) + R^3 = 0.$$

Setting the imaginary part of the equation above to zero, we get

$$R'(\tau) = 0.$$

To satisfy the initial condition $A_0(0) = 1$, we must put

$$R(\tau) = 1/2.$$

Setting the real part of the equation above to zero, we get

$$\theta' = 1/8.$$

Thus

$$\theta = -\pi/2 + \tau/8.$$

We get

$$A_0(\tau) = -i\frac{e^{i\tau/8}}{2}.$$

Thus

$$x_0(t) = \sin\left(1 + \frac{\epsilon}{8}\right) t.$$

We note that the differential equation (10.50) is invariant under $t \to -t$, and hence the solution is periodic. Thus the solution can also be obtained with the method of renormalized perturbation. The renormalized perturbation series for the solution is good for all times. If we approximate ω by the first two terms in its perturbation series, the resulting lowest-order term in the renormalized perturbation series of the solution is the same as $x_0(t, \tau)$ yielded by the two-scale method, and is good only for values of t up to $O(\epsilon^{-1})$.

We emphasize that there is no need to apply singular perturbation to problems in which there are no secular terms. We give two examples below.

Problem for the Reader

Determine for yourself if there is a need to employ the two-scale method to solve the following problems:

a. $\ddot{x} + \epsilon e^{-t}\dot{x}^5 + x = 0,$

b. $\ddot{x} + \epsilon e^{-\epsilon t}x^3 + x = 0.$

Solution

Applying regular perturbation to (a), we find that there are no secular terms in the differential equation satisfied by $x_n(t)$. For example, the differential equation satisfied by $x_1(t)$ is

$$\ddot{x}_1 + x_1 = -e^{-t}\dot{x}_0^5.$$

The source term on the right side of the equation has no secular terms. Thus $x_1(t)$ is bounded at all times. This is also true for $x_n(t)$. There is no need to use the two-scale method for this problem.

Next we turn our attention to the equation of case (b). We expand $e^{-\epsilon t}$ as a Taylor series of ϵt and apply regular perturbation. The differential equation satisfied by $x_1(t)$ is

$$\ddot{x}_1 + x_1 = -x_0^3.$$

There are secular terms of the form $e^{\pm it}$ coming from the source term $-x_0^3(t)$ in this equation. Thus $x_1(t)$ has $te^{\pm it}$ terms.

The differential equation satisfied by $x_2(t)$ is

$$\ddot{x}_2 + x_2 = tx_0^3 - 3x_0^2 x_1.$$

The source terms on the right side of the equation above have secular terms of the form $te^{\pm it}$. Thus $x_2(t)$ has $t^2 e^{\pm it}$ terms. Therefore the two-scale method is needed.

One may wonder how to treat this problem with one of the coefficients of the differential equation proportional to $e^{-\epsilon t}$. For example, do we expand $e^{-\epsilon t}$ into a Maclaurin series of ϵt, as we would have done in regular perturbation? The answer is "no." The point is that $\tau = \epsilon t$ is considered $O(1)$ in the two-scale method. Thus we express this exponential as $e^{-\tau}$ and treat it as a function of τ and not a function of t as we apply the two-scale method. (See homework problem 5.) We remark that while t appears explicitly in this differential equation, which means that it is not autonomous, the two-scale method is applicable.

As a final demonstration of the two-scale method, consider the equation

$$\ddot{x} + e^{-\epsilon t} x = 0,$$

with the initial conditions

$$x(0) = 1 \text{ and } \dot{x}(0) = 0.$$

This is the equation for the motion of a harmonic oscillator with a spring constant that wears out slowly like $e^{-\epsilon t}$. This problem is hence not autonomous.

The differential equation is linear and can be solved by the WKB method. (See homework problem 6b of Chapter 7.) We shall here solve this problem with the two-scale method.

When ϵ is small, we have

$$e^{-\epsilon t} = 1 - \epsilon t + \epsilon^2 t^2/2 + \cdots .$$

This is a useful expansion only if

$$\epsilon t \ll 1. \tag{10.51}$$

Applying the method of regular perturbation, we express the differential equation as

$$\ddot{x}_0 + \epsilon \ddot{x}_1 + \cdots = -(1 - \epsilon t + \epsilon^2 t^2/2 + \cdots)(x_0 + \epsilon x_1 + \cdots). \tag{10.52}$$

The zeroth-order solution $x_0(t)$ is, as usual, the motion of a harmonic oscillator. But the secular terms in the equation for $x_1(t)$ are as large as $te^{\pm it}$, with the factor t coming from the $-\epsilon t$ term in the expansion of $e^{-\epsilon t}$. Because of this factor t, the leading terms of $\epsilon x_1(t)$ are $\epsilon t^2 e^{\pm it}$ terms. Thus regular perturbation is a good approximation only if $\epsilon t^2 \ll 1$, or

$$t \ll \epsilon^{-1/2}, \tag{10.53}$$

which is a region smaller than the region of (10.51).

Going to higher orders, we find that the leading terms of $\epsilon^n x_n(t)$ are $\epsilon^n t^{2n} e^{\pm it}$ terms. Thus the sum of leading terms is in the form

$$A(\rho)e^{it} + A^*(\rho)e^{-it}, \tag{10.54}$$

where the long-time variable ρ is

$$\rho \equiv \sqrt{\epsilon} t, \tag{10.55}$$

not the variable $\tau = \epsilon t$.

To determine $A(\rho)$, we apply the two-scale method with ρ the long-time variable. Then (10.52) becomes

$$\left(\frac{\partial^2}{\partial t^2} + 2\sqrt{\epsilon}\frac{\partial^2}{\partial t \partial \rho} + \epsilon\frac{\partial^2}{\partial \rho^2}\right)\left[x_0(t,\rho) + \sqrt{\epsilon}x_1(t,\rho) + \cdots\right]$$

$$= -\left(1 - \sqrt{\epsilon}\rho + \cdots\right)\left[x_0(t,\rho) + \sqrt{\epsilon}x_1(t,\rho) + \cdots\right].$$

From the equation for $x_0(t,\rho)$, we find that $x_0(t,\rho)$ is in the form (10.54). To determine $A(\rho)$, we write down the differential equation for $x_1(t,\rho)$. We find that the secular terms in this equation are

$$-2i(A'e^{it} - A^{*\prime}e^{-it}) + \rho(Ae^{it} + A^*e^{-it}).$$

Eliminating the secular terms above, we get

$$-2iA' + \rho A = 0.$$

Solving this differential equation for A, we get

$$A(\rho) = \frac{e^{-i\rho^2/4}}{2}.$$

Thus we have

$$x_0 = \frac{e^{i(t-\epsilon t^2/4)}}{2} + c.c.$$

This approximate solution holds in the region

$$t = O(\epsilon^{-1/2}),$$

which is a somewhat smaller region than that of $O(\epsilon^{-1})$, where the approximate solution of the two-scale method usually holds.

In order to obtain an approximate solution of the differential equation that holds in a larger region of t, let us first review why the nth-order leading terms of regular perturbation for this problem have the magnitude $\epsilon^n t^{2n}$ and not $\epsilon^n t^n$. These leading terms owe their existence to the terms of the order of $\epsilon^m t^m x$, $m = 1, 2, \cdots$, in the differential equation, which are obtained as we expand $e^{-\epsilon t}$ in this equation into a Taylor series. In order to get a better approximation, we must eliminate such terms from the equation.

This is done by making a transformation of the independent variable from t to T. We have

$$\frac{dx}{dt} = \frac{dT}{dt}\frac{dx}{dT}$$

and

$$\frac{d^2x}{dt^2} = \left(\frac{dT}{dt}\right)^2 \frac{d^2x}{dT^2} + \frac{d^2T}{dt^2}\frac{dx}{dT}.$$

Then the differential equation becomes

$$\left(\frac{dT}{dt}\right)^2 \frac{d^2x}{dT^2} + \frac{d^2T}{dt^2}\frac{dx}{dT} + e^{-\epsilon t}x = 0. \tag{10.56}$$

We choose

$$\frac{dT}{dt} = e^{-\epsilon t/2},$$

or

$$T = \frac{1 - e^{-\epsilon t/2}}{\epsilon/2}.$$

Then eq. (10.55) becomes

$$\frac{d^2x}{dT^2} + x = \frac{\epsilon e^{\epsilon t/2}}{2}\frac{dx}{dT}. \tag{10.57}$$

This equation is that governing the motion of a harmonic oscillator, with T the time variable, driven by a small, slowly varying force. It is not autonomous. Indeed, in this equation, there is a short-time variable, T, and a long-time variable, $\tau = \epsilon t$. Thus we apply the two-scale method to this equation. We get

$$\frac{d}{dT} = \frac{\partial}{\partial T} + \frac{d\tau}{dT}\frac{\partial}{\partial \tau} = \frac{\partial}{\partial T} + \epsilon e^{\tau/2}\frac{\partial}{\partial \tau}$$

and

$$\frac{d^2}{dT^2} = \frac{\partial^2}{\partial T^2} + 2\epsilon e^{\tau/2}\frac{\partial^2}{\partial \tau \partial T} + \epsilon^2 e^{\tau}\frac{\partial^2}{\partial \tau^2}.$$

Therefore, (10.57) becomes

$$\left(\frac{\partial^2}{\partial T^2} + 1 + 2\epsilon e^{\tau/2}\frac{\partial^2}{\partial \tau \partial T} + \epsilon^2 e^{\tau}\frac{\partial^2}{\partial \tau^2}\right)x = \frac{\epsilon e^{\epsilon t/2}}{2}\left(\frac{\partial}{\partial T} + \epsilon e^{\tau/2}\frac{\partial}{\partial \tau}\right)x.$$

As usual, the equation for $x_0(T,\tau)$ gives

$$x_0(T,\tau) = A(\tau)e^{iT} + c.c.$$

The function $A(\tau)$ is determined by setting the secular terms in the differential equation for $x_1(T,\tau)$ to zero. We get

$$2e^{\tau/2}iA' = ie^{\tau/2}A/2,$$

which yields

$$A(\tau) = e^{\tau/4}/2.$$

Thus we have

$$x_0(T,\tau) = e^{\tau/4}\cos T. \tag{10.58}$$

This approximation is the same as that obtained with the WKB method. Its validity is predicated on the coefficient of dx/dT in (10.57) being small, or

$$\frac{1}{2}\epsilon e^{\epsilon t/2} \ll 1.$$

D. The Renormalized Two-Scale Method

In the preceding section, we've shown how to apply the two-scale method to problems of small nonlinear oscillations. As we have discussed, this method is basically a method of summing the leading terms. Since nonleading terms may become large when t is much larger than ϵ^{-1}, the solution yielded by the two-scale method often fails when the value of t is much larger than ϵ^{-1}. In this section, we show how to obtain approximate solutions that are valid for all times.

We wish to first make clear that it is not possible to do this for all problems of small nonlinear oscillations. If the motion is unstable and grows to be very large, the nonlinear force, although proportional to the small parameter ϵ, may grow to the point that it cannot be regarded as a perturbation. When this happens, all perturbation methods fail.

As an example of unstable motions, consider the motion governed by

$$\ddot{x} + x = \epsilon \dot{x}^3,$$

which is obtained from (10.28) with $t \to -t$ and with ω_0 set to unity. Let the initial conditions continue to be $x(0) = 1$ and $\dot{x}(0) = 0$. Then the lowest-order solution yielded by the two-scale method for this problem is obtained from (10.47) with t and τ replaced by $-t$ and $-\tau$, respectively, and with ω_0 set to unity. Thus we have

$$x_0(t, \tau) = \frac{\cos t}{\sqrt{1 - 3\tau/4}}.$$

This solution blows up at $\tau = 4/3$.

To see precisely at which point the two-scale method fails, we note that the nonlinear force cannot be treated as a perturbation when x_0 is so large that $\epsilon \dot{x}_0^3$ is comparable to x_0, which happens when

$$\frac{4}{3} - \tau = O(\epsilon).$$

For problems in which the motion is stable, the two-scale method fails when the motion has more than two time scales and t reaches the third time scale. As we shall see, many such problems can be solved by unifying the two-scale method with the method of renormalized perturbation. The approximate solution obtained is valid for all times. This unification of two methods of singular perturbation will be called the renormalized two-scale method.

Consider as the first example the solution of the equation

$$\ddot{x} + x = -\epsilon(\dot{x} - \dot{x}^3/3), \tag{10.59}$$

satisfying the initial conditions

$$x(0) = 2a > 0, \ \dot{x}(0) = 0. \tag{10.60}$$

By multiplying equation (10.59) with $\dot{x}$, we get

$$\frac{dE}{dt} = -\epsilon(\dot{x}^2 - \dot{x}^4/3), \tag{10.61}$$

where

$$E = \frac{\dot{x}^2 + x^2}{2}.$$

We see from (10.61) that the force $-\epsilon \dot{x}$ decreases the energy of the motion, while the force $\epsilon \dot{x}^3/3$ increases the energy of the motion. If $\dot{x}^2$ is always less than 3, the process described by (10.59) is dissipative. Therefore, if a does not exceed a certain critical value, we expect $E(t)$ to decrease as t increases. As we shall see, the critical value is given by (10.74) below.

If a satisfies (10.74) and the energy $E(t)$ decreases to zero as $t \to \infty$, the magnitude of x becomes very small when t is large. Then at very large t, the cubic force $\epsilon\dot{x}^3/3$ becomes negligible compared to the linear force $\epsilon\dot{x}$, and (10.59) is approximately

$$\ddot{x} + x \simeq -\epsilon\dot{x}.$$

The solution of the equation above is

$$x = ce^{-\epsilon t/2}e^{i\omega t} + c.c.,$$

where

$$\omega^2 = 1 - \epsilon^2/4, \tag{10.62}$$

with c an arbitrary constant.

The solution above is the approximate solution of (10.59) for t very large, while the two-scale method is capable of yielding an approximate solution of (10.59) for t not much larger than ϵ^{-1}. To obtain an approximate solution of (10.59) valid for all times, we must incorporate the features of both solutions.

Since the angular frequency at very large t is $\sqrt{1-\epsilon^2/4}$, not unity, it is appropriate to make a frequency renormalization before we apply the two-scale method. Thus we write (10.59) as

$$\ddot{x} + \omega^2 x = -\epsilon(\dot{x} - \dot{x}^3/3) - \epsilon^2 x/4, \tag{10.63}$$

where the last term in (10.63) is the counter term of frequency renormalization. We note that while the counter term for the pendulum problem is an infinite series of ϵ, the counter term in the present problems has only an ϵ^2 term that is easily determined.

Problem for the Reader

Apply the renormalized two-scale method to (10.63).

Solution

We put

$$x(t) = x_0(t,\tau) + \epsilon x_1(t,\tau) + \epsilon^2 x_2(t,\tau) + \cdots ,$$

where $\tau = \epsilon t$. As before, we have

$$x_0(t,\tau) = A_0(\tau)e^{i\omega t} + c.c. \tag{10.64}$$

The function $A_0(\tau)$ is determined by the requirement that the differential equation satisfied by x_1 have no secular terms. This equation is

$$\left(\frac{\partial^2}{\partial t^2} + \omega^2\right) x_1 = -2\frac{\partial^2 x_0}{\partial t \partial \tau} - \frac{\partial x_0}{\partial t} + \left(\frac{\partial x_0}{\partial t}\right)^3 /3. \tag{10.65}$$

Setting the sum of secular terms in (10.65) to zero, we get

$$-2i\omega A_0' - i\omega A_0 + i\omega^3 A_0^2 A_0^* = 0. \tag{10.66}$$

Let

$$A_0 = Re^{i\theta}; \tag{10.67}$$

then (10.66) is

$$R' + Ri\theta' + R/2 - \omega^2 R^3/2 = 0. \tag{10.68}$$

Setting the real part and the imaginary part of this equation to zero, we get

$$R' + R/2 - \omega^2 R^3/2 = 0 \tag{10.69}$$

and

$$\theta' = 0. \tag{10.70}$$

Solving the equations above, we get, after some algebra,

$$A_0(\tau) = R(\tau) = \frac{a}{\sqrt{a^2\omega^2 + (1 - a^2\omega^2)e^{\tau}}} \tag{10.71}$$

and

$$\theta(\tau) = 0. \tag{10.72}$$

In the above, the initial conditions have been incorporated. We therefore have

$$x_0(t,\tau) = \frac{2a\cos\omega t}{\sqrt{a^2\omega^2 + (1-a^2\omega^2)e^\tau}}. \tag{10.73}$$

The function $R(\tau)$ vanishes like $e^{-\epsilon t/2}$ as $\tau \to \infty$, provided that

$$1 - a^2\omega^2 > 0,$$

or

$$a < 1/\omega. \tag{10.74}$$

We shall show that under the condition (10.74), the perturbation series of the renormalized two-scale method is good for all times.

With the secular terms eliminated, we have from (10.65) that

$$x_1(t,\tau) = A_1(\tau)e^{i\omega t} - \frac{\omega R^3(\tau)e^{3i\omega t}}{24i} + c.c. \tag{10.75}$$

The function $A_1(\tau)$ is determined by requiring that the differential equation satisfied by $x_2(t,\tau)$ secular terms. This equation is

$$\begin{aligned}\left(\frac{\partial^2}{\partial t^2} + \omega^2\right)x_2 = {} & -2\frac{\partial^2 x_1}{\partial t\partial\tau} - \frac{\partial^2 x_0}{\partial\tau^2} - \frac{\partial x_1}{\partial t} - \frac{\partial x_0}{\partial\tau} \\ & + \left(\frac{\partial x_0}{\partial t}\right)^2\left(\frac{\partial x_1}{\partial t} + \frac{\partial x_0}{\partial\tau}\right) - x_0/4,\end{aligned} \tag{10.76}$$

where the last term is the counter term of frequency renormalization. Setting the secular terms in (10.76) to zero, we get

$$\begin{aligned}& -2i\omega A_1' - i\omega A_1 + i\omega^3R^2A_1^* + 2i\omega^3R^2A_1 = \\ & R'' + R' - \omega^2R^2R' - \frac{\omega^4R^5}{8} + \frac{R}{4}.\end{aligned} \tag{10.77}$$

We shall verify that $A_1(\tau)$ is of the same order of magnitude as $A_0(\tau)$ for all values of τ.

Equation (10.77) is a linear first-order differential equation for $A_1(\tau)$ and can be solved in a closed form, as we have discussed in Chapter 1. As all coefficients as well as the source terms of this equation are finite for all τ, the solution of such a linear equation is of the order of unity if τ is of the order of unity. Thus $A_0(\tau)$ and $A_1(\tau)$ are of the same order of magnitude when τ is of the order of unity. It remains to check whether this is still true when τ is large. When τ is large, $R(\tau)$ is small compared to unity and the equation above is approximately

$$-2i\omega A_1' - i\omega A_1 \simeq R'' + R' + R/4.$$

Since $R(\tau)$ is approximately equal to a constant times $e^{-\tau/2}$ when τ is large, the terms on the right side of the equation cancel completely and we get

$$-2i\omega A_1' - i\omega A_1 \simeq 0. \tag{10.78}$$

The solution of this equation is

$$A_1(\tau) = de^{-\tau/2},$$

where d is a constant to be determined from the initial conditions. We see that $A_1(\tau)$ is of the same order of magnitude as $A_0(\tau)$ when τ is large. Thus $A_1(\tau)$ and $A_0(\tau)$ are of the same order of magnitude for all times, and so are $x_1(t, \tau)$ and $x_0(t, \tau)$.

We remark that had we not made a frequency renormalization, (10.78) would have been

$$-2i\omega A_1' - i\omega A_1 \simeq R'' + R'.$$

The terms in the right side of this equation as $\tau \to \infty$ do not cancel completely and the sum of these two terms is asymptotically equal to a nonzero constant times $e^{-\tau/2}$. Since $e^{-\tau/2}$ is the complementary solution of the equation, the particular solution of this equation has a term proportional to $i\tau e^{-\tau/2}$. The existence of such a term means that had we not made a frequency renormalization, $\epsilon x_1(t, \tau)$ would have been comparable to $x_0(t, \tau)$ when $\epsilon\tau$ is of the order of unity. Thus the two-scale method fails for $t = O(\epsilon^{-2})$.

That there is the time scale of ϵ^{-2} can also be seen from (10.62), which gives

$$\omega = 1 - \frac{\epsilon^2}{8} + \frac{\epsilon^4}{128} - \cdots.$$

When t is of the order of ϵ^{-2},

$$e^{i\omega t} \simeq e^{it} e^{-i\epsilon^2 t/8}.$$

In the two-scale method without frequency renormalization, one makes the expansion

$$e^{-i\epsilon^2 t/8} = e^{-i\epsilon\tau/8} = 1 - i\epsilon\tau/8 + \cdots.$$

When τ is of the order of $8\epsilon^{-1}$, truncating this series, as the two-scale method would do, is not a good approximation. To make things worse, there are, in addition, the time scales of $\epsilon^{-4}, \epsilon^{-6} \cdots$. Therefore, even if one is able to take care of the third time scale, there are other time scales, indeed an infinite number of them, which we must handle. But we now see that all these time scales are due to the fact that ω, the renormalized angular frequency of the motion, differs from unity. If we define the time scale by ω^{-1}, then the solution has only two time scales: ω^{-1} and ϵ^{-1}. The solution obtained with the frequency renormalization (10.62) is a good approximation for all t.

We may similarly show that all $x_n(t, \tau)$ obtained with the renormalized two-scale method are of the order of $e^{-\tau/2}$ as τ goes to infinity. (See homework problem 4.) Thus $\epsilon^n x_n(t, \tau)$ is much smaller than $\epsilon^{n-1} x_{n-1}(t, \tau)$ for all t.

For completeness, I will say a few words about why the process is dissipative when a satisfies (10.74). We note that, very roughly, the solution of the present problem is

$$x \simeq 2a \cos t,$$

at least for t not very large. Thus we have

$$\dot{x} \simeq -2a \sin t.$$

The average value of $\dot{x}^2$ over one period of the motion is

$$\int_0^{2\pi} 4a^2 \sin^2 t \frac{dt}{2\pi} = 2a^2,$$

and the average value of $\dot{x}^4$ over one period of the motion is

$$\int_0^{2\pi} (2a)^4 \sin^4 t \frac{dt}{2\pi} = 6a^4.$$

Substituting these values into (10.61), we find that the value of $E(t)$ averaged over one period decreases if

$$a < 1,$$

which agrees with (10.74) in the lowest order.

If $a > 1/\omega$, $E(t)$ is an increasing function of t and the nonlinear force is not negligible at very large times. We remark that $x_0(t, \tau)$ obtained is still a good approximation when τ is not too large. From (10.71) we have, when $a > 1/\omega$, $x_0(t, \tau)$ blowing up at

$$\tau_0 = \ln \frac{a^2}{a^2\omega^2 - 1}. \tag{10.79}$$

Therefore, as τ approaches τ_0, $R(\tau)$ becomes very large. Now x_1 has third harmonic terms that are of the order of $R^3(\tau)$. When R is large, R^3 is even larger. Therefore, if $a > 1/\omega$, ϵx_1 becomes comparable to x_0 as τ approaches τ_0 and the two-scale method fails.

The renormalized two-scale method is also applicable to nonlinear problems in which $x(t)$ does not approach zero as $t \to \infty$. Two well-known examples of such problems are the Rayleigh equation and the van der Pol equation. We will leave the van der Pol equation as a homework problem and discuss here how to solve the Rayleigh equation approximately for all t.

Let

$$\ddot{x} + \omega_0^2 x = \epsilon(\dot{x} - \dot{x}^3/3), \tag{10.80}$$

and let the initial conditions for the solution of this equation be

$$x(0) = 2a > 0, \ \dot{x}(0) = 0.$$

From (10.83), we have

$$\frac{d}{dt}\frac{(\dot{x}^2 + \omega_0^2 x^2)}{2} = \dot{x}^2(1 - \dot{x}^2/3).$$

From the equation above, we see that dE/dt is positive if $\dot{x}$ is small and the magnitude of the motion grows. But dE/dt is negative if $\dot{x}$ is large and the magnitude of the motion shrinks. Thus it is not surprising that the motion approaches a limit circle as $t \to \infty$ no matter what the initial conditions are, as we shall soon see. Since we expect the period of such a limiting circular motion to be different from $2\pi/\omega_0$, we make the frequency renormalization

$$\omega_0^2 = \omega^2(1 + b_1\epsilon + b_2\epsilon^2 + \cdots), \tag{10.81}$$

where b_n, $n = 1, 2, \cdots$, are real constants to be determined. Equation (10.80) becomes

$$\ddot{x} + \omega^2 x = \epsilon(\dot{x} - \dot{x}^3/3) - \omega^2(b_1\epsilon + b_2\epsilon^2 + \cdots)x. \tag{10.82}$$

Problem for the Reader

Apply the two-scale method to (10.82).

Solution

The lowest-order term $x_0(t, \tau)$ is given by

$$x_0(t, \tau) = A_0(\tau)e^{i\omega t} + c.c.$$

The function $A_0(\tau)$ is determined by the requirement that the sum of secular terms in the equation satisfied by $x_1(t, \tau)$ vanishes. This equation is

$$\left(\frac{\partial^2}{\partial t^2} + \omega^2\right) x_1 = -2\frac{\partial^2 x_0}{\partial t \partial \tau} + \frac{\partial x_0}{\partial t} - \left(\frac{\partial x_0}{\partial t}\right)^3 /3 - \omega^2 b_1 x_0. \qquad (10.83)$$

By requiring that the sum of secular terms in (10.83) vanish, we get

$$-2i\omega A_0' + i\omega A_0 - i\omega^3 A_0^2 A_0^* - \omega^2 b_1 A_0 = 0.$$

As before, we write

$$A_0(\tau) = R(\tau)e^{i\theta(\upsilon)}$$

and substitute it into the differential equation above. We get, after removing the common factor $\omega e^{i\theta}$ from all the terms of the resulting equation,

$$-2i\left(\frac{d}{d\tau} + i\theta'\right) R + iR - i\omega^2 R^3 - \omega b_1 R = 0.$$

Equating the imaginary part as well as the real part of this equation to zero, we get the differential equations

$$2R' - R + \omega^2 R^3 = 0 \qquad (10.84)$$

and

$$2\theta' - \omega b_1 = 0. \qquad (10.85)$$

Solving (10.84), we get

$$R(\tau) = \frac{a}{\sqrt{a^2\omega^2 + (1 - a^2\omega^2)e^{-\tau}}}. \qquad (10.86)$$

Note that the function $R(\tau)$ approaches the constant value ω^{-1} exponentially as τ approaches infinity, independent of the value of a.

The solution of (10.85) is

$$\theta_1(\tau) = \frac{\omega b_1}{2}\tau + \theta_1(0).$$

Requiring that $\theta_1(\tau)$ not become large as τ becomes large, we choose

$$b_1 = 0.$$

Then (10.85) together with the initial conditions give

$$\theta(\tau) = 0.$$

Thus $A_0(\tau)$ approaches the finite limit $1/\omega$ as $\tau \to \infty$. Indeed, the difference of $A_0(\tau)$ with its limiting value vanishes exponentially as $\tau \to \infty$. Therefore, the lowest-order term of x is given by

$$x_0(t, \tau) = \frac{2a \cos \omega t}{\sqrt{a^2\omega^2 + (1 - a^2\omega^2)e^{-\tau}}}, \tag{10.87}$$

which approaches

$$\frac{2 \cos \omega t}{\omega},$$

independent of the value of a.

The equation for x_{n+1} for $n \geq 1$ is

$$\left(\frac{\partial^2}{\partial t^2} + \omega^2\right) x_{n+1} + 2\frac{\partial^2 x_n}{\partial t \partial \tau} - \frac{\partial x_n}{\partial t} + \frac{\partial^2 x_{n-1}}{\partial \tau^2} - \frac{\partial x_{n-1}}{\partial \tau} + \omega^2 \sum b_m x_{n-m+1}$$
$$= -\frac{1}{3} \sum \left(\frac{\partial x_{n-k-m}}{\partial t} + \frac{\partial x_{n-k-m-1}}{\partial \tau}\right) \left(\frac{\partial x_k}{\partial t} + \frac{\partial x_{k-1}}{\partial \tau}\right) \left(\frac{\partial x_m}{\partial t} + \frac{\partial x_{m-1}}{\partial \tau}\right), \tag{10.88}$$

where by definition $x_{-1} = x_{-2} = \cdots = 0$. Let

$$x_n(x, \tau) = A_n(\tau)e^{i\omega t} + c.c. + \text{higher harmonic terms.}$$

We shall prove that $A_n(\tau)$ approaches a constant exponentially as τ approaches infinity.

This is done by induction. We assume that $A_m(\infty)$, $m < n$, is finite, and that $A_m(\tau) - A_m(\infty)$ vanishes exponentially as $\tau \to \infty$, both of which have been established for $m = 1$. By setting the secular terms of (10.88) to zero, we get a linear inhomogeneous differential equation for $A_n(\tau)$, $n = 1, 2, \cdots$. When τ is very large, we approximate $A_0(\tau)$ by ω^{-1} and the equation for $A_n(\tau)$ becomes

$$2\frac{d}{d\tau}A_n + A_n + A_n^* = S_n, \tag{10.89}$$

where the left side of (10.89) comes from the second and the third term on the left side of (10.88) as well as the sum on the right side of (10.88). Also, S_n depends only on A_m, $m < n$, which, by assumption, approaches a constant exponentially as τ goes to infinity. Let

$$A_n(\tau) = R_n(\tau) + iI_n(\tau),$$

where $R_n(\tau)$ and $I_n(\tau)$ are real. Then we have from (10.89) that

$$\frac{d}{d\tau}R_n + R_n = c_n, \tag{10.90}$$

where c_n approaches a constant exponentially as τ goes to infinity. We may then argue that $R_n(\tau)$ approaches c_n exponentially as $\tau \to \infty$.

Next we note that the term $\omega^2 b_{n+1} x_0(t, \tau)$ in the sum on the left side of (10.88) contributes to the differential equation for I_n. We choose the constant b_{n+1} so that the right side of the equation for I_n vanishes as $\tau \to \infty$. Thus

$$\frac{d}{d\tau}I_n \approx 0, \tau \gg 1. \tag{10.91}$$

The terms neglected in (10.91) are exponentially small as $\tau \to \infty$. Therefore, $I_n(\tau)$ approaches a constant as $\tau \to \infty$. Since both the real part and the imaginary part of $A_n(\tau)$ approach constants as $\tau \to \infty$, all $A_n(\tau)$ are of the same order of magnitude for all τ.

E. The Renormalization Group

Another method to treat small nonlinear oscillations is that of renormalization group. The method of renormalization group has been applied to diverse branches of physics. The techniques employed and even the basis on which the method is applied vary significantly from one branch to another. For example, in Yang-Mills theories, the justification of the method is the proven renormalizability of the scattering amplitudes. On the other hand, in the treatment of critical phenomena in condensed matter physics, that the iterated Hamiltonian has a fixed point under a prescribed renormalization procedure is an assumption. There is, nevertheless, a common ground of these applications, and that is the theme of expressing the dynamics of the system with parameters that depend on the point of normalization.

Here we shall only demonstrate how to apply the idea of renormalization group to problems of small nonlinear oscillations. Let there be a nonlinear differential equation for $x(t)$. The initial conditions are $x(0) = a$ and $\dot{x}(0) = b$. In the two-scale method, the goal is to find an approximation of $x(t)$ good for all values of t up to the order of ϵ^{-1}. We may achieve the same result with the method of renormalization group.

In the approach of renormalization group, one calculates the solution with regular perturbation and expresses the arbitrary constants of the solution not by a and b, but by the (unknown) initial values $x(t_0)$ and $\dot{x}(t_0)$, where t_0 can be any point. The idea is that the solution $x(t)$ is independent of t_0, and hence its partial derivative with respect to t_0 is zero. This gives us a first-order differential equation that yields an approximation of $x(t_0)$ good for all values of t_0 up to $O(\epsilon^{-1})$.

To see how this works out in detail, we revisit the equation

$$\ddot{x} + x = -\epsilon \dot{x}^3, \tag{10.92}$$

satisfying the given initial values $x(0) = a$ and $\dot{x}(0) = b$. The equation above is equal to (10.28) with ω_0 set to be unity.

In order to employ the method of renormalization group, one needs some input from the differential equation (10.92). This input is the solution obtained with regular perturbation. Let

$$x(t) = x_0(t) + \epsilon x_1(t) + \cdots .$$

The initial conditions for $x_n(t)$ and $\dot{x}_n(t)$ are

$$x_0(t_0) = x(t_0), \tag{10.93}$$

$$x_1(t_0) = x_2(t_0) = \cdots = 0, \tag{10.94}$$

$$\dot{x}_0(t_0) = \dot{x}(t_0), \tag{10.95}$$

and

$$\dot{x}_1(t_0) = \dot{x}_2(t_0) = \cdots = 0. \tag{10.96}$$

As we know, this regular perturbation series is a good approximation of the solution only if $(t - t_0)$ is not large.

The lowest-order term $x_0(t)$ is easily found to be

$$x_0(t) = A_0(t_0)e^{it} + A_0^*(t_0)e^{-it}, \tag{10.97}$$

where $A_0(t_0)$ is determined from the initial conditions (10.93) and (10.95). From the initial condition (10.93), we have

$$A_0(t_0)e^{it_0} + A_0^*(t_0)e^{-it_0} = x(t_0); \tag{10.98}$$

from the initial condition (10.95), we have

$$iA_0(t_0)e^{it_0} - iA_0^*(t_0)e^{-it_0} = \dot{x}(t_0).$$

Thus we get

$$A_0(t_0) = e^{-it_0}\frac{x(t_0) - i\dot{x}(t_0)}{2}.$$

Now

$$\frac{\partial}{\partial t_0}x(t) = 0.$$

Since

$$x(t) = x_0(t) + O(\epsilon),$$

we have

$$\frac{\partial}{\partial t_0}x_0 = O(\epsilon).$$

With (10.97), we get

$$A_0'(t_0)e^{it} + A_0^{*\prime}(t_0)e^{-it} = O(\epsilon),$$

where

$$A_0'(t_0) \equiv \frac{dA_0(t_0)}{dt_0}.$$

Since the equation above holds as a function of t, the coefficient of e^{it} and that of e^{-it} must be individually of the order of ϵ. Thus we have

$$A_0'(t_0) = O(\epsilon).$$

This shows that $A_0(t_0)$ varies slowly with t_0; in particular, if $\epsilon = 0$, $A_0(t_0)$ is independent of t_0.

To see more precisely how $A_0(t_0)$ varies with t_0, we calculate the next-order term $x_1(t)$ of regular perturbation. Since we have already made similar calculations in this chapter, we will omit the details here. We get

$$x_1(t) = A_1(t_0)e^{it} - \frac{3}{2}A_0^2(t_0)A_0^*(t_0)(t - t_0)e^{it} - \frac{i}{8}A_0^3(t_0)e^{3it} + c.c., \quad (10.99)$$

where the sum of $A_1(t_0)e^{it}$ and its complex conjugate is the complementary solution, and the sum of the rest of the terms is a particular solution. The solution $x_0(t) + \epsilon x_1(t)$ approximates $x(t)$ up to the order of ϵ, provided that the values of $(t - t_0)$ are much smaller than ϵ^{-1}.

The initial conditions enable us to express $A_1(t_0)$ in terms of $A_0(t_0)$. By (10.94) we have

$$A_1(t_0)e^{it_0} + \frac{i}{8}A_0^3(t_0)e^{3it_0} + c.c. = 0;$$

by (10.96) we have

$$iA_1(t_0)e^{it_0} - \frac{3}{2}A_0^2(t_0)A_0^*(t_0)e^{it_0} + \frac{3}{8}A_0^3(t_0)e^{3it_0} + c.c. = 0.$$

One may solve the two algebraic equations above and get an expression for $A_1(t_0)$ in terms of $A_0(t_0)$. This will not be needed here.

The partial derivative of $x_0(t) + \epsilon x_1(t)$ with respect to t_0 is $O(\epsilon^2)$. Thus we get from (10.97) and (10.99) that

$$A'(t_0)e^{it} + \frac{3}{2}\epsilon A^2(t_0)A^*(t_0)e^{it} + c.c. = O(\epsilon^2) \tag{10.100}$$

where

$$A(t_0) = A_0(t_0) + \epsilon A_1(t_0).$$

An example of the unexhibited $O(\epsilon^2)$ terms in (10.100) is

$$-\frac{i}{8}\epsilon e^{3it}\frac{d}{dt_0}A_0^3(t_0),$$

which is of the order of ϵ^2 as $A_0'(t_0)$ is of the order of ϵ. Since (10.100) holds as a function of t, the coefficient of e^{it} in (10.100) is zero. Thus we get

$$2\frac{d}{dt_0}A(t_0) = -3\epsilon A^2(t_0)A^*(t_0) + O(\epsilon^2), \tag{10.101}$$

which is the same as (10.45) if we ignore the $O(\epsilon^2)$ terms, identify ϵt_0 as τ, and set ω_0 to unity.

By (10.98), we get an approximation of $x(t_0)$ for all values of t_0 up to $O(\epsilon^{-1})$, which agrees with the approximate solution $x_0(t, \tau)$ of the two-scale method. The third time scale of the solution is revealed as we calculate $x_2(t)$ with regular perturbation and apply the argument of renormalization group to the sum $x_0 + \epsilon x_1 + \epsilon^2 x_2$. It has not been possible to use this approach to yield an approximate solution valid for all t.

References

For the renormalized two-scale method, see H. Cheng, *Studies in Applied Mathematics* 113: 381–387 (2004).

For more discussion of the renormalization group, see Lin-Yuan Chen, Nigel Goldenfeld, and Y. Oono, *Physical Review E* 54: 376 (1996).

Homework Problems for This Chapter

Solutions to the Homework Problems can be found at www.lubanpress.com.

1. Let

$$\ddot{x} + \omega_0^2 x = \epsilon x^2,$$

with the initial conditions

$$x(0) = 0 \text{ and } \dot{x}(0) = 1.$$

a. Use the method of regular perturbation and find the solution up to and including the term of the order of ϵ. Are there secular terms in the equation for $x_1(t)$? Argue that regular perturbation still holds when $t = O(\epsilon^{-1})$.

b. Find the equation satisfied by $x_2(t)$. Show that regular perturbation fails when $t = O(\epsilon^{-2})$.

c. Use the method of counter terms and obtain the lowest-order correction to the angular frequency of oscillation.

d. Argue that the secular terms of any order can be eliminated by the frequency counter terms, and that the solution yielded by renormalized perturbation is a good approximation for all t.

2. Consider the solution of

$$\ddot{x} + (1 + \epsilon \dot{x}^2)x = 0,$$

satisfying

$$x(0) = 1, \dot{x}(0) = 0.$$

a. Show that regular perturbation fails when $t = O(\epsilon^{-1})$.

b. Introduce counter terms and express the angular frequency ω of the system as

$$\omega^2(1 + a_1\epsilon + a_2\epsilon^2 + \cdots) = 1.$$

Express the solution $x(t)$ as the renormalized perturbation series $x_0(t) + \epsilon x_1(t) + \cdots$ and obtain the differential equation for $x_0(t)$ and that for $x_1(t)$.

c. Show that one may eliminate the secular terms in the equation satisfied by $x_1(t)$ if one chooses

$$a_1 = -\frac{1}{4\omega^2}.$$

Determine all the constants in the expression for $x_0(t)$.

d. Obtain the differential equation satisfied by $x_2(t)$ and show that one may eliminate the secular terms of this equation by choosing a real a_2.

3. Consider the differential equation in problem 2. Argue that with the use of frequency counter terms, you can eliminate the secular terms in any order and hence the renormalized perturbation series is good for all times. Argue more generally that the same conclusion is reached if the term $\dot{x}^2 x$ in the differential equation is replaced by $P(\dot{x})x$, where $P(\dot{x})$ is a polynomial of $\dot{x}$ with only even powers of $\dot{x}$.

4. Consider the differential equation

$$(1 + \epsilon e^{-\epsilon t}\dot{y}^2)\ddot{y} + y = 0, \text{ with } y(0) = 0, \dot{y}(0) = 1.$$

a. Solve with regular perturbation up to and including terms of the order of ϵ. Discuss the region of t in which this perturbation series is meaningful.

b. Solve this problem with the two-scale method and obtain $y_0(t, \tau)$. What is the asymptotic form of $y_0(t, \tau)$ as $\tau \to \infty$? Can you derive this asymptotic form directly from the differential equation without using perturbation expansion?

c. Determine $y_1(t, \tau)$. For what values of t is $\epsilon y_1(t, \tau)$ small compared to $y_0(t, \tau)$?

d. Argue that $y_0(t, \tau)$ is a good approximation of the solution for all t.

5. Repeat problem 4 with the differential equation

$$\ddot{y} + y = -\epsilon e^{-\epsilon t} y^3, \epsilon \ll 1, \quad y(0) = 0, \dot{y}(0) = 1.$$

6. Apply the two-scale method and find the approximate solution of

$$\ddot{x} + x = -\epsilon(\dot{x}^3 + 2x^3),$$

with

$$x(0) = 0 \text{ and } \dot{x}(0) = 1.$$

Calculate the lowest-order term $x_0(t, \tau)$ of the solution. What is the asymptotic form of $x_0(t, \tau)$ in the limit $\tau \to \infty$? For what values of t do you expect $x_0(t, \tau)$ be a good approximation of the solution?

7. Consider the differential equation

$$\frac{d^2y}{dt^2} + \epsilon(1 + y^2)\frac{d}{dt}y + y = 0,$$

with $\dot{y}(0) = 1, y(0) = 0$. Find the lowest-order approximate solution of this problem with the use of the two-scale method. What is the asymptotic form of this solution as $\tau \to \infty$? Do you expect this solution be a good approximation when $t = O(\epsilon^{-2})$?

8. Use the renormalized two-scale method and obtain the lowest-order solution for the differential equation in problem 7 that is a good approximation for all t.

9. Consider the solution of the van der Pol equation

$$\frac{d^2x}{dt^2} + \epsilon(x^2 - 1)\frac{dx}{dt} + x = 0$$

satisfying the initial condition

$$x(0) = 2a, x^{\cdot}(0) = 0.$$

a. Derive the equation

$$\frac{d}{dt}\frac{\dot{x}^2 + x^2}{2} = -\epsilon(x^2 - 1)\dot{x}^2.$$

Using this equation, argue that the solution of the van der Pol equation is approximately given by $x(t) = 2\cos t$ as $t \to \infty$, no matter what the initial conditions are.

b. Give a qualitative argument that in order to obtain an approximate solution good for all times, one must make a frequency renormalization.

c. Show with the two-scale method that $x_0(t,\tau) = A_0(\tau)e^{i\omega t} + A_0(\tau)e^{-i\omega t}$, and that $A_0(\tau)$ approaches 1 as $\tau \to \infty$.

d. Consider the differential equation satisfied by $x_2(t,\tau)$. Show that you will need to make a frequency renormalization in order to eliminate the secular terms in this differential equation. Find $x_1(t,\tau)$ and show that $\epsilon x_1(t,\tau)$ is much smaller than $x_0(t,\tau)$ at all times.

e. Let

$$x_n(t,\tau) = A_n(\tau)e^{i\omega t} + A_n^*(\tau)e^{-i\omega t} + \text{higher harmonic terms.}$$

Show that by setting to zero the sum of secular terms in the differential equation satisfied by $x_{n+1}(t)$, we get

$$\frac{dA_n}{d\tau} + \frac{1}{2}(A_n + A_n^*) = S_n,$$

where S_n depends only on A_m with $m < n$.

f. Prove by induction that by making frequency renormalization of the order of ϵ^n, $A_n(\tau)$ approaches a constant as $\tau \to \infty$. Argue that $\epsilon^n x_n(t,\tau)$ for any n is then much smaller than $\epsilon^{n-1}x_{n-1}(t,\tau)$ at all times.

Appendix of Useful Formulae

Chapter 1

Two useful formulae involving the differential operator D are

$$D(uv) = u\left(D + \frac{u'}{u}\right)v$$

and

$$(D - a)(D + a) = D^2 - a^2 + a'.$$

The general solution of the first-order ODE

$$y' + p(x)y = q(x)$$

is

$$y(x) = e^{-P(x)}\left(\int_0^x e^{P(x')}q(x')dx' + c\right),$$

where

$$P(x) = \int_0^x p(x')dx'$$

with c an arbitrary constant. The lower limit of integration in the integrals above may be chosen to be any constant other than 0.

Chapter 2

The Euler formula is

$$e^{iz} = \cos z + i \sin z.$$

From the Euler formula, we get

$$\cos z = \frac{e^{iz} + e^{-iz}}{2}$$

and

$$\sin z = \frac{e^{iz} - e^{-iz}}{2i}.$$

The Maclaurin series for $\sin z$ and $\cos z$ are

$$\sin z = z - z^3/3! + z^5/5! + \cdots$$

and

$$\cos z = 1 - z^2/2! + z^4/4! + \cdots,$$

which converge for all z. We also have

$$\frac{1}{1-z} = 1 + z + z^2 + \cdots,$$

which converges for $|z| < 1$.

If $f(z)$ has a pole of order n at z_0, the residue of $f(z)$ at z_0 is

$$Res(z_0) = \lim_{z \to z_0} \frac{1}{(n-1)!} \frac{d^{n-1}}{dz^{n-1}} \left[(z - z_0)^n f(z) \right].$$

If z_0 is a simple pole of $f(z)$, and if $f(z) = N(z)/D(z)$, where $N(z)$ and $D(z)$ are analytic at z_0 with $D(z)$ having a simple zero at z_0, a good way to compute the residue of $f(z)$ at z_0 is to use the formula

$$\text{Res}(z_0) = \frac{N(z_0)}{D'(z_0)}.$$

The Plemelj formulae are

$$\int_a^b \frac{r(x')}{x' - (x \pm i\epsilon)} dx = P \int_a^b \frac{r(x')}{x' - x} dx \pm i\pi r(x)$$

where $a < x < b$.

The Fourier transform of $F(x)$, which is defined for $-\infty < x < \infty$, is

$$\tilde{F}(k) \equiv \int_{-\infty}^{\infty} e^{-ikx} F(x) dx.$$

And Fourier's inversion formula is

$$F(x) = \int_{-\infty}^{\infty} e^{ikx} \tilde{F}(k) \frac{dk}{2\pi}.$$

Also, let

$$A(k) = \int_{-\infty}^{\infty} \cos kx F(x) dx, \ B(k) = \int_{-\infty}^{\infty} \sin kx F(x) dx,$$

then

$$F(x) = \int_{-\infty}^{\infty} \left[A(k) \cos kx + B(k) \sin kx\right] \frac{dk}{2\pi}.$$

The Fourier coefficient of $f(\theta)$, which is defined for $-\pi \leq \theta \leq \pi$, is

$$a_n \equiv \frac{1}{2\pi} \int_{-\pi}^{\pi} e^{-in\theta} f(\theta) d\theta.$$

The function $f(\theta)$ can be expressed by the Fourier series

$$f(\theta) = \sum_{n=-\infty}^{\infty} a_n e^{in\theta}.$$

The Fourier coefficient of $df/d\theta$ is equal to

$$\frac{1}{2\pi} \int_{-\pi}^{\pi} e^{-in\theta} \frac{df}{d\theta} d\theta = (-1)^n \frac{f(\pi) - f(-\pi)}{2\pi} + ina_n.$$

Thus the Fourier transform of the derivative of $f(\theta)$ is equal to ina_n only if $f(\pi) = f(-\pi)$. In the above, we have assumed that $f(\theta)$ is continuous. If $f(\theta)$ is discontinuous at θ_0, there is an additional term

$$e^{-in\theta_0} \frac{f(\theta_0^-) - f(\theta_0^+)}{2\pi}$$

on the right-side of the equation above.

Another form of the Fourier series for $f(\theta)$ is

$$f(\theta) = a_0 + \sum_{n=1}^{\infty} \left[A_n \cos n\theta + B_n \sin n\theta\right],$$

where

$$A_n = \frac{1}{\pi}\int_{-\pi}^{\pi} f(\theta)\cos n\theta d\theta,\, B_n = \frac{1}{\pi}\int_{-\pi}^{\pi} f(\theta)\sin n\theta d\theta.$$

The Laplace transform of $f(x)$, $x \geq 0$, is

$$L(s) = \int_0^{\infty} e^{-sx} f(x)dx.$$

The Laplace inversion formula is

$$f(x) = \int_{-i\infty}^{i\infty} e^{sx} L(s)\frac{ds}{2\pi i}.$$

The Laplace transform of $f'(x)$ is

$$-f(0) + sL(s).$$

And the Laplace transform of $f''(x)$ is

$$-f'(0) - sf(0) + s^2 L(s).$$

Chapter 3

The equations determining the characterisitc curves for the quasi-linear PDE

$$a(x, y, u)u_x + b(x, y, u)u_y = c(x, y, u)$$

are

$$\frac{dx}{a(x, y, u)} = \frac{dy}{b(x, y, u)} = \frac{du}{c(x, y, u)}.$$

The equations determining the characterisitc curves for the PDE

$$F(x, y, u, q, p) = 0,$$

with

$$p \equiv u_x, q \equiv u_y$$

are

$$\frac{dx}{F_p} = \frac{dy}{F_q} = \frac{du}{pF_p + qF_q} = -\frac{dp}{F_x + F_u p} = -\frac{dq}{F_y + F_u q}.$$

Chapter 4

The Laplace equation is

$$\nabla^2 u = 0.$$

The heat equation is

$$\frac{\partial u}{\partial t} = \nabla^2 u.$$

The wave equation is

$$\nabla^2 u - \frac{\partial^2}{\partial t^2} u = 0.$$

The Fourier transform of the Guassian is

$$\int_{-\infty}^{\infty} e^{ikx - ax^2} dx = \sqrt{\frac{\pi}{a}} e^{-k^2/4a}.$$

Chapter 5

The Laplace equation in the polar coordinates is:

$$\left(\frac{\partial^2}{\partial r^2} + \frac{1}{r}\frac{\partial}{\partial r} + \frac{1}{r^2}\frac{\partial^2}{\partial \theta^2} \right) u = 0.$$

The Schrödinger equation for a particle in a potential V is

$$i \frac{h}{2\pi} \frac{\partial}{\partial t} \Psi = -\frac{1}{2m} \left(\frac{h}{2\pi} \right)^2 \nabla^2 \Psi + V\Psi,$$

where $h = 6.6238 \times 10^{-27}$ erg-sec and m is the particle mass.

Chapter 6

Airy's differential equation is

$$\frac{d^2 y}{dx^2} - xy = 0.$$

Bessel's differential equation is

$$x^2\frac{d^2y}{dx^2}+x\frac{dy}{dx}+(x^2-p^2)y=0,$$

where p is a constant.

The parabolic cylinder equation is

$$\frac{d^2y}{dx^2}+(\nu+1/2-z^2/4)y=0,$$

where ν is a constant.

The general solution of the Airy equation is

$$y(x)=C_1\sum_m\frac{x^{3m}}{3^{2m}m!\Gamma\left(m+\frac{2}{3}\right)}+C_2\sum_m\frac{x^{3m+1}}{3^{2m}m!\Gamma\left(m+\frac{4}{3}\right)},$$

where the Gamma function $\Gamma(z)$ is defined by the integral

$$\Gamma(z)\equiv\int_0^\infty e^{-t}t^{z-1}\,dt.$$

The gamma function satisfies the identity

$$z(z-1)(z-2)\cdots(z-n+1)=\Gamma(z+1)/\Gamma(z-n+1).$$

One of the independent solutions of the Airy equation is the Airy function $A_i(x)$ which is given by

$$Ai(x)=3^{-2/3}\sum_m\frac{x^{3m}}{3^{2m}m!\Gamma\left(m+\frac{2}{3}\right)}-3^{-4/3}\sum_m\frac{x^{3m+1}}{3^{2m}m!\Gamma\left(m+\frac{4}{3}\right)}.$$

This solution of the Airy equation has the integral representation

$$Ai(x)=\frac{1}{\pi}\int_0^\infty\cos(xt+t^3/3)dt.$$

The function $Ai(x)$ vanishes exponentially as $x\to\infty$, and is oscillatory as $x\to-\infty$. Specifically, we have

$$\begin{aligned}Ai(x)&\approx\frac{\exp(-2x^{3/2}/3)}{2\sqrt{\pi}x^{1/4}}, && x\to\infty,\\ &\approx\frac{\sin(2\left|x\right|^{3/2}/3+\pi/4)}{\sqrt{\pi}\left|x\right|^{1/4}}, && x\to-\infty.\end{aligned}$$

Two of the independent solutions of the Bessel equation are

$$J_p(x) = \sum_{m \geq 0} (-1)^m \frac{\left(\frac{x}{2}\right)^{2m+p}}{\Gamma(m+p+1)\Gamma(m+1)}$$

and

$$Y_p(x) = \frac{J_p(x)\cos \pi p - J_{-p}(x)}{\sin \pi p}.$$

If $p = n$, we have, by l' Hopital's rule,

$$Y_n(x) = \pi^{-1} \left[\frac{\partial J_p(x)}{\partial p} - (-1)^n \frac{\partial J_{-p}(x)}{\partial p} \right]_{p=n}.$$

The asymptotic forms for $J_p(x)$ and $Y_p(x)$ when x is very large are

$$J_p(x) \approx \sqrt{\frac{2}{\pi x}} \cos(x - \pi/4 - p\pi/2),\ x >> 1,$$

and

$$Y_p(x) \approx \sqrt{\frac{2}{\pi x}} \sin(x - \pi/4 - p\pi/2),\ x >> 1.$$

A solution of the hyperbolic cylinder equation is denoted as $D_\nu(x)$. The functions $D_\nu(-x), D_{-\nu-1}(ix)$ and $D_{-\nu-1}(-ix)$ are also solutions of this equation. The function $D_\nu(x)$ vanishes exponentially as $x \to \infty$, and blows up exponentially as $x \to -\infty$. Specifically, we have

$$\begin{aligned} D_\nu(x) &\approx x^\nu \exp(-x^2/4), \quad x \to \infty, \\ &\approx \frac{\sqrt{2\pi}}{\Gamma(-\nu)} |x|^{-\nu-1} \exp(x^2/4), \quad x \to -\infty. \end{aligned}$$

Chapter 7

The WKB approximate solutions of the differential equation

$$y'' + \lambda^2 P^2 y = 0,\ \lambda >> 1,$$

are

$$\frac{1}{\sqrt{P(x)}} e^{\pm i\lambda \int^x P(x')dx'}.$$

Chapter 8

If $v(x)$ is minimum at a with $v'(a) > 0$ and if $h(a) \neq 0$, the contribution of the lower endpoint a to the integral

$$\int_a^b e^{-\lambda v(x)} h(x) dx, \quad \lambda >> 1,$$

is

$$\frac{e^{-\lambda v(a)}}{\lambda v'(a)} h(a).$$

If $v(x)$ is minimum at b with $v'(b) < 0$ and if $h(b) \neq 0$, the contribution of the upper endpoint b to the integral is

$$-\frac{e^{-\lambda v(b)}}{\lambda v'(b)} h(b).$$

If $v(x)$ is minimum at an interior point x_0, where $v'(x_0) = 0$ and $v''(x_0) > 0$, and if $h(x_0) \neq 0$, the contribution of x_0 to the integral is

$$\sqrt{\frac{2\pi}{\lambda v''(x_0)}} e^{-\lambda v(x_0)} h(x_0).$$

The endpoint contributions to the integral

$$\int_a^b e^{i\lambda u(x)} h(x) dx, \quad \lambda >> 1$$

are

$$\frac{e^{i\lambda u(b)}}{i\lambda u'(b)} h(b) - \frac{e^{i\lambda u(a)}}{i\lambda u'(a)} h(a),$$

provided that $u'(a)$, $u'(b)$, $h(a)$ and $h(b)$ do not vanish. If $u'(x)$ vanishes at an interior point x_0, there is a contribution from the point of stationary phase x_0. In the case of $u''(x_0) > 0$ and $h(x_0) \neq 0$, this contribution is

$$e^{i\pi/4} \sqrt{\frac{2\pi}{\lambda u''(x_0)}} e^{i\lambda u(x_0)} h(x_0);$$

and in the case of $u''(x_0) < 0$ and $h(x_0) \neq 0$, this contribution is

$$e^{-i\pi/4} \sqrt{\frac{2\pi}{\lambda \left| u''(x_0) \right|}} e^{i\lambda u(x_0)} h(x_0).$$

Chapter 9

We give here the approximate solution of the differential equation

$$\epsilon \frac{d^2y}{dx^2} + a(x)\frac{dy}{dx} + b(x)y = 0,\ 0 < x < 1,\ \epsilon << 1,$$

satisfying the boundary conditions

$$y(0) = A,\ y(1) = B.$$

If $a(x) > 0$, the solution is

$$y_{uniform}(x) = y_{out}(x) + [A - y_{out}(0)] \exp\left[-\frac{a(0)x}{\epsilon}\right],$$

where

$$y_{out}(x) = B \exp\left[\int_x^1 \frac{b(x')}{a(x')} dx'\right]$$

And if $a(x) < 0$, the solution is

$$y_{uniform}(x) = y_{out}(x) + [B - y_{out}(1)] \exp\left[\frac{a(1)(1-x)}{\epsilon}\right],$$

where

$$y_{out}(x) = A \exp\left[-\int_0^x \frac{b(x')}{a(x')} dx'\right].$$

Bibliography

Chapter 1

1. The use of qualitative methods for solving ordinary differential equations can be found in the classic text V. I. Arnold, *Ordinary Differential Equations*, M.I.T. Press, Cambridge, MA, 1978.

2. An advanced text on the subject of geometrical methods applied to ordinary differential equations is V. I. Arnold, M. Levi, and J. Szucs, *Geometrical Methods of Ordinary Differential Equations*, Springer-Verlag, New York, NY, 1996.

3. The theory of ordinary differential equations from the viewpoint of dynamical systems is given in D. K. Arrowsmith and C. M. Place, *An Introduction to Dynamical Systems*, Cambridge University Press, New York, NY, 1990.

4. A book on the quantitative and qualitative behavior of solutions to ordinary differential equations, such as the related stability issues, is R. E. Bellman, *Stability Theory of Differential Equations*, Dover, NY, 1969.

5. An introductory text on the theory of ordinary differential equations is G. Birkhoff and G.-C. Rotta, *Ordinary Differential Equations*, John Wiley, New York, NY, 4th Edition, 1989.

6. An introductory text on differential equations is W. E. Boyce and R. C. DiPrima, *Elementary Differential Equations*, John Wiley and Sons, New York, NY, 1969.

7. An introductory text on the qualitative theory of ordinary differential equations is F. Brauer and J. A. Nohel, *The Qualitative Theory of Ordinary Differential Equations: An Introduction*, Dover, NY, 1989.

8. Many solved and unsolved exercises and problems on, and applications of, ordinary differential equations and difference equations can be found in

G. F. Carrier and C. E. Pearson, *Ordinary Differential Equations*, Society for Industrial and Applied Mathematics (SIAM), Philadelphia, PA, 1991.

9. A comprehensive introduction to ordinary differential equations can be found in E. A. Coddington, *Introduction to Ordinary Differential Equations*, Dover, NY, 1989.

10. A thorough treatment of ordinary differential equations with emphasis on analytical concepts and methods can be found in the classic text E. A. Coddington and N. Levinson, *Theory of Ordinary Differential Equations*, McGraw-Hill, New York, NY, 1955.

11. A classic, but quite advanced, text on the theory of ordinary differential equations and dynamical systems is J. Guckenheimer and P. Holmes, *Nonlinear Oscillations, Dynamical Systems, and Bifurcations of Vector Fields*, Springer-Verlag, New York, NY, 1983.

12. An account of ordinary differential equations in the complex plane can be found in E. Hille, *Ordinary Differential Equations in the Complex Domain*, John Wiley, New York, NY, 1976.

13. For the interested reader, connections of ordinary differential equations and dynamical systems are given in M. W. Hirch and S. Smale, *Differential Equations, Dynamical Systems, and Linear Algebra,* Academic Press, New York, NY, 1974.

14. An introductory text on the theory and applications of ordinary differential equations is H. Hochstadt, *Differential Equations*, Dover, NY, 1975.

15. A classic text on the theory of ordinary differential equations, with some emphasis on nonlinear equations and behavior of solutions in the complex plane, is E. L. Ince, *Ordinary Differential Equations*, Dover, NY, 1956.

16. An account of the theory of certain nonlinear differential equations can be found in D. Jordan and P. Smith, *Nonlinear Ordinary Differential Equations*, Clarendon Press, New York, NY, 2nd Edition, 1987.

17. An introduction to the theory of ordinary differential equations from the viewpoint of physical applications can be found in P. M. Morse and H. Feshbach, *Methods of Theoretical Physics, Part I*, McGraw-Hill, New York, NY, 1953.

18. The operator method (based on use of *D*) described in the main text is found in H. T. H. Piaggio, *Differential Equations*, John S. Swift Co., Inc., Chicago, IL.

Chapter 2

1. An introduction to the theory of complex variables with various modern applications can be found in M. J. Ablowitz and A. S. Fokas, *Complex Variables: Introduction and Applications*, Cambridge University Press, New York, 2nd Edition, 2003.

2. An advanced text on applications of functions of a complex variable, especially contour integration, conformal mapping and integral transforms, is G. F. Carrier, M. Krook, and C. E. Pearson, *Functions of A Complex Variable: Theory and Technique*, Hod Books, Ithaca, NY, 1983.

3. A classic text on the theory of Fourier series and integrals is H. S. Carslaw, *Introduction to the Theory of Fourier's Series and Integrals*, Dover, NY, 3rd Edition, 1950.

4. A basic treatment of the theory of functions of one and several variables can be found in H. Cartan, *Elementary Theory of Analytic Functions of One or Several Complex Variables*, Dover, NY, 1973.

5. A classic reference on elements of functions of a complex variable is R. V. Churchill and J. W. Brown, *Complex Variables and Applications*, McGraw-Hill, New York, NY, 6th Edition, 1995.

6. A specialized text on multivalued functions and Riemann surfaces is H. Cohn, *Conformal Mapping on Riemann Surfaces*, McGraw-Hill, New York, NY, 1967.

7. Elements and applications of the theory of complex variables can be found in J. W. Dettman, *Applied Complex Variables*, Macmillan, New York, NY, 1965.

8. An advanced exposition of Fourier series and integrals can be found in H. Dym and H. P. McKean, *Fourier Series and Integrals*, Academic Press, New York, 1972.

9. An elementary introduction to complex analysis and Fourier series, with many unsolved exercises and their answers, can be found in the text by F. B. Hildebrand, *Advanced Calculus for Applications*, Prentice Hall, Englewood-Cliffs, NJ, 2nd Edition, 1976.

10. A classic, somewhat advanced, account of the theory of analytic functions, with theorems and proofs and many exercises with solutions, can be found in K. Knopp, *Theory of Functions, Parts I, II and III*, Dover, NY, 1996.

11. A comprehensive introduction to the subject of complex variables from the viewpoint of applications can be found in P. M. Morse and H. Feshbach, *Methods of Theoretical Physics, Part I*, McGraw-Hill, New York, NY, 1953.

12. An advanced, comprehensive exposition of analytic functions can be found in G. Sansone and J. Gerretsen, *Lectures on the Theory of Functions of a Complex Variable*, P. Noordhoff, Groningen, The Netherlands, 1960.

13. An introduction to the theory of functions of a complex variable with various applications can be found in R. A. Silverman, *Complex Analysis with Applications*, Prentice-Hall, Englewood Cliffs, NJ, 1973.

14. A classic advanced text on the theory of analytic functions, with theorems and proofs, is E. D. Tichmarsh, *Theory of Functions*, Oxford University Press, Oxford, UK, 2nd Edition, 1976.

15. A classic advanced text on the theory of Fourier integrals is E. D. Tichmarsh, *Introduction to the Theory of Fourier Integrals*, Chelsea, New York, NY, 3rd Edition, 1988.

16. A classic text on elements of functions of a complex variable, special functions, and Fourier series and transforms is E. T. Whittaker and G. N. Watson, *A Course of Modern Analysis*, Cambridge University Press, Cambridge, UK, 4th Edition, 2002.

17. An extensive introduction to complex analysis, with many exercises, can be found in D. Wunsch , *Complex Variables with Applications*, Addison Wesley, Boston, MA, 2nd Edition, 1993.

Chapter 3

1. An introduction to first-order partial differential equations with applications can be found in G. F. Carrier and C. E. Pearson, *Partial Differential Equations: Theory and Technique*, Academic Press, Boston, MA, 2nd Edition, 1988.

2. The general theory of first-order partial differential equations is described thoroughly in C. Chester, *Techniques in Partial Differential Equations*, McGraw-Hill, New York, NY, 1971.

3. A reference on complete integrals of first-order partial differential equations is the monumental text by R. Courant and D. Hilbert, *Methods of Mathematical Physics, Vol. 2*, Wiley InterScience, New York, NY, 1953.

4. A classic text on shock waves originating from first-order partial differential equations is R. Courant and K. O. Friedrichs, *Supersonic Flow and Shock Waves*, Springer-Verlag, New York, NY, 1976.

5. A text on theory and applications of first-order partial differential equations to traffic flow and gas dynamics is L. Debnath, *Nonlinear Partial Differential Equations for Scientists and Engineers*, Birkhaüser, Boston, MA, 1997.

6. The theory of first-order partial differential equations is presented in a rigorous but comprehensive way in the advanced text by L. C. Evans, *Partial Differential Equations*, American Mathematical Society (AMS), Providence, RI, 1998.

7. Advanced theory and applications of first-order partial differential equations are given in P. R. Garabedian, *Partial Differential Equations*, American Mathematical Society (AMS)/Chelsea, Providence, RI, 1998.

8. An elementary introduction to first-order partial differential equations with many exercises can be found in H. Levine, *Partial Differential Equations*, American Mathematical Society (AMS), Providence, RI, 1997.

9. Applications of first-order partial differential equations and nonlinear waves to traffic flow, gas dynamics and water waves, with special emphasis on shock waves, can be found in G. B. Whitham, *Linear and Nonlinear Waves*, John Wiley, New York, NY, 1999.

Chapter 4

1. A mathematical introduction to the linear partial differential equations of theoretical physics can be found in the classic text by H. Bateman, *Partial Differential Equations of Mathematical Physics*, Dover, New York, NY, 1944.

2. Solution techniques for and applications of the Laplace and heat equations can be found in G. F. Carrier, M. Krook, and C. E. Pearson, *Functions of a Complex Variable: Theory and Technique*, Hod Books, Ithaca, NY, 1983.

3. The elementary theory of the Laplace, wave and heat equations with examples is described in G. F. Carrier and C. E. Pearson, *Partial Differential Equations: Theory and Technique*, Academic Press, Boston, MA, 2nd Edition, 1988.

4. The theory of the Laplace, wave and heat equations, with emphasis on conditions for well-posed problems, are developed rigorously in E. DiBenedetto, *Partial Differential Equations*, Birkhaüser, Boston, MA, 1995.

5. The theory of second-order linear partial differential equations is developed rigorously with emphasis on solution spaces in L. C. Evans, *Partial Differential Equations*, American Mathematical Society (AMS), Providence, RI, 1998.

6. The classification and uses in gas dynamics and fluid mechanics of linear and quasi-linear partial differential equations are described in the classic text by P. R. Garabedian, *Partial Differential Equations*, American Mathematical Society (AMS)/Chelsea, Providence, RI, 1998.

7. The rigorous theory of elliptic, second-order partial equations and associated boundary-value problems is described in D. Gilbarg and N. S. Trudinger, *Elliptic Partial Differential Equations of Second Order*, Springer-Verlag, New York, NY, 2nd Edition, 1983.

8. The theory of linear partial differential equations is developed in the classic text F. John, *Partial Differential Equations*, Springer-Verlag, New York, NY, 4th Edition, 1995.

9. The theory of second- and higher-order hyperbolic partial differential equations in connection to systems of conservation laws is presented in P. Lax, *Hyperbolic Systems of Conservation Laws and the Mathematical Theory of Shock Waves*, Society for Industrial and Applied Mathematics (SIAM), Philadelphia, PA, 1973.

10. An elementary introduction to second-order partial differential equations, with many examples, can be found in H. Levine, *Partial Differential Equations*, American Mathematical Society (AMS), Providence, RI, 1997.

11. The rigorous theory of parabolic, second-order partial differential equations can be found in G. M. Lieberman, *Second Order Parabolic Partial Differential Equations*, World Scientific, Singapore, 1996.

12. A comprehensive introduction to the Laplace, Poisson, wave, diffusion and Schrödinger equations from the viewpoint of physical applications can be

found in P. M. Morse and H. Feshbach, *Methods of Theoretical Physics, Part II*, McGraw-Hill, New York, NY, 1953.

13. An elementary introduction to models of wave and diffusion phenomena is found in F. Y. M. Wan, *Mathematical Models and Their Analysis*, Harper & Row, New York, NY, 1989.

Chapter 5

1. Variants of the method of separation of variables for partial differential equations can be found in L. Debnath, *Nonlinear Partial Differential Equations*, Birkhaüser, Boston, MA, 1997.

2. Special solution techniques for linear and nonlinear partial differential equations are described rigorously in the advanced text L. C. Evans, *Partial Differential Equations*, American Mathematical Society (AMS), Providence, RI, 1998.

3. An elementary introduction to the separation of variables and eigenfunction expansions for linear partial differential equations, with many exercises and their answers, can be found in F. B. Hildebrand, *Advanced Calculus for Applications*, Prentice Hall, Englewood Cliffs, NJ, 2nd Edition, 1976.

4. The method of variable separation for the Laplace and Poisson equations, in the physical context of electrostatics, can be found in J. D. Jackson, *Classical Electrodynamics*, Wiley, New York, NY, 1999.

5. Advanced applications of variable separation and other techniques are described in J. Kevorkian, *Partial Differential Equations: Analytical Solution Techniques*, Chapman & Hall, London, 1996.

6. The methods of separation of variables and eigenfunction expansions, and other solution techniques are described in an elementary way with many examples in H. Levine, *Partial Differential Equations*, American Mathematical Society (AMS), Providence, RI, 1997.

7. Applications of variable separation to Schrödinger's equation, in the physical context of quantum mechanics, are described in E. Merzbacher, *Quantum Mechanics*, Wiley, New York, NY, 1998.

8. The separation of variables for the Laplace, Poisson, diffusion and Schrödinger equations in various coordinate systems is described in the classic text P. M. Morse and H. Feshbach, *Methods of Theoretical Physics, Part II*, McGraw-Hill, New York, NY, 1953.

9. The theory of boundary value problems and the associated Green's functions is developed from an applied perspective in I. Stakgold, *Green's Functions and Boundary Value Problems*, Wiley InterScience, New York, NY, 2nd Edition, 1998.

Chapter 6

1. A comprehensive reference for the Frobenius method is G. B. Arfken, *Mathematical Methods for Physicists*, Academic Press, Orlando, FL, 1985.

2. An introduction to the theory of singular points of ordinary differential equations from an applied perspective is described in C. M. Bender and S. A. Orszag, *Advanced Mathematical Methods for Scientists and Engineers*, McGraw-Hill, New York, NY, 1978; Springer-Verlag, New York, NY, 1999.

3. Extensive formulas for series expansions of special functions, along with comments on their derivations, can be found in A. Erdelyi, W. Magnus, F. Oberhettinger, and F. G. Tricomi, *Higher Transcendental Functions, Vols. 1, 2, 3*, McGraw-Hill, New York, NY, 1953.

4. One of the first papers on the Frobenius method is G. F. Frobenius, Ueber die Integration der linearen Differentialgleichungen durch Reihen, *Journal für die reine und angewandte Mathematik*. Vol. 76, pp. 214–235, 1873.

5. An elementary introduction to the Frobenius method can be found in F. B. Hildebrand, *Advanced Calculus for Applications*, Prentice Hall, Englewood Cliffs, NJ, 2nd Edition, 1976.

6. An advanced theory for singular points of ordinary differential equations can be found in E. L. Ince, *Ordinary Differential Equations*, Dover, NY, 1956.

7. An elementary introduction to the Frobenius method is described in D. A. McQuarrie, *Mathematical Methods for Scientists and Engineers*, University Science Books, Sausalito, CA, 2003.

8. The theory of singular points of Bessel and Bessel-related functions, along with the associated power series expansions, can be found in the classic text G. N. Watson, *A Treatise on the Theory of Bessel Functions*, Cambridge University Press, Cambridge, UK, 1966.

Chapter 7

1. An introduction to the WKB method with examples and comparisons with numerical solutions can be found in C. M. Bender and S. A. Orszag, *Advanced Mathematical Methods for Scientists and Engineers*, McGraw-Hill, New York, NY, 1978; Springer-Verlag, New York, NY, 1999.

2. One of the earlier papers related to the WKB method appears to be L. Brillouin, Remarques sur la méchanique ondulatoire, *Journal de physique et le radium* Vol. 7, pp. 353–368, 1926.

3. Some advanced applications of the WKB method can be found in G. F. Carrier, M. Krook, and C. E. Pearson, *Functions of A Complex Variable: Theory and Technique*, Hod Books, Ithaca, NY, 1983.

4. A classic text for the WKB method is N. Fröman and P. O. Fröman, *JWKB Approximation*, North-Holland, Amsterdam, The Netherlands, 1965.

5. A thorough, modern introduction to techniques related to the WKB method can be found in N. Fröman and P. O. Fröman, *Physical Problems Solved by the Phase-Integral Method*, Cambridge University Press, Cambridge, UK, 2002.

6. Applications of the WKB method can be found in E. J. Hinch, *Perturbation Methods*, Cambridge University Press, Cambridge, UK, 1995.

7. An introductory account of the WKB method can be found in M. H. Holmes, *Introduction to Perturbation Methods*, Springer-Verlag, New York, NY, 1995.

8. One of the earlier papers related to the WKB method appears to be H. A. Kramers, Wellenmechanik und halbzahlige Quantisierung, *Zeitschrift für Physik* Vol. 38, pp. 828–840, 1926.

9. An advanced theory for the WKB method can be found in V. P. Manslov, *The Complex WKB Method for Nonlinear Equations I: Linear Theory*, translated from the Russian by M. A. Shishkova and A. B. Sossinsky, Birkhaüser, Boston, MA, 1994.

10. Applications of the WKB method to non-relativistic quantum mechanics are described in E. Merzbacher, *Quantum Mechanics*, Wiley, New York, NY, 1998.

11. A comprehensive introduction to the WKB method can be found in P. M. Morse and H. Feshbach, *Methods of Theoretical Physics, Part II*, McGraw-Hill, New York, NY, 1953.

12. The application of WKB-related methods to the scattering of scalar (acoustic and quantum mechanical) fields at high frequencies can be found in S. I. Rubinow and T. T. Wu, "First Correction to the Geometric-optics Scattering Cross Section from Cylinders and Spheres," *J. Appl. Physics* Vol. 27, pp. 1032–1039, 1956. For a more systematic treatment see T. T. Wu, "High-frequency Scattering," *Phys. Rev.* Vol. 104, pp. 1201–1212, 1956.

13. Asymptotic expansions for ordinary differential equations are discussed generally in the classic text W. Wasow, *Asymptotic Expansions for Ordinary Differential Equations*, Dover, New York, NY, 2002.

14. The theory of turning points for linear differential equations is described in W. R. Wasow, *Linear Turning Point Theory*, Springer-Verlag, New York, NY, 1985.

15. One of the earlier papers associated with the WKB method appears to be G. Wentzel, "Eine Verallgemeinerung der Quantenbedingungen für die Zwecke der Wellenmechanik," *Zeitschrift für Physik* Vol. 38, pp. 518–529, 1926.

Chapter 8

1. Extensive tables with asymptotic expansions of special functions can be found in M. Abramowitz and I. A. Stegun, *Handbook of Mathematical Functions*, Dover, New York, NY, 1964.

2. An account of asymptotic methods for integrals with examples and comparisons with numerical evaluations can be found in C. M. Bender and S.

A. Orszag, *Advanced Mathematical Methods for Scientists and Engineers*, McGraw-Hill, New York, NY, 1978; Springer-Verlag, New York, NY, 1999.

3. Various techniques for the asymptotic evaluation of integrals are described in N. Bleistein and R. A. Handelsman, *Asymptotic Expansion of Integrals*, Dover, New York, NY, 1986.

4. Some advanced problems in asymptotic expansions can be found in G. F. Carrier, M. Krook, and C. E. Pearson, *Functions of A Complex Variable: Theory and Technique*, Hod Books, Ithaca, NY, 1983.

5. The use of the Mellin transform technique in the evaluation of Feynman integrals is presented in H. Cheng and T. T. Wu, *Expanding Protons: Scattering at High Energies*, M.I.T. Press, Cambridge, MA, 1987.

6. A classic text on asymptotic expansions is E. T. Copson, *Asymptotic Expansions*, Cambridge University Press, Cambridge, UK, 1967.

7. Asymptotic methods are discussed generally in N. G. de Bruijn, *Asymptotic Methods in Analysis*, North-Holland, Amsterdam, The Netherlands, 1958.

8. A general and concise discussion of asymptotic expansions can be found in A. Erdelyi, *Asymptotic Expansions*, Dover, New York, NY, 1956.

9. Extensive asymptotic expansions of integrals of special functions, along with comments on their derivations, can be found in A. Erdelyi, W. Magnus, F. Oberhettinger, and F. G. Tricomi, *Higher Transcendental Functions, Vols. 1, 2, 3*, McGraw-Hill, New York, NY, 1953.

10. The nature of a class of asymptotic expansions as divergent series is discussed rigorously in the classic text G. H. Hardy, *Divergent Series*, Oxford University Press, New York, NY, 1949.

11. A discussion of Watson's lemma in the asymptotic evaluation of integrals can be found in E. J. Hinch, *Perturbation Methods*, Cambridge University Press, Cambridge, UK, 1995.

12. An introduction to asymptotic series and methods can be found in P. M. Morse and H. Feshbach, *Methods of Theoretical Physics, Part I*, McGraw-Hill, New York, NY, 1953.

13. A detailed discussion on the theory of asymptotic expansions is provided in F. W. J. Olver, *Asymptotics and Special Functions*, Academic Press, New York, NY, 1974.

14. Modern applications of the Mellin transform technique to the evaluation of integrals are described in the expository paper R. J. Sasiela and J. D. Shelton, Mellin "Transform Methods Applied to Integral Evaluation: Taylor Series and Asymptotic Approximations," *J. Math. Phys.* Vol. 34, pp. 2572–2617, 1993.

Chapter 9

1. An account of boundary layer theory, including comparisons with numerical solutions, can be found in C. M. Bender and S. A. Orszag, *Advanced Mathematical Methods for Scientists and Engineers*, McGraw-Hill, New York, NY, 1978; Springer-Verlag, New York, NY, 1999.

2. The nature of series expansions stemming from singular perturbations is discussed in the paper J. P. Boyd, "The Devil's Invention: Asymptotic, Superasymptotic and Hyperasymptotic Series," *Acta Applicandae Mathematicae* Vol. 56, pp. 1–98, 1999.

3. Applications of boundary layer theory to ordinary differential equations are described in G. F. Carrier and C. E. Pearson, *Ordinary Differential Equations*, Society for Industrial and Applied Mathematics (SIAM), Philadelphia, PA, 1991.

4. Examples of boundary layers in boundary value problems for elliptic partial differential equations can be found in G. F. Carrier and C. E. Pearson, *Partial Differential Equations: Theory and Technique*, Academic Press, Boston, MA, 2nd Edition, 1988.

5. A mathematical, concise introduction to boundary layers for the Navier-Stokes equations (in fluid mechanics) is provided in A. J. Chorin and J. E. Marsden, *A Mathematical Introduction to Fluid Mechanics*, Springer-Verlag, New York, NY, 1990.

6. A seminal research paper on the theory of matched asymptotics is L. E. Fraenkel, "On the Method of Matched Asymptotic Expansions, Parts I, II, and III," *Proc. Camb. Philos. Soc.* Vol. 65, pp. 209–284, 1969.

7. A comprehensive treatment of boundary layers with representative examples can be found in E. J. Hinch, *Perturbation Methods*, Cambridge University Press, Cambridge, UK, 1995.

8. An introduction to boundary layers is provided in M. H. Holmes, *Introduction to Perturbation Methods*, Springer-Verlag, New York, NY, 1995.

9. A general discussion on boundary layer theory with applications is provided in J. Kevorkian and J. D. Cole, *Perturbation Methods in Applied Mathematics*, Springer-Verlag, New York, NY, 1981.

10. A review article on singular perturbations is P. A. Lagerstrom and R. G. Casten, "Basic Concepts Underlying Singular Perturbation Techniques," *SIAM Rev.* Vol. 14, pp. 63–120, 1972.

11. A general discussion of singular perturbation theory is provided in S. A. Lomov, *Introduction to the General Theory of Singular Perturbation*, American Mathematical Society (AMS), Providence, RI, 1992.

12. An application of boundary layer theory to the evolution of crystal surfaces is described in D. Margetis, M. J. Aziz, and H. A. Stone, "Continuum Approach to Self-Similarity and Scaling in Morphological Relaxation of a Crystal with a Facet," *Phys. Rev. B* Vol. 71, art. 165432, pp. 1–22, 2005.

13. Boundary layers are discussed extensively in A. H. Nayfeh, *Perturbation Methods*, John Wiley and Sons, New York, NY, 1973.

14. A general discussion of singular perturbations is given in R. E. O'Malley, *Introduction to Singular Perturbations*, Academic Press, New York, NY, 1974.

15. Boundary layer theory and other methods are described in D. R. Smith, *Singular Perturbation Theory*, Cambridge University Press, Cambridge, UK, 1985.

16. Applications of boundary layer theory to fluid mechanics are described in the classic text M. Van Dyke, *Perturbation Methods in Fluid Mechanics*, Academic Press, New York, NY, 1964.

17. The theory of turning points for linear differential equations is described formally in W. R. Wasow, *Linear Turning Point Theory*, Springer-Verlag, New York, NY, 1985.

Chapter 10

1. Elements of multiple-scale analysis with examples and comparisons with numerical solutions are described in C. M. Bender and S. A. Orszag, *Advanced Mathematical Methods for Scientists and Engineers*, McGraw-Hill, New York, NY, 1978; Springer-Verlag, New York, NY, 1999.

2. A research paper on the application of the renormalization group to singular perturbation problems is L.-Y. Chen, N. Goldenfeld, and Y. Oono, "Renormalization Group and Singular Perturbations: Multiple Scales, Boundary Layers, and Reductive Perturbation Theory," *Phys. Rev. E* Vol. 54, pp. 376–394, 1996.

3. The renormalized two-scale method with counterterms to treat nonlinear motions is given in H. Cheng, "The Renormalized Two-scale Method," *Stud. Applied Math.* Vol. 113, pp. 381–387, 2004.

4. An application of renormalization theory to the abelian-Higgs quantum field theory with chiral fermions is described in H. Cheng and S. P. Li, "How to Treat γ_5," *International J. Mod. Phys.*, Vol. 13, pp. 2991–3050, 1998.

5. A general introduction to renormalization in perturbation theory for quantum fields can be found in J. Collins, *Renormalization*, Cambridge University Press, Cambridge, UK, 1998.

6. A thorough introduction to the renormalization group approach in statistical physics is given in N. D. Goldenfeld, *Lectures on Phase Transitions and the Renormalization Group*, Addison-Wesley, Reading, MA, 1992; in particular, see Chapter 10.

7. An extension of the renormalization group approach to multiscale problems in materials is described in N. Goldenfeld, B. P. Athreya, and J. A. Dantzig, "Renormalization Group Approach to Multiscale Simulation of Polycrystalline Materials Using the Phase Field Crystal Model," *Phys. Rev. E* Vol. 72, art. 020601, pp. 1–4, 2005.

8. An introduction to the method of multiple scales with examples can be found in J. Kevorkian and J. D. Cole, *Perturbation Methods in Applied Mathematics*, Springer-Verlag, New York, NY, 1981.

9. Applications of multiple-scale analysis to chemical problems are described in Y. Kuramoto, *Chemical Oscillations, Waves, and Turbulence*, Springer-Verlag, Berlin, 1984.

10. A research paper on application of two-scale analysis in fluid mechanics is A. C. Newell and J. A. Whitehead, "Finite Bandwidth, Finite Amplitude Convection," *J. Fluid Mech.* Vol. 38, pp. 279–303, 1969.

11. An introduction to multiple-scale analysis is provided in A. H. Nayfeh, *Perturbation Methods*, John Wiley and Sons, New York, NY, 1973.

12. An introduction to the related method of reconstitution is described in A. J. Roberts, "An Introduction to the Technique of Reconstitution," *SIAM (Society for Industrial and Applied Mathematics) J. Math. Anal.* Vol. 16, pp. 1243–1257, 1985.

13. Theory and examples of multiple-scale analysis in vibrations are presented in J. J. Stocker, *Nonlinear Vibrations*, Wiley InterScience, New York, NY, 1950.

14. A paper on the "reductive perturbation method" related to two- or multiple-scale problems is T. Taniuti and C. C. Wei, "Reductive Perturbation Method in Nonlinear Wave Propagation. I," *J. Phys. Soc. Japan* Vol. 24, pp. 941–946, 1968.

15. Other aspects of multiple-scale problems are discussed in S. L. Woodruff, "The Use of an Invariance Condition in the Solution of Multiple-scale Singular Perturbation Problems-Ordinary Differential Equations," *Stud. Appl. Math.* Vol. 90, pp. 225–248, 1993.

Index

U

V

W

Z